VII Conference on Low Temperature Plasma in the Processes of Functional Coating Preparation 2015

Journal of Physics: Conference Series Volume 669

Kazan, Russia
4 - 7 November 2015

ISBN: 978-1-5108-1960-3
ISSN: 1742-6588

VII Conference on Low Temperature Plasma in the Processes of Functional Coating Preparation 2015

Journal of Physics: Conference Series Volume 669

Kazan, Russia
4 - 7 November 2015

IOP
Institute of Physics

Printed from e-media with permission by:

Curran Associates, Inc.
57 Morehouse Lane
Red Hook, NY 12571

CURRAN ASSOCIATES INC. proceedings.com

Some format issues inherent in the e-media version may also appear in this print version.

Printed by Curran Associates, Inc. (2016)

For permission requests, please contact the Institute of Physics
at the address below.

Institute of Physics
Dirac House, Temple Back
Bristol BS1 6BE UK

Phone: 44 1 17 929 7481
Fax: 44 1 17 920 0979

techtracking@iop.org

Additional copies of this publication are available from:

Curran Associates, Inc.
57 Morehouse Lane
Red Hook, NY 12571 USA
Phone: 845-758-0400
Fax: 845-758-2634
Email: curran@proceedings.com
Web: www.proceedings.com

Table of contents

Volume 669

VII Conference on Low temperature Plasma in the Processes of Functional Coating Preparation
4–7 November 2015, Kazan, Russia

Accepted papers received: 20 November 2015
Published online: 14 January 2016

Preface

Papers

Journal of Physics: Conference Series **669** (2016) 011001 doi:10.1088/1742-6596/669/1/011001

The VII All-Russian (with international participation) Scientific Technical Conference "Low-temperature plasma during the deposition of functional coatings" took place from 4–7 November 2015 at the Academy of Sciences of the Republic of Tatarstan and the Kazan Federal University.

The conference was attended by over 150 people from Russia and abroad. The participants proposed a wide range of issues affecting the theoretical and experimental aspects of the problems of the physics of low-temperature plasma.

We heard the reports of experts from leading universities and research organizations in the field of plasma physics: Moscow State University, St. Petersburg State University, MEPhI, Tomsk Polytechnic University, Institute of High Current Electronics SB RAS, etc.

A series of works were devoted to the study of thin films obtained by low-temperature plasma. This year, work dedicated to the related field of heat mass transfer in multiphase media and low-temperature plasma was also presented. Of special interest were reports on the exploration of gas discharges with liquid electrolytic electrodes and the study of dusty plasmas.

Kashapov Nail, D.Sc., professor (Kazan Federal University)

Journal of Physics: Conference Series **669** (2016) 011002 doi:10.1088/1742-6596/669/1/011002

Peer review statement

All papers published in this volume of *Journal of Physics: Conference Series* have been peer reviewed through processes administered by the proceedings Editors. Reviews were conducted by expert referees to the professional and scientific standards expected of a proceedings journal published by IOP Publishing.

Way of stability increase of the glow discharge

T Ya Asadullin I G Galeyev and B A Timerkayev

Kazan National Research Technical University named by A.N.Tupolev
420111 Russia Kazan, K.Marx str., 10

gpd_kstu@rambler.ru

Abstract. Instabilities of a glow discharge significantly limit the opportunities for use of nonequilibrium plasma of a glow discharge [1]. The ionization overheat instability is the main type of glow discharge instabilities. It causes the loss of a non-equilibrium and put the glow discharge to a contracted state. The way of stabilization of glow discharge in transverse gas flow by the use of adaptive control system is considered in the work. The system controls voltage (resistance) of cathodic sections of the discharge chamber. Such fast advancing influence prevents the cathode spot shrinking and subsequent transition to contracted state. The controlling algorithm can be realized both following type and dynamic non-equilibrium type.

The essentially nonequilibrium glow discharge plasma has many various technology application, including cleaning and modification of surfaces. The glow discharge with a transverse direction of a gas flow is more often used in most cases of practical application of a glow discharge for surfaces modification. In this case it is possible to use chambers with various geometry, to apply various ways of preionizing and so on.

In all cases of practical application of a glow discharge it is required to provide the stable homogeneous space mode of glowing at the considerable values of electric current and specific energy supply. Increase of these discharge characteristics is limited due to problem of development of instabilities of the discharge [1]. First of all it is the ionization overheating instability [2] leading to transition of the discharge to contracted state. In order to prevent developing of instabilities of glow discharge in transverse gas flow the partitioned cooled cathodic blocks are widely used. Usually cathodic sections in the cathodic block are separated from each other by dielectric layers. In this work possibility of use of system of dynamic management of potentials (resistance) of sections of the cathode for stabilization of a glow discharge in a transverse gas flow is considered.

There are various ways of glow discharge stabilization which effectiveness depends on many factors. The common principle of work of all techniques of stabilization is the prevention of formation of conditions for emergence of instabilities. For this purpose it can be used: ballast resistors, twisting of the gas flow, various ways of preionizing, sectioning of the cathode, influence by the magnetic field and so on. Ways on prevention of development of instabilities can be divided into two types:
1) the influences uniforming the discharge characteristics in space of the chamber, for example, preionizing and twisting of a gas flow;
2) the influences faster than instabilities development, for example, influence by the rotating magnetic field.

Here the way relating to the second type is considered.

The reference time of ionization overheating instability development is about 10^{-3} s. Therefore, stabilizing influence has to eliminate the conditions for ionization overheating instability development for significantly smaller time.

As it was told above, the glow discharge in transverse gas flow is considered. The used cathodic block consists of the small copper sections dielectrically separated from each other. Sections are cooled by distilled water. The anode is the copper plate. At rather low values of pressure, gas flow rate and electric current the discharge glows is in diffuse mode, and the cathode spot fulfills all surface of cathodic sections. With increase of discharge parameters the cathode spot area is shrinking usually down to the stream. Then, at further increase of discharge parameters, discharge contracts, and cathode spot is shrinking to the point on the cathode surface with preferred conditions.

In most cases conditions for development ionization overheating instabilities are realized near the cathode area where there are processes of intensive ionization and considerable thermal emission. Therefore the stabilizing effect in the considered way is made through management of the cathodic sections currents.

The essence of a method consists in carrying out the continuous monitoring of dynamics of change of values of currents through separate cathodic sections. At the stable diffuse glow mode the law of constancy of a current density on the cathode is usually carried out. If the glow mode will become not stable, and the glow will be changing to contracted form of the discharge, it will be followed by appearing of the increasing nonuniformity of current distribution. It is necessary to manage fast and optimally the ballast resistance values to make cathode spot spread along the cathode surface.

It is evident that because of quite small reference time of development ionization overheating instabilities (10^{-3} c) the stabilization system has to be automated.

The system of stabilization consists of the block of measuring resistors, a chain of coordination, the switchboard with the multiplexer, the analog-to-digital converter (ADC), the computer or the microcontroller, the control unit of potentials of cathodic sections.

Current through each section is defined by a voltage drop on measuring resistors with a small resistance. After passing of a chain of coordination the transformed signals from each section sequentially are connected by means of the switchboard to ADC. Then data about voltage (current) of each section come on processing to the computer. The computer processes the data. Values of currents of the next sections and dynamics of change of currents values in all sections are compared. After the data processing the managing signals are formed and sent to the system of cathodic section potentials management. The cathodic section potentials management circuit consists of the digital-to-analog converter, the demultiplexer and switching elements which resistance can be changed depending on a managing signal.

The hardware of system of stabilization is supplemented with the corresponding software which operates processes:
- of the measurement of parameters of the experimental setup,
- of the measurement results transfer to the computer,
- of the analysis of the experimental data flow,
- of the development of the operating teams,
- of the realization monitoring and so on.

The role of the software consists in providing dynamically supported labile equilibrium of the discharge near the limiting (critical) values of currents through cathodic sections. On the one hand the system of stabilization has to provide the steady diffuse mode of combustion of the discharge, and with another not to give the chance to reduce values of currents to minimum because of an overshoot. Depending on installations of the operator the system can provide glowing of the discharge in the different modes on the relation of proximity to the critical mode. Use of the system of preionization which is built in the pilot unit considerably expands range of adjustment of the limiting currents which size is limited to system effectiveness of electrodes cooling.

Except the basic purpose for stabilization of the discharge this system with the disconnected chain of regulation can be used for carrying out measurements of speed of development of instabilities in the discharge and current distributions on a cathode surface.

References

[1] Raizer Yu P 1991 Gas Discharge Physics, Springer, Berlin, Heidelberg, N.Y.
[2] D. Staack, B. Farouk, A. Gutsol, A. Fridman. J. Appl. Phys. 106, 013303 (2009).

Low pressure glow discharge in a system with hollow electrode at floating potential

N A Babinov

Department of Electron Devices and Systems, Saint Petersburg Electrotechnical University "LETI", Saint Petersburg, Russia

babinovnikita@gmail.com

Abstract. This article describes the research of the low pressure gas discharge in a system with hollow electrode at the floating potential. The main characteristic features of the discharge distinguishing it from the glow discharge with hollow cathode are described. The studied type of discharge has good perspective to use in the plasma emission systems allowing to reach high current efficiency of the ion sources.

1. Introduction

In the present time devices intended to create beams of fast ions generated in the plasma volume are intensively used for technological processes of surface treatment, coatings deposition and modification. This devices are also known as plasma emission systems. The most common method for plasma generation is to use gas discharges of different types such as arc discharge, glow discharge, inductively coupled discharge etc. The choice of this method is caused by its relatively simplicity, high efficiency and the fact that a gas discharge plasma is well studied.

At the same time, existing in the recent years tendency to reduce an operating pressure during processes of coatings deposition and treatment. This tendency is due to intention to reduce the flow of polluting substances. Moreover increasing requirements to beam quality and uniformity demand study of new gas discharge types that are able to provide the generation of high-density and uniform plasma at the low gas pressure.

One of the promising gas discharge types that are able to provide necessary parameters is a glow discharge with electron electrostatic trapping discussed in details in [1]. At the expense of electron multiply oscillations in the electrostatic trap area the electron path length in the gas discharge area is increased many times. This leads to increasing the proportion of energy that are spent by fast cathode-emitted electrons for gas ionization. This allows significant reducing of the system operation pressure down to the 0.1 and even lower in some cases.

A large number of different emission systems that use the glow discharge with electron electrostatic trapping (in particular, glow discharge with hollow cathode) are known [2, 3]. Such systems operate on direct current and are able to generate ions beams with high density and large cross-section.

At the same time, the systems based on glow discharge with electron electrostatic trapping have some disadvantages. One of the main disadvantages is low current efficiency (the ratio of beam current to the discharge current) of such systems. It is caused by the fact that to ensure the effect of electron electrostatic trapping it is necessary to create a closed cathode structure while the extraction grid is placed on the one side of the system; the grid area is much smaller than cathode area. Considering that ion current density is practically independent of electrode potential and is equal for grid and cathode; this leads to the fact that extracting beam current is much smaller than discharge current.

There are ion sources designs in which the current efficiency is increased by, for example, magnetic fields using [4]. But even in this case current efficiency in 20–30 % is considered high.

For increasing of the current efficiency of plasma emission systems it may be helpful to use the glow discharge with hollow cathode at the floating potential. Due to the zero total current to the isolated electrode in this case all the discharge current is closed through the cathode that plays the role of an extraction grid as well. Thank to this the current efficiency of such system can reach the 90–95 %.

In this work the argon gas discharge in the system consisting of three electrodes: hollow cylindrical isolated electrode that was closed from the one side; cathode formed in the shape of disc and placed at the open side of isolated electrode and anode placed outside of hollow electrode.

To ensure the absence of a parasitic discharge in the system and the charges draining from the isolated electrode it was surrounded by the grounded screen placed at the range of about 5 mm. It ensure the absence of parasitic discharge even when the potential of the isolated electrode was greater than 1 kV.

2. Description of experiments

It is known [5, 6] that floating potential of the electrode placed at the path of the electron beam can nonlinearly depend on the energy of the beam electrons. It is associated with the influence of the secondary electron emission phenomenon.

The studied system has the similarities with described in [5, 6] systems in that the isolated electrode is placed at the path of high-energy electrons, in our case they are emitted from the cathode and accelerated in the cathode potential drop.

The significant distinction of the systems is that due to the large area of the isolated electrode, its potential significantly affects the gas discharge plasma configuration, its density and electron temperature; on the other hand the floating potential is affected by the parameters of gas discharge plasma. Moreover, in this work the main attention is given to the discharge with voltage higher than 1 kV and with low current density.

During the experiments the anode of the system was grounded and the cathode voltage was applied to the negative high voltage power supply. The discharge current and voltage as well as potential of the isolated electrode were measured by the analogue and digital instruments; leakage current was not greater than 100 uA.

The measurements were carried out during the burning of gas discharge in argon at the pressure from 3 to 100 Pa. Due to the discharge instability and big influence of the random error, each experiment series was carried out from 3 to 5 times.

One of the experiment series is shown in a figure 1. It can be seen there are two main forms of gas discharge: with high and low potential at the isolated electrode. Transition from one from to another occurs when the discharge current exceeds certain value. Also there is the phenomenon of hysteresis since the revers transition doesn't occur.

The further experiments allowed to state some assumptions:

1. The low pressure glow discharge with isolated electrode has 2 or 3 forms, significantly that have different values of floating potential.

2. The transition from one form to another occurs when the current exceed certain value, in these moments the visual changes are visible, moreover the discharge voltage changes abruptly.

3. Depending on the form of discharge the voltage of the isolated electrode can take both value of –20…–40 V, that are typical for floating potential of plasma and the very large negative values close to the cathode potential. This phenomena was not observed in [6] where the floating potential always was significantly lower than electron accelerated voltage.

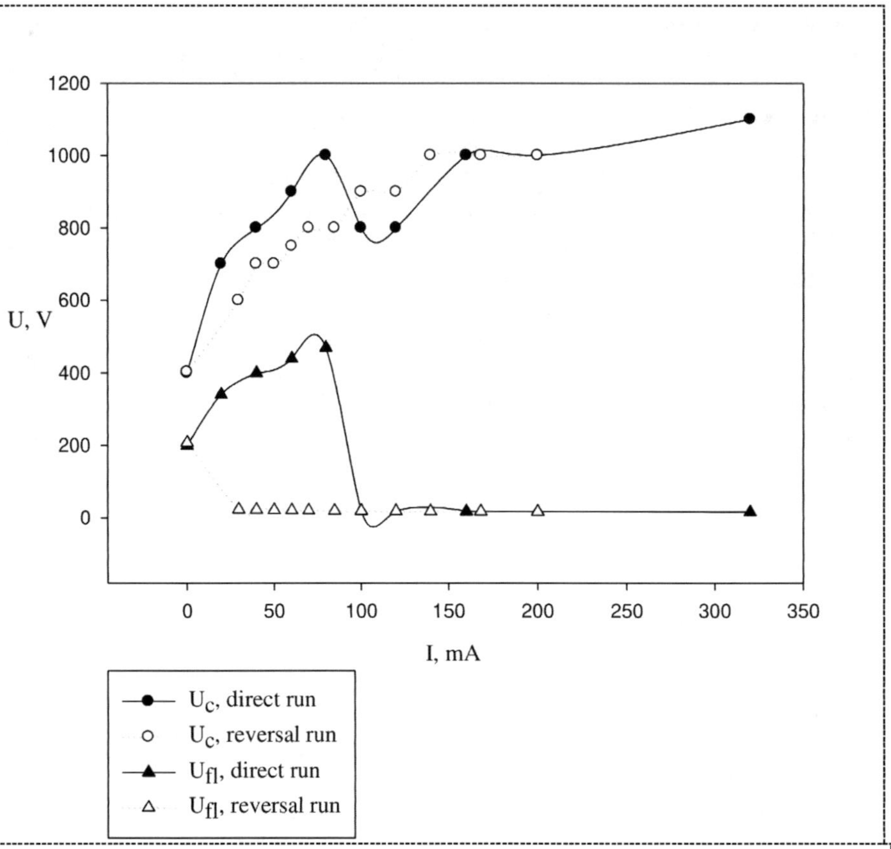

Figure 1. VAC and floating potential at the pressure of 5 Pa

The strongest differences between the discharge forms are observed when the pressure is low, not greater than 10…20 Pa. When the gas pressure is greater, the differences are slowly erased and then are completely disappear at the pressure of about 50 Pa.

3. Conclusion

The obtained experimental data allows to say the glow discharge with hollow isolated large electrode has some has the characteristic features that distinguish it from other types of gas discharge.

Gas discharge in the described system has much in common with well-studied hollow cathode discharge, although there are significant distinctions. First of all it seems the important role in the occurring processes is played by the secondary electron-electron emission that doesn't play a big role in the discharge with hollow cathode. Based on the works [5, 6] it can be assumed that it is the secondary electron-electron emission which causes the existence of several forms of gas discharge.

At the same time, the physic processes determining the gas discharge burning are not fully clear. First of all, to compile a satisfactory theory of the discharge with hollow isolated electrode it is necessary to continue experimental work and obtain new data that will allows accurately determining the existing patterns of such gas discharge.

Acknowledgments

This work was performed as a part of the Russian Federation government project assignment №8.2456.2014/K. The research was supported by Department of Electron Devices and Systems of Saint Petersburg Electrotechnical University "LETI".

Journal of Physics: Conference Series **669** (2016) 012002 doi:10.1088/1742-6596/669/1/012002

References

[1] S N Grygoryev, A S Metel 2007 *Modification of surface by glow discharge with electron electrostating trapping* (Moscow, Moscow State University, 452 p.)
[2] S N Grygoryev, A S Metel A N Isaykov and Yu A Melnik 2005 Har *Hardening Technologies and coatings.* issue 9, pp. 36–40.
[3] Ichiki K., Hatakeyama M. 2008. *J. Phys. D: Appl. Phys.* Vol. 41, p. 024003
[4] S N Grygoryev, A S Metel, Yu A Melnik and V V Panin RU pathent № 2373603 Published 20.11.2009
[5] V Ya Martens 1996 *Journal of tech. physics* Vol. 66, issue 6, pp. 70–76
[6] N V Gavrilov, A I Menshakov, A S Kamenetskih 2013 *Journal of tech. physics* Vol. 83 issue 1, pp. 74–79

Hydrogen trapping in graphite materials in various conditions

L B Begrambekov, A A Ayrapetov, Ya A Sadovskiy, P A Shigin

National Research Nuclear University MEPhI, 31 Kashirskoe shosse, Moscow, 115409, Russia

E-mail : lbb@plasma.mephi.ru

Abstract. The paper presents a short revue of the study of not been investigated specific features of hydrogen trapping in graphite occurring as the results of inelastic interaction of impinging ions with carbon materials (potential trapping). The influence of potential mechanism on different regularities of hydrogen trapping are considered. Among them there is trapping dependence on irradiating ion current density, ion energy, irradiation fluence, temperature of the investigated samples and oxygen addition in the plasma.

1. Introduction

The problem of hydrogen trapping in the graphites, graphite materials and carbon films was the subject of interest during last decades. The main reason of it was an application of graphites and graphite material as the plasma facing elements for number of fusion devices [1-3]. The main sides of carbon material behavior under ion and plasma irradiation were evaluated due to investigations made in fusion devices and in experimental stands. In particular, serious attention was paid on study of features of hydrogen trapping and retention in the graphite materials [4-7]. The authors of the paper studded of not been investigated specific features of hydrogen trapping in graphite occurring as the result of potential interaction of impinging ions with carbon materials and carbon films [8-11]. The paper presents a short revue of the main results of these works.

2. Deuterium trapping in carbon materials exposed to D plasma

2.1. Dependence of trapping on irradiating ion energy

The experiments were performed in a device [8] intended for irradiation of samples with plasma ions and for measurement of the amount of retained gases by the method of Thermal Desorption Spectrometry (TDS). The temperature ramp up during TDS investigation was equal to 5 K/s. Deuterium (D) plasma was ignited in a three electrode system composed of heated cathode, preliminary anode and anode. The specimens of carbon fiber composite (CFC) and pyrolitic graphite (PG) were investigated. The temperature of the samples during irradiation was 470 ± 10 K and fluence was 5×10^{23} at/m^2. The results were basically similar for both materials. The paper presents results for CFC and notice the differences between two materials, if it is needed.

The energy dependence of the amount of deuterium released as D_2 molecules from the CFC (carbon fiber composite) irradiated with D ions at various energies (Fig. 1) demonstrates that the deuterium retention decreases when the ion energy reduces from 1000 to 200 eV/D. The main peaks in D_2 thermal desorption spectra for the samples irradiated with deuterium ions at relatively high energies (from 500 to 1000 eV/D) locate in the temperature ranges 900–1000K and 1050–1150 K with shoulders between 1200 and 1300 K.

The authors of the paper [9] basing on the analysis of the papers dealing with measurements of TDS spectra concluded that temperature of TDS peaks is relevant to the irradiation conditions and reflects to some extent the structure of the near surface layer of irradiated material. In particular, they associate TDS peaks at 900-1000 K with traps created by keV-ions in the stopping zone, i.e. with the near surface layer damaged by implanted ions. It was also shown that the high temperature peaks and/or shoulders relate to hydrogen release from traps in non-damaged layers behind the stopping zone. It can be seen (fig. 2 b and c) that both parts of TDS spectra of high-energy irradiated samples (HEIS) increase with the ion energy. Thus, one can say that the deuterium trapping in HEIS occurs both in and behind the stopping zone and depends on the ion kinetic energy. This type of trapping can be named as "kinetic trapping".

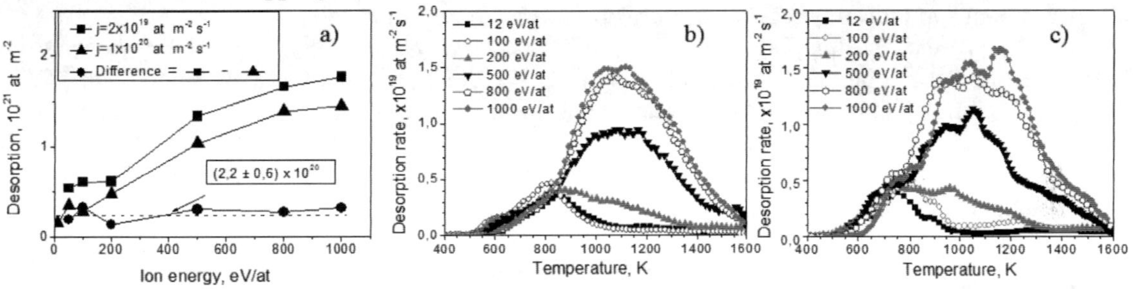

Figure 1. Total deuterium desorption from CFC (a), spectra of thermal desorption of PG (b) and CFC (c) irradiated with ion flux density 10^{20} at/m^{2}s.

At ion energies below 200 eV/D, the decrease of the deuterium retention becomes inessential. A noticeable desorption from the CFC samples is observed even from the samples irradiated under floating potential [8]. Low energy ions cannot create traps through knock out collision with carbon atoms. Conclusion is made that low energy ions are captured at active centers created due to inelastic interaction with the surface carbon atoms. The active centers can also initiate trapping of D atoms from the layer of surface sorption. This type of trapping can be called as "potential trapping".

The main TDS peak for the low-energy irradiated samples (LEIS) locates in the temperature range 700-850 K. The difference in the TDS peak positions for the LEIS and HEIS indicates that traps which are responsible for deuterium accumulation in the LEIS differ from those in the HEIS. Efficiency of low energy trap formation and filling seems to be at its maximum, when impinging ions stop in the near surface region, and decreases, as the ion penetration depth increases. Potential trapping constitutes presumable part of entire deuterium trapping at energies below 200 eV/D.

TDS peak and shoulders at the temperatures 500-600 K exist in the TDS spectra of both LEIS and HEIS. The same peak was found in TDS spectra of unirradiated CFC exposed to the air. Thus, one can assume that relevant traps are also formed by the mechanism of "potential trapping".

The TDS peaks and shoulders in the temperature ranges 1050–1150 K and 1200-1400 exist in TDS spectra of hydrogen retained in CFC not subjected to any plasma irradiation, i.e. they relevant to the traps in the not damaged CFC structure. Thus, we believe that some part of implanted D atoms diffuses and are trapped in the not damaged area behind the ion range. This mechanism is not active for the dense PG structure. As a result, a peak in the temperature range 1200-1400 exists in TDS spectra for the CFC and absent in similar spectra for the PG.

Irradiation of the CFC samples with D ions at low ion flux of 2×10^{19} D/m^{2}s to a fluence of 5×10^{23} D/m^2 leads to higher D retention as compared to the irradiation at high ion flux of 1×10^{20} D/m^{2}s to the same fluence. This fact is observed for ion energies in the range from 10 to 1000 eV/D. The time/ion flux dependence of D trapping (fig. 1) shows that the increase of irradiation time results in growth of all the parts of TDS spectra. Trapping of impinging fast ions by kinetic mechanism does not depend on duration of the ion irradiation, whereas the probability of the surface penetration and trapping is time dependent process. These finding shows that 1) potential trapping constitutes the main part of deuterium trapping at energies below 200 eV/D; 2) the fast ions provide potential trapping of D atoms from the layer of surface sorption layer, it seems to happen at expense of the energy of their in elastic collisions with the surface carbon; 3) the atoms penetrating surface by potential mechanism then can diffuse and be trapped in the traps created by fast ions in the ion stopping zone or in the not damaged area behind it.

Moreover, it is obviously, that number of deuterium traps created in the ion stopping zone of HEIS increases with increase of ion energy. But the amount of deuterium additionally trapped in longer experiments is practically the same for ions with different energies. Thus, one can believe, that dissociation of deuterium molecules and penetration of deuterium atoms into graphite is the limiting stage for particle diffusion and filling the traps in the stopping zone and in the bulk.

2.2. Dependence of trapping on the temperature of irradiated sample

It is seen (fig.2) that the trapping in LEIS steadily decreases with the elevation of the temperature during the sample irradiation. Contrary, the trapping in HEIS drops sharply between 500 and 650 K [10]. Fig. 2 c shows that in this temperature range the trapping is decreased in both low and high temperature traps, i.e. in the traps filled by both potential and kinetic mechanisms. This fact can also be explain by assumption that some fraction of particles penetrating the surface by potential mechanism is trapped in the kinetic traps.

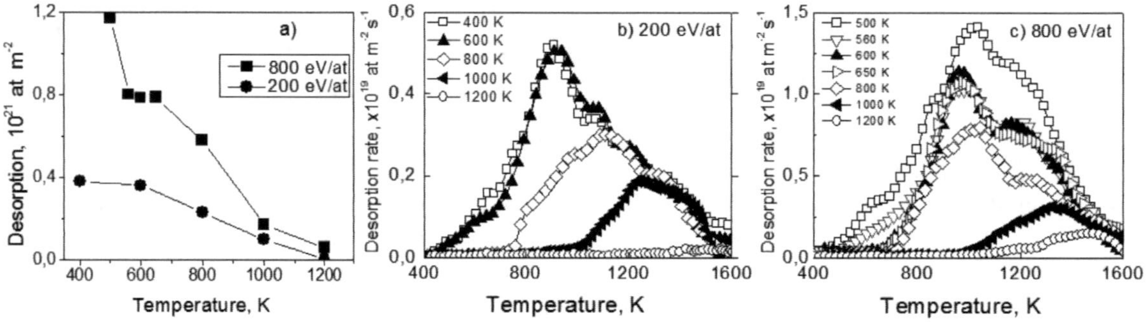

Figure 2. Temperature dependence of D_2 thermal desorption from CFC (a), spectra of thermal desorption of CFC irradiated by ions with energy 200 eV/at. (b), and 800 eV/at. (c).

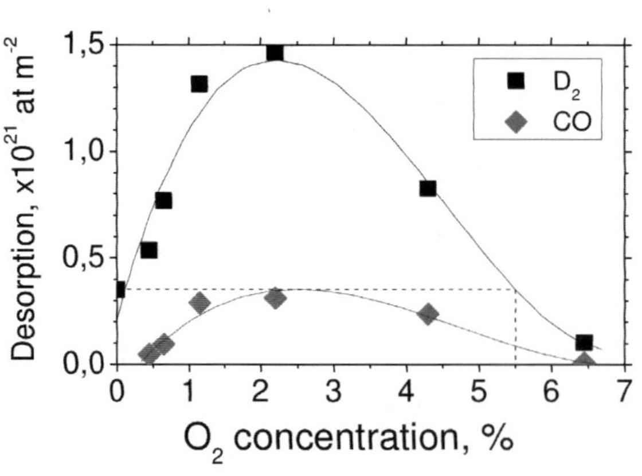

Figure 3. Deuterium and CO retention in CFC in dependence on oxygen concentration in plasma. (Ion energy is 50 eV/at., ion flux density is 10^{20} at/m^{2}s and fluence is 5×10^{23} at/m^2)

2.3. Influence of oxygen addition to a deuterium plasma on D trapping

It is seen (fig.3 a) that trapping of both oxygen and deuterium increases with the increase of oxygen concentration in plasma. They reach maximums when oxygen concentration elevates up to ≈2%, decreases with further growth of oxygen concentration, and matches the trapping in CFC irradiated in D_2 plasma, when oxygen concentration equals 5.5%. These results could be a consequence of two opposite influences on deuterium trapping. On one hand, oxygen interaction with surface activates formation of active centers and deuterium trapping. One other hand, oxygen sorbed on the surface more than deuterium, thus, hardening reaching the surface by deuterium. As the result, the increase of the oxygen concentration above certain limit leads to decrease of deuterium trapping.

Trapping of deuterium results in modification of graphite structure. It seems to be the reason of oxygen penetration and trapping and finally the reason of enhancement of oxygen retention. That is why change of deuterium trapping eventually leads to similar change of oxygen trapping.

3. Conclusion

The main results of the work can be summarized as follows. Deuterium trapping in- and desorption from CFC and PG samples irradiated with plasma ions was examined with thermal desorption spectrometry. It has been found that: 1) the deuterium trapping takes place even when energy of deuterium ions approaches zero; 2) for ion irradiation to equal fluences, the deuterium retention is higher at lower ion flux, i.e. the deuterium retention increases with the irradiation time.

Deuterium ions at energies above 200 eV/D are captured in the traps created due to their elastic collisions with carbon atoms. This process may be named as “kinetic trapping”.

The low energy ions are captured at active centers created due to inelastic interaction with the surface carbon atoms. The active centers created by both low- and high energy ions can also initiate trapping of D atoms from the layer of the surface sorption. The term “potential trapping” is proposed for this type of trapping. Potential trapping constitutes presumable part of entire deuterium trapping at energies below 200 eV/D. Under high energy irradiation the atoms penetrating surface through potential mechanism can fill the traps formed by fast ions in the ion stopping zone.

Time dependence of trapping under both low- and high energy irradiation is provided by trapping through potential mechanism.

Deuterium and oxygen trapping in the oxygen added plasma with the increase of oxygen concentration in plasma up to ≈2% and decrease with further growth of oxygen concentration.

References

[1] Y. Hirohata, T. Tanabe, Y. Oya et.al., JNM, 363–365 (2007),P.854-861
[2] Y. Corre, R Dejarnac, J Gardarein, et al., JNM, 463 (2015), P. 832-836,
[3] T. Eich, B. Sieglin, A. Scarabosio et al., JNM, 438 Supplement, (2013), P.S72-S77
[4] V. Philipps, R.Neu, J.Rapp, U.Samm, M. Tokar, T.Tanabe, M.Rubel, (2000) Plasma Physics and Controlled Fusion, 42 (12 SUPPL. B), pp. B293-B310.
[5] H. Atsumi, T.Tanabe, T. Shikama, (2015) Fusion Science and Technology, 67 (2), pp. 245-249.
[6] J. Küppers, The hydrogen surface chemistry of carbon as a plasma facing material, Surface Science Reports, Volume 22, Issues 7–8, 1995, P. 249-321
[7] L.L. Snead and M. Ferraris, 4.18 - Carbon as a Fusion Plasma-Facing Material, In Comprehensive Nuclear Materials, edited by Rudy J.M.Konings, Elsevier,2012, Pages 583-620.
[8] A. Airapetov, L. Begrambekov, C. Brosset, et al, JNM, 390–391 (2009), P.589-592
[9] A. Airapetov, L. Begrambekov, S.Vergazov, et al. Journal of Surface Investigation: X-Ray, Synchrotron and Neutron Techniques. 2010. T. 4. № 4. P. 567-571.
[10] A. Airapetov, L. Begrambekov, S.Vergazov, et al. (2010) Bulletin of the Russian Academy of Sciences: Physics, 74 (2), pp. 227-232.
[11] L. Begrambekov, A. Ayrapetov, V. Ermakov. Nuclear Instruments and Methods in Physics Research Section B: Beam Interactions with Materials and Atoms, 315 (2013), P. 110-116

Additive manufactured Ti6Al4V scaffolds with the RF-magnetron sputter deposited hydroxyapatite coating

E Chudinova[1], M Surmeneva[1], A Koptioug[2], P Scoglund[2], R Surmenev[1]

[1]Tomsk Polytechnic University, Institute of Physics and Technologies, 30, Lenina ave., Tomsk 634050, Russia
[2]Mid Sweden University, Akademigatan 1, Östersund SE-831 25, Sweden

Abstract. Present paper reports on the results of surface modification of the additively manufactured porous Ti6Al4V scaffolds. Radio frequency (RF) magnetron sputtering was used to modify the surface of the alloy via deposition of the biocompatible hydroxyapatite (HA) coating. The surface morphology, chemical and phase composition of the HA-coated alloy were studied. It was revealed that RF magnetron sputtering allows preparing a homogeneous HA coating onto the entire surface of scaffolds.

1. Introduction
Modern additive manufacturing (AM) is already using a wide variety of very different materials including metals. One of the key advantages of AM technologies is the freedom of three-dimensional shapes of manufactured components. One of the application areas where the advantages of additive manufacturing can be successfully used is biomedicine. In particular this technology allows for better operation planning and manufacturing functionalised implants specifically fit to the particular clinical case [1-3]. Metallic additively manufactured implants can also successfully replicate the complex microstructure of the substituted bones by integrating porous sections into monolithic implants improving the implant integration process and its long term stability in the body [3-5]. In many cases, when materials used for the metallic implants manufacturing are bioinert additional surface coatings can significantly improve the osseointegration [6-9]. But achieving homogenous coatings of the surfaces for the components with a complex three-dimensional structure, and especially the ones containing porous and 3D-lattice structures is quite challenging. Present
work is devoted to the investigation of the morphology and phase composition of the coatings based on hydroxyapatite (HA) deposited on Ti64 scaffolds by RF-magnetron sputtering.

2. Materials and methods
Test samples containing solid and porous sections (made as 3D-lattices with regular lattice cells) were manufactured from titanium alloy Ti6Al4V by the Electron Beam Melting technology in ARCAM A2 EBM® machine (Arcam AB, Mölndal, Sweden) [10, 11] using manufacturing parameter settings recommended by the machine manufacturer.
Hydroxyapatite (HA) with a specific stechiometric composition $Ca_{10}(PO_4)_6(OH)_2$ was selected as a material for coating.

A HA powder was prepared in a mechanochemical activation process and then used as a precursor to prepare a target for magnetron-based sputtering. Critical sputtering parameters such as radiofrequency power (400 W), argon gas pressure (working pressure 0.4 Pa, base pressure 10^{-4} Pa) and the distance between the target and substrate (40 mm) were kept constant in all experiments. The

HA coatings were deposited for 8 hours which resulted in the coating thickness of 700 nm. Sputtering equipment was described in detail earlier [12].

The phase composition and the structure of the coatings were identified by X-ray diffraction (XRD-7000, Shimadzu, Japan) with CuKα radiation (λ=0.154 nm) in the 2Θ range from 10° to 60° with a scan speed of 2.0°/min, sampling pitch of 0.03°, preset time of 5.0 sec at 30 kV and 30 mA. The average crystallite size was determined using the Scherrer's equation from the broadening of the diffraction peaks taking into account the instrumental broadening with the help of PowderCell 2.4 software. An instrumental broadening of 0.14 in 2θ was determined by the full width at half maximum of a calibration silicon powder. The database patterns #9-0432 and #44-1294 (ICDD database) were used for the pure HA and Ti respectively. The micro scale morphology of the samples was observed with SEM (CamScan MV2300 SEM, Great Britain) and field-emission SEM (FESEM, Quanta-200, FEI, USA) equipped with an Energy Dispersive X-ray (EDX) system.

3. Results and discussion

The RF-magnetron sputtering of the HA coating on the different rough surfaces offers the opportunity to create conformal coatings on the substrates that have specific surface features, without significantly changing the underlying microtopography [12, 13]. But the majority of the studies were performed with relatively planar sample surfaces. But it was unclear if the magnetron coating of the structures with multi-layer lattice structures backed by the solid parts suggested for the new generation implants [4, 5] would be at all effective. One of the serious concerns was due to the fact that electric fields in the conductive lattices drops dramatically when moving from outer to inner layers, which can completely prevent the coating of the deeper structure layers with the electrochemical and plasma-based methods.

Typical microscopy images of the surface of the lattice parts of the samples fabricated by EBM® before and after deposition of biocompatible HA coatings are shown in figure 1. The figure 1a shows 'the view from the top' of few cellular structure layers backed by the solid section of 'as manufactured' sample before coating. In the case of implants solid inner core of the structures should provide the mechanical integrity, and outer lattice layers should improve the biocompatibility promote the bone ingrowth for better implant fixation in the body. In present experiments low density regular lattice with diamond-shaped cells 5 × 5 mm and starts thickness of 0.35 mm were used.

As illustrated by the images in figure 1 resulting HA coating is quite conformal and well replicates the original surface features of the lattice with some smoothing of the smallest features in their microstructure. The same time the surface of the coating has its own microroughness, which can also be beneficial for the osseointegration.

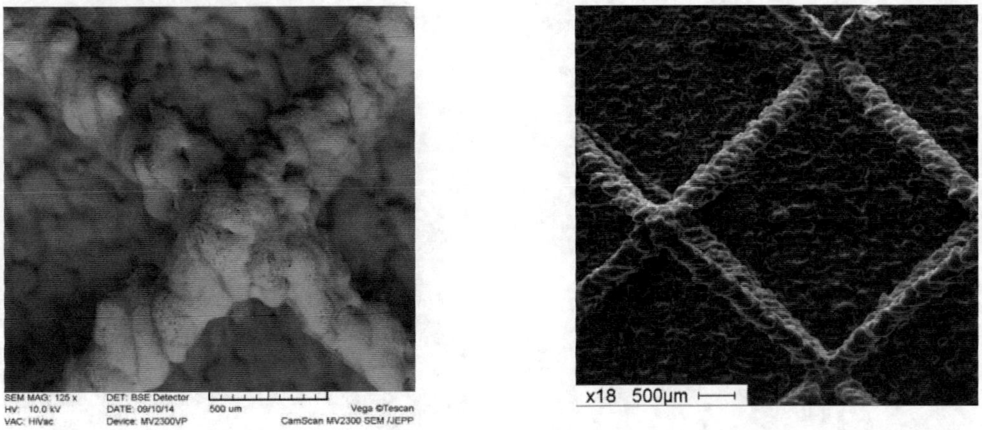

a b

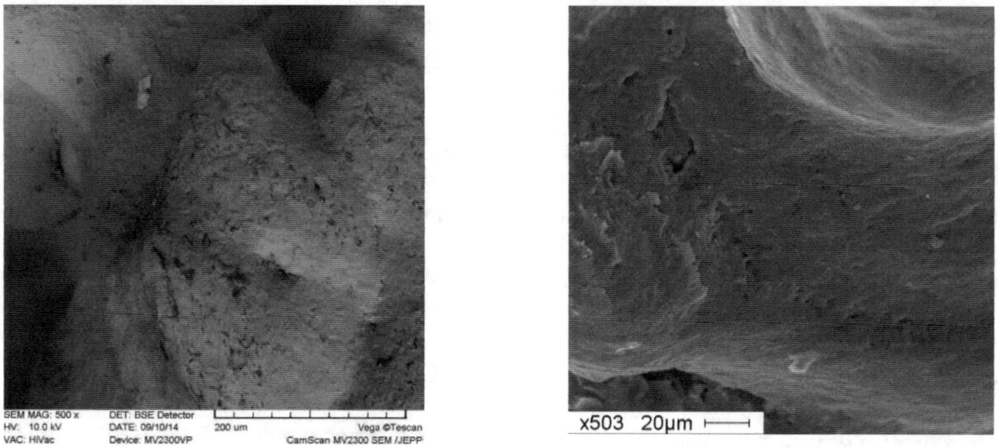

c d

Figure 1. SEM image of Ti64 mesh composite before (a, c) and after the coating (b, d)

Coated samples were investigated to identify the properties of the HA coating on the struts of the lattice scaffolds (figure 2). The boundary of the HA coating layer can be easily visible on the broken struts of the lattice structure. Cutting and polishing of such structures for the analysis of the coatings is quite challenging, as many of easily available methods either contaminate the samples or overheat them. Thus careful mechanical braking provides easy alternative for the express analysis of the lattice structure elements.

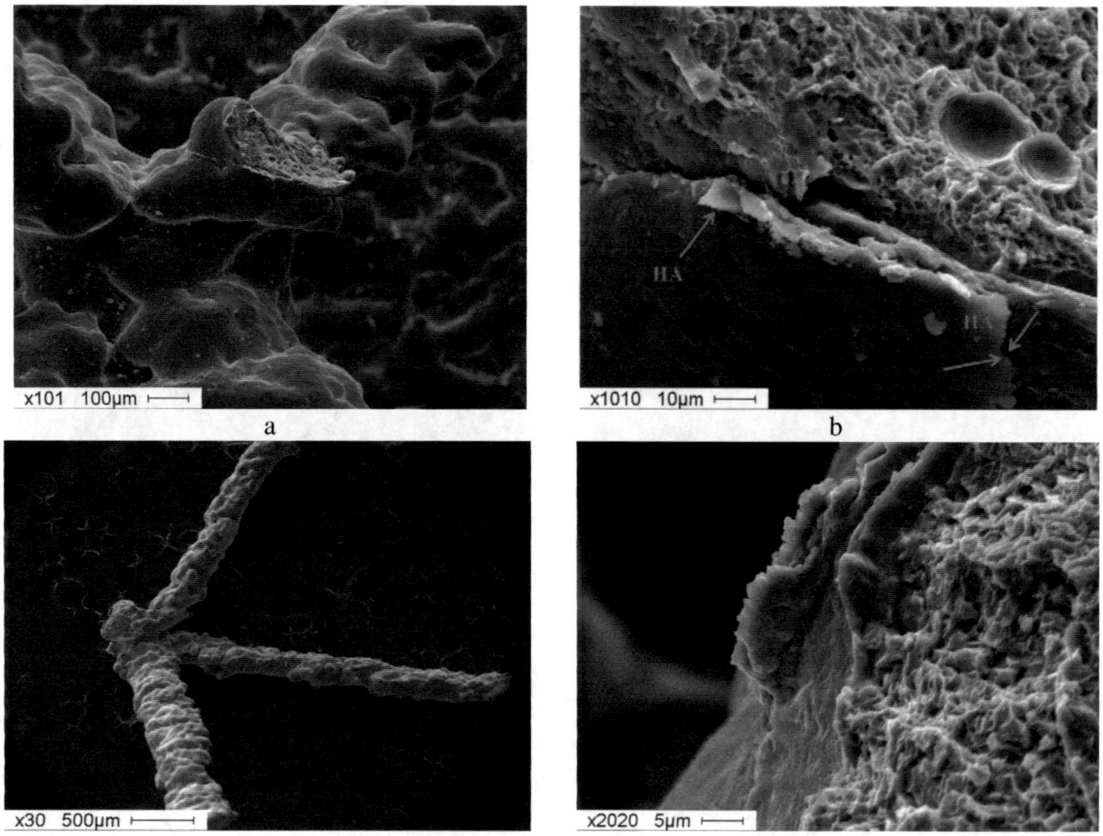

a b

c d

Figure 2. SEM image of the HA coated Ti6Al4V 3D-lattice structures prepared by additive manufacturing in EBM® technology. The HA coating is shown by arrows.

A typical EDX spectrum of the HA-coated scaffold is shown in figure 3. The study showed the presence of elemental composition on the surface of Ca, P, C and O. The high concentration of carbon on the coated surfaces appears after cutting and most probably results from the sample contamination.

Since the EDX-technique is a semi-quantitative method, it is not possible to reveal a true Ca/P ratio for the coating. However, the calculated value of the Ca/P ratio in the case of the coated substrates was 1.0± 0.1. According to the XPS results presented in our previous study, the Ca/P ratio for nanocrystalline HA coatings deposited via the RF magnetron sputtering technique with comparable process parameters was reported in the range of 1.65–1.86 [14, 15].

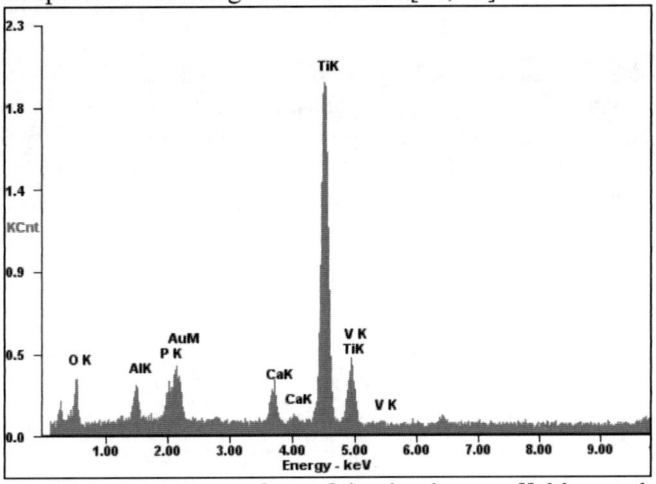

Figure 3. EDX spectrum on surface of the titanium scaffold coated with HA

The structure and phase composition are important characteristics having a decisive influence on the functional properties of the coatings.

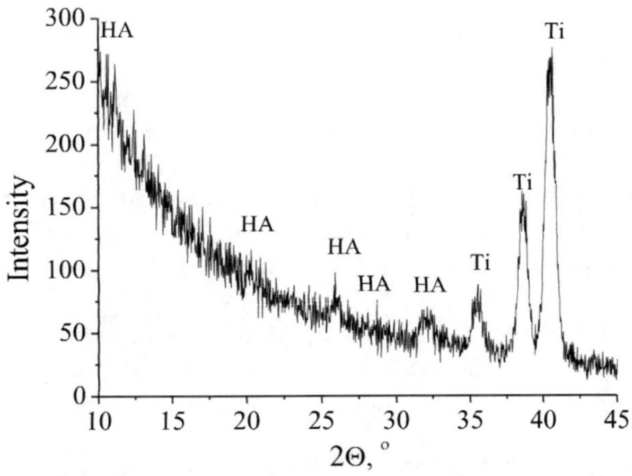

Figure 4. XRD pattern of the HA-coated titanium alloy prepared via additive manufacturing

The typical XRD-pattern of the HA coating fabricated via RF-magnetron sputtering on the surface of Ti64 alloy prepared via additive manufacturing is shown in figure. 4. The peaks at 25.9° (002), 31.8° (211), 32.2° (112), 32.9° (300) correspond to the diffraction pattern of the HA with hexagonal crystalline structure. The most intense peak in the HA spectrum is at 31.8°(211), but the second most intense one at 25.9° (002) seems to have higher relative intensity as compared to the database values. Such phenomenon was also observed previously [16]. The parameters of the lattice a = 9.4042 and c = 6.8875 Å correspond to the HA hexagonal arrangement (space group P63/m). The deposition mechanism of HA in magnetron-based method can be described as the formation of amorphous calcium phosphate clusters, their conversion into HA nanodomains and the crystallization of the grain domains with a preferential orientation along the HA [002] direction [17]. The magnetron sputtering plasma parameters have a great influence on the physicochemical and mechanical properties of the calcium phosphate films, and can be adjusted to produce single-phase HA films generated at a high deposition rate and high thermal stability [18].

4. Conclusions

Results of present studies indicate that additively manufactured metallic 3D-lattice structures can be effectively coated with hydroxyapatite using magnetron-based deposition methods. HA coating of the lattice elements is homogeneous and quite conformal but has certain smoothing effect hiding microstructure of the original metallic surfaces. The same time surface of the HA coatings has its own microroughness. Structural analysis of coatings indicates the hexagonal structure of HA, though the relative intensities of the corresponding peaks in the XRD spectra show some difference with the database values. Further studies should be carried out to identify the efficiency of magnetron-based method for the HA coating of dense 3D-lattice scaffolds, and the possibility to control the coating layer surface microroughness.

Acknowledgments

The authors would like to express their gratitude to Mr. T. M. Mukhametkaliyev for the help with the fabrication of thin HA films and to Mr. M. S. Syrtanov for the assistance with XRD measurements. This research was supported by the Russian Science Foundation grant Nr.15-13-00043.

References

[1] Ryan G, Pandit A and Apatsidis D P 2006 Fabrication methods of porous metals for use in orthopaedic applications *Biomaterials.* **27** 2651-2670.

[2] Dérand P, Rännar L E and Hirsch J M 2012 Imaging, virtual planning, design, and production of patient-specific implants and clinical validation in craniomaxillofacial surgery *Craniomaxillofacial trauma & reconstruction.* **5** (3) 137.

[3] Koptyug A, Rännar L, Bäckström M, Fager Franzén S and Dérand P 2013 Additive manufacturing technology applications targeting practical surgery *International Journal of Life Science and Medical Research.* **3** 15-24.

[4] Murr L E, Gaytan S M, Medina F, Lopez H, Martinez E , Machado B I, Hernandez D H, Martinez L, Lopez M I, Wicker R B, Bracke J 2010 Next-generation biomedical implants using additive manufacturing of complex cellular and functional mesh arrays *Philosophical Transactions of the Royal Society.* **368** 1999-2032.

[5] Murr L E, Gaytan S M, Medina F, Martinez E, Martinez J L, Hernandez D H, Machado B I, Ramirez D A, Wicker R B 2010 Characterization of Ti-6Al-4V open cellular foams fabricated by additive manufacturing using electron beam melting *Materials Science and Engineering*: A. **527** 1861-1868.

[6] Surmenev R A, Surmeneva M A, Ivanova A A 2014 Significance of calcium phosphate coatings for the enhancement of new bone osteogenesis–A review *Acta biomaterialia.* **10** (2) 557-579.

[7] Nakano T. et al. 2011 Formation of new bone with preferentially oriented biological apatite crystals using a novel cylindrical implant containing anisotropic open pores fabricated by the electron beam melting (EBM) method *ISIJ international.* **51** (2) 262-268.

[8] Koptyug A. et al.2014 Additive Manufacturing for Medical and Biomedical Applications: Advances and Challenges *Materials Science Forum.* **783** 1286-1291.

[9] Dorozhkin S V and Epple M 2002 Biological and Medical Significance of Calcium Phosphates *Angewandte Chemie International Edition* **41**(17) 3130-3146.
[10] Surmeneva M , Surmenev R, Nikonova Y, Selezneva I, Ivanova A, Putlyaev V, Prymak O, Epple M 2014 Fabrication, ultra-structure characterization and in vitro studies of RF magnetron sputter deposited nano-hydroxyapatite thin films for biomedical applications *Applied Surface Science.* **317** 172-180.
[11] Elias C. N. et al. 2008 Biomedical applications of titanium and its alloys *Jom.* **60** (3) 46-49.
[12] Surmenev R. et al. 2011 The influence of the deposition parameters on the properties of an rf-magnetron-deposited nanostructured calcium phosphate coating and a possible growth mechanism *Surface and coatings technology* **205** (12) 3600-3606.
[13] Surmeneva M A, Kleinhans C, Vacun G, Kluger P J, Schönhaar V, Müller M, Hein S B, Wittmar A, Ulbricht M, Prymak O, Oehr C, Surmenev R A 2015 Nano-hydroxyapatite-coated metal-ceramic composite of iron-tricalcium phosphate: Improving the surface wettability, adhesion and proliferation of mesenchymal stem cells in vitro *Colloids and Surfaces B: Biointerfaces.* **135** 386-393.
[14] Surmeneva M A, Mukhametkaliyev T M, Tyurin A I, Teresov A D, Koval N N, Pirozhkova T S, Shuvarin I A, Shuklinov A V, Zhigachev A O, Oehr C, Surmenev R A 2015 Effect of silicate doping on the structure and mechanical properties of thin nanostructured RF magnetron sputter-deposited hydroxyapatite films *Surface and Coatings Technology.* **275** 176–184.
[15] Ivanova A A, Surmeneva M A, Surmenev R A, Depla D 2015 Influence of deposition conditions on the composition, texture and microstructure of RF-magnetron sputter-deposited hydroxyapatite thin films *Thin Solid Films.***591** 368-374
[16] Surmeneva M, Surmenev R, Nikonova Y, Selezneva I, Ivanova A, Putlyaev V, Prymak O, Epple M 2014 Fabrication, ultra-structure characterization and in vitro studies of RF magnetron sputter deposited nano-hydroxyapatite thin films for biomedical applications *Applied Surface Science.* **317** 172-180.
[17] López E O, et al. 2015 Growth of crystalline hydroxyapatite thin films at room temperature by tuning the energy of the RF-magnetron sputtering plasma *ACS applied materials & interfaces.* **5**(19) 9435-9445.
[18] López E O, et al. 2015 Nanoscale analysis of calcium phosphate films obtained by RF magnetron sputtering during the initial stages of deposition *Surface and Coatings Technology.* **279** 16-24.

Characteristics of the electric arch and stream of plasma in the channel with porous cooling

G. Yu. Dautov[1], R. R. Khairetdinova[2], I. G. Dautov[2]
[1]Kazan National Research Technical University named after A. N. Tupolev – KAI, Kazan, Russia
[2]Kazan Federal University, Kazan, Russia

E-mail: Ramilya.19@mail.ru

Abstract. We study the characteristics of the arc plasma torch with a porous wall. The increase in mass flow of gas through the porous wall leads to an increase in thermal efficiency of the plasma torch. Compared it with the characteristics of the plasma torch with interbay gas supply.

Saving energy with plasma generation by improving thermal efficiency of the plasma torch continues to be an urgent task.
The formula for thermal efficiency:

$$\eta = 1 - \frac{q_1+q_2+q_3}{IU}, \quad (1)$$

where q_1 - heat loss through the cathode, q_2 - heat loss through the anode, q_3 - the loss of heat through the wall of the arc chamber, I-force power, U - the arc voltage.
In works [1-8] have demonstrated the possibility of reducing q_3 through the use of transpiration cooling of the wall of the arc chamber. Formula (1) shows that it improves the efficiency of the plasma torch. This work is devoted to study the efficiency of the plasma torch and characteristics of electric arc in the channel with a porous cooling. The diameter of the arc chamber is 1 cm, the wall of the arc chamber was made of copper with porosity of 30%. Part of the plasma gas-air was supplied into the arc chamber through a porous wall. The speed of plasma on the axis of the arc chamber was directed from the cathode to the anode, the pressure at the outlet of the cathode was equal to 10^5 Pa. The mass flow rate of the gas through the surface per unit of length of arc in one second camera is denoted m. Experiments have shown that long-term stable operation of the plasma torch at I=100-200 A is possible only if $m \geq 0.02$ kg s^{-1} m^{-1}.
As can be seen from Fig.1 (graphs 1 and 2), the efficiency of the plasma torch are investigated very slowly decreases with increasing current and is determined mainly by the intensity of inflow of gas through a porous wall. A comparison of charts 1 and 2 shows a significant increase in the efficiency of the plasmotron with increasing m. At the same figure presents data for the plasma torch assembly with gas flow [9] (curve 3). From the comparison graphs, it follows that the plasma torch with a porous cooling provides the greatest η value. A high value of η in a porous cooling is achieved by changing

the heat flow to the wall of the arc chamber and cooling stabilizing the walls of the injected gas. It should be emphasized that even at a very high intensity for the gas efficiency of the plasma torch with a cross-sectional flow remains less than the efficiency of the plasma torch with a porous cooling, since the presence of water-cooled sections does not allow to eliminate losses due to radiation. In porous cooling, the heat flow returns to the arc chamber.

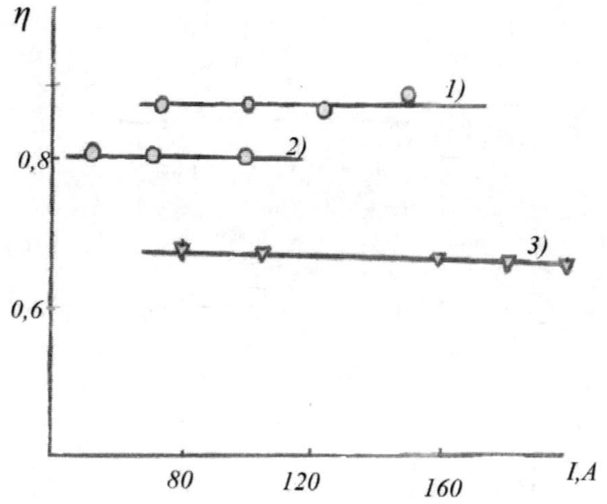

Figure 1. The dependence of the thermal efficiency of the plasma torch from a current when the length of the positive column of 7.2 cm, d = 1 cm. 1) m = 0.03 kg s^{-1} m^{-1}; 2) m = 0.02 kg s^{-1} m^{-1}; 3) [9]

Intensive cooling of the positive column of flowing through the surface of relatively cold gas leads to a narrowing of the profile and a certain growth temperature on the axis of the arc column. This is evident from comparing Fig.2 a) and b) which presents graphs of the temperature distribution of plasma along the radius of the positive arc column stabilized in a channel of the plasma torch, respectively, with gas and without it.

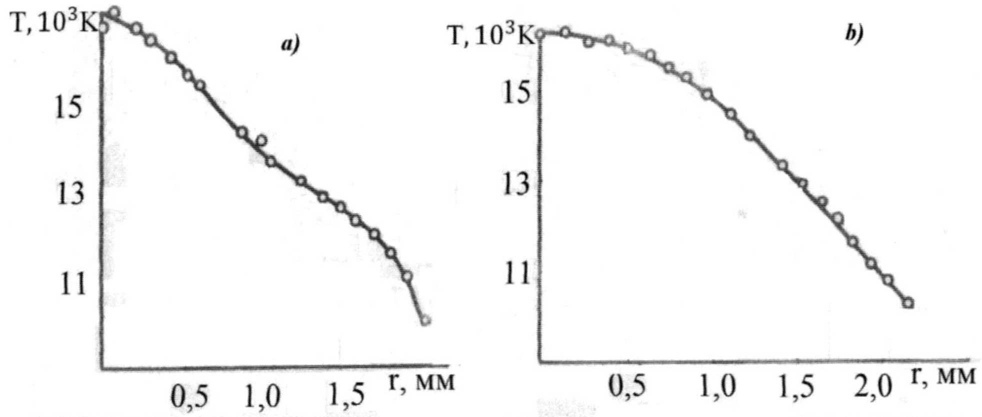

Figure 2. Distribution of plasma temperature at a distance x = 5.2 cm from the cathode spot at I = 100 A. a) the partitioned water-cooled plasma torch with the interelectrode insert, G(x)= 2 $Gauss^{-1}$, m = 0; b)porous cooling, G(x) = 2 $Gauss^{-1}$, m = 0.2 g cm^{-1} s^{-1}

The experiments showed that the mean mass enthalpy of plasma at the exit of the plasma torch h_c increases linearly with increasing current. With increasing air flow rate through the porous wall of the h_c is reduced, and the electric field strength increases.

References

[1] Yu. V. Kurochkin, A. V. Pustogarov, « Research of plasma torches with the flow of the working fluid through a porous interelectrode insert. Experimental research of plasma torches », Novosibirsk. Science. 1977 (p. 82-103)

[2] J. Anderson, E. Eckert, « Porous cooling stable electric arc heater. Rocket technology and cosmonautics», 1967, №4 (p.113-122)

[3] Yu. V. Kurochkin, A. V. Pustogarov, V. V. Ukolov « Electric arc discharge in a porous channel with high pressure gas. Thermophysics of high temperatures », 1978, 16, №1 (p.195-197)

[4] J. Kheberlayn, E. Pfander « Porous cooling of the walls of the chamber with a stable arc of high power. Heat transfer », 1972, №2 (p.17-25)

[5] Dautov G, Dautov I, Fayrushin I and Kashapov N 2013 *J. Phys.: Conf. Series* **479** 012001

[6] Dautov G, Kashapov N, Fayrushin I and Dautov I 2013 *J. Phys.: Conf. Series* **479** 012014

[7] Dautov G, Fayrushin I, Kashapov N and Dautov I 2014 *J. Phys.: Conf. Series* **567** 012008

[8] Fayrushin I, Kashapov N and Dautov I 2014 *J. Phys.: Conf. Series* **567** 012009

[9] G. M. Mustafin «The features of stable arc in a channel with distributed gas flow », 1968, № 4 (p.127-131)

Evaluation of the effectiveness of the use of arc plasma generator with distributed parameters for the deposition of functional coatings at low pressures.

N. F. Kashapov[1], G. Yu. Dautov[2], I. G. Dautov[1,2], G. R. Zakirova[1].
Kazan Federal University[1],
Kazan National Research Technical University named after A. N. Tupolev[2]

E-mail: gulnaraz93@mail.ru

Abstract. The estimation of the mean mass enthalpy at the outlet of the plasmatron with distributed parameters. Non-equilibrium plasma at low pressures allows to combine the processes of plasma treatment and sputtering. It is necessary to consider the influence of electrode erosion.

Arc plasma torch with a sectioned inter-electrode insert, wherein the anode is part of an expanding conical nozzle, ensures that at pressures of 130 to 1300 Pa, the temperature of the electron gas up to 50 000 K and allows you to change the size of the plasma plume in a vacuum chamber. During the conducted experiments, the dependences of electron temperature T_e, their concentration n_e and the velocity of the plasma jet along its axis. A description of the setup used for the measurements and the results are shown in the paper 1. Further studies were conducted to assess the average mass enthalpy h_c of the gas stream flowing from the nozzle anode, the heat flux distribution q along the axis of the plasma torch is uniform along the length of the channel the distribution of consumption by section and defining resource of continuous operation of the plasma torch.

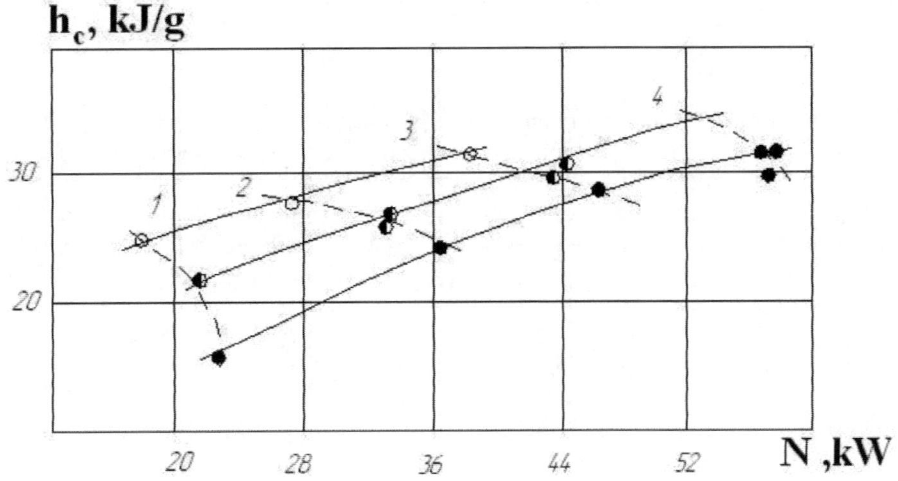

○- $G = 0{,}2 \cdot 10^{-3} kg \cdot s^{-1}$ ◐- $G = 0{,}6 \cdot 10^{-3} kg \cdot s^{-1}$

●- $G = 0{,}8 \cdot 10^{-3} kg \cdot s^{-1}$

Fig.1. The dependence of h_c from N.

In Fig. 1 shows the dependence of the enthalpy at the exit of the anode nozzle set power plasmatron N at constant gas flow rate G. Here the solid lines show the dependence of hc on the strength of the current I in the discharge channel at constant cost plasma-forming gas G, the dotted curves 1,2,3,4 represent the change in h_c when I=const(I_1=90 A, I_2=140 A, I_3= 180 A, I_4= 220 A).

As can be seen from the figure, the plasma torch provides a high value of the enthalpy at the exit of the anode nozzle, therefore, the powdered cathode material can be delivered into the zone of the substrate in the molten state. In combination with high flow rate that ensures "spletovoe" formation of functional coatings on the substrate [2-6].

To study the effect of electrode erosion on the life of the plasma torch used, the cathode and the anode, made of titanium alloy VT1. Were they weighing after 1, 3 and 16 hours. The plasma gases (Ar and air) was supplied through the screw grooves of the cathode node and the tangential opening between the sections of the interelectrode inserts. At a given power optimal length the interelectrode inserts leads to a decrease I and decrease the rate of erosion of the electrodes. The movement of the anode spot prevents the destruction of the surface of the anode at high values of heat flux q. The distribution of q along the channel of the plasma torch shown in Fig.2.

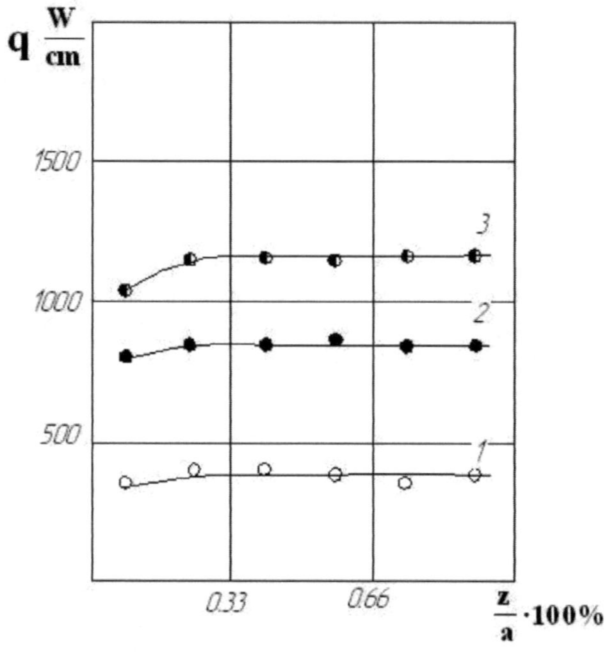

Fig.2. The distribution of q along the channel of a plasmatron with a distributed flow rate $G=0.8\cdot10^{-3}$ $kg\cdot s^{-1}$;$G_o=0,2\cdot10^{-3}$ $kg\cdot s^{-1}$; $a=13\cdot10^{-2}$ m. The curves 1, 2 and 3 correspond to currents of 90, 140, 180 A.

The total gas flow was $0.8\cdot10^{-3}$ $kg\cdot s^{-1}$, using the cathode unit was supplied $G_0=0,2\cdot10^{-3}$ $kg\cdot s^{-1}$, the channel length $a=13\cdot10^{-2}$ m. Small energy losses on the walls of the plasma torch and the cathode with a relatively low consumption of plasma gases provide an efficient heating of the spray material of the cathode.

In Fig.3 shows the dependence of the rate of erosion of the cathode Ψ from the arc current. A significant part of the erosion products deposited on the anode.

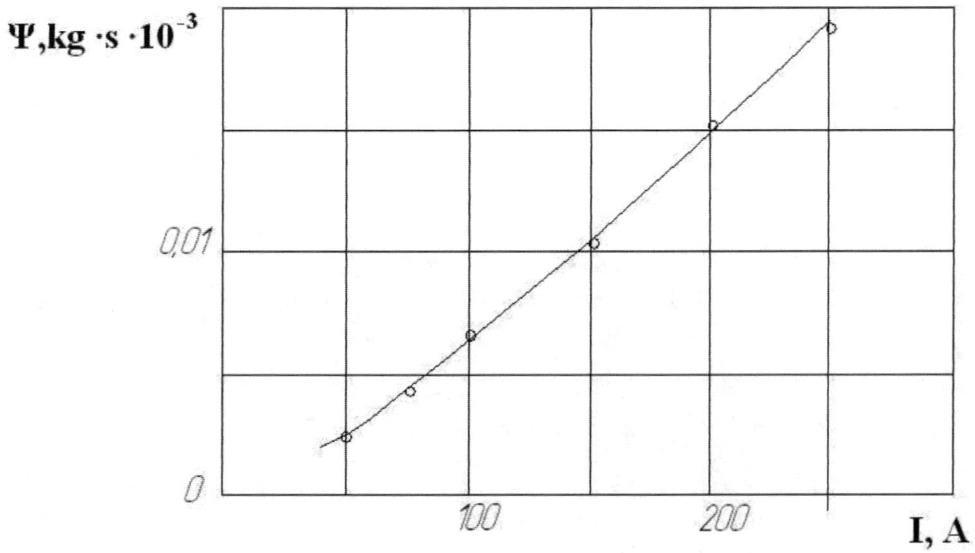

Fig.3. The dependence of the rate of erosion of the cathode arc current.

When operating with the air resource of the plasma torch in a mode of stable values of parameters of the plasma jet without replacement of the electrodes is 6 to 8 hours, with Ar up to 60 hours.

Conclusions

Plasmatron with distributed parameters with a suitable choice of electrodes and the mode of operation is an effective means of plasma spraying coatings at low pressures. Further research is needed obtained with it coatings, as their functional properties and their dependence on parameters of the plasmatron. You want to measure the maximum coating thickness, the speed of their growth, adhesion and porosity.

References

[1] Dautov, I. G., I. I. Fairushin, Tonight O. A., Mustafina, A. N. Characteristics of the plasmatron with distributed parameters to obtain the arc plasma at low pressures. All-Russian conference FNTP-2014, vol. 1, pp. 75-78 (2014).
[2] Generation of low temperature plasma and plasma technologies. Problems and prospects..P.Solonenko, A. V. Smirnov, A. A. Mikhalchenko, E. V. Thermophysics impact melt droplet onto the base by plasma spraying. Novosibirsk, Nauka,2004, pp. 253-277
[3] Dautov G, Dautov I, Fayrushin I and Kashapov N 2013 *J. Phys.: Conf. Series* **479** 012001
[4] Dautov G, Kashapov N, Fayrushin I and Dautov I 2013 *J. Phys.: Conf. Series* **479** 012014
[5] Dautov G, Fayrushin I, Kashapov N and Dautov I 2014 *J. Phys.: Conf. Series* **567** 012008
[6] Fayrushin I, Kashapov N and Dautov I 2014 *J. Phys.: Conf. Series* **567** 012009

The influence of ripple current on characteristics of electric arc, stable in a cylindrical plasmatron channel.

G Y Dautov[1], N F Kashapov[2], G R Zakirova[2], I G Dautov[1,2].
Kazan National Research Technical University named after A. N. Tupolev[1]
Kazan Federal University[2]

E-mail: gulnaraz93@mail.ru

Abstract. The paper presents the results of experimental studies of nonstationary electric arc, stable partitioned in a cylindrical plasmatron channel. The experimental data obtained in the range of electric current I=50-200A, I_m=1-7 A, when the gas flow rate $G=2\cdot10^{-4}$ $kg\cdot s^{-1}$, the frequency of alternating current f=300-20000Hz.

Research shows that the arc in the plasmatron channel DC has nonstationary properties, the degree of manifestation of which depends on the ripple voltage of the power source, and processes, due to the shunting arc [1-4]. Therefore, for a more complete understanding of physical processes of interaction of the arc with the gas flow, calculation of energy characteristics of the plasma torch and the stability of the system arc the power source must take into account the nonstationarity of the discharge.
The paper presents the results of experimental studies of nonstationary electric arc, stable partitioned in a cylindrical plasmatron channel.

Transient behavior was carried out by imposing on the arc direct current alternating voltage of a predetermined amplitude and frequency source connected in parallel to the main power source. The direct and alternating currents were separated by capacitive and inductive resistances. The experiments were carried out in the range of electric current I=50-200A, I_m=1 to 7 A, the gas flow rate $G=2\cdot10^{-4}$ $kg\cdot s^{-1}$, alternating current frequency f=300-20000 Hz. The radius of the arc chamber of the plasma torch was 0.5 cm, the interelectrode distance l=14cm. Here I_m is the root-mean-square of the alternating current. As the plasma gas used was air.

In Fig.1 shows the distribution of heat flow through a unit length of the channel along q_o arc chamber in a stationary mode, the x axis is directed along the axis of the arc chamber from the cathode toward the anode, the cathode spot is in the cross-section x=0.

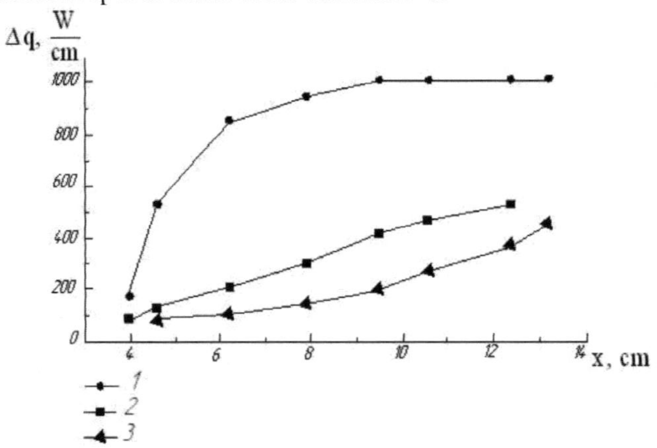

Fig.1. The heat flux distribution along the channel at I=60A. Graphs 1,2 and 3 correspond to the expenditures of air $G=6\cdot10^{-4}$ $kg\cdot s^{-1}$, $G=4\cdot10^{-4}$ $kg\cdot s^{-1}$ and $G=2\cdot10^{-4}kg\cdot s^{-1}$.

In the direction of the bulk temperature of the plasma and its temperature at the channel wall increases and therefore q_o along the x-axis increases. With decreasing gas flow temperature increases along the axis x increases and accordingly, increases in q_o.
Present heat flow in modulated mode in the form

$$q_i = q_0 + \Delta q \tag{1}$$

Figure 2 shows the graphs Δq from H.

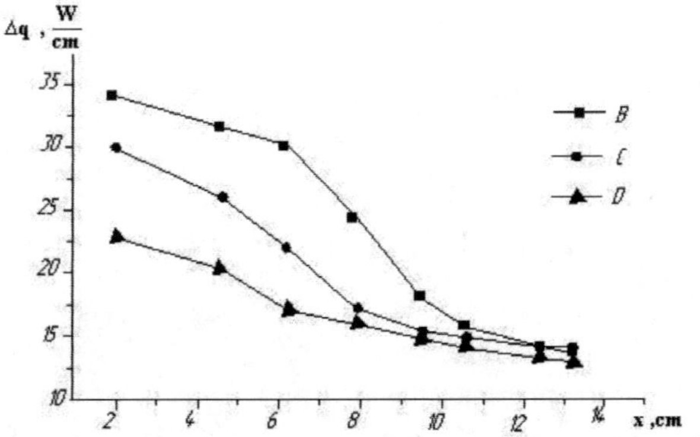

Fig.2. The distribution of heat flow Δq along the canal at I_o=60A, I_m=6A, f=103Hz. Graphs B, C, D correspond to airflow rates of $2{\cdot}10^{-4}$kg·s^{-1}, $4{\cdot}10^{-4}$ kg·s^{-1} and $6{\cdot}10^{-4}$ kg·s^{-1}.

With the growth of fluctuations of the electrical parameters of the arc voltage of the arc increases, thus increase and heat flow. As can be seen from Fig.2, the greatest increase in the heat flux Δq is observed in the initial and transition sections of the positive column. To limit the area effect of small perturbations on the distribution of heat flow is very weak. One of the possible reasons for this as follows. In the initial section of the channel the radius of the arc column is small and, therefore, even a small current fluctuations lead to noticeable fluctuations in the radius of the arc and the emergence of enforced radial convection. Since the heat flux q_o on the initial site is small, the heat transfer enhancement due to the modulation current plays a prominent role. With distance from the initial portion of the radius of the pillar increases, the relative change of the radius of the post when the modulation current decreases correspondingly and the contribution of forced convection to the heat flux on the channel wall.

To explain this experimental fact must also take into account the nature of the change in bulk temperature of the flow along the channel. As is well known the bulk temperature of the gas increases in the direction of the gas flow. At a relatively low temperature at the initial section of the channel modulation of the current, apparently, leads to significant temperature fluctuations, consequently, the electrical conductivity of plasma. While there may be significant fluctuations in the field strength, which leads to a higher power Joule dissipation. In combination with the above-noted fact of heat transfer enhancement increased Joule dissipation can also be the reason for the increase Δq. At high temperatures, which occurs at the maximum phase channel, the relative fluctuations of temperature, therefore, conductivity and field strength will be small. Therefore, the increase of the Joule dissipation capacity here is negligible. Minor and the increment Δq.

Experiments have shown that changing the frequency of the disturbing current from 300 to 20000 Hz virtually no effect on heat loss through the wall of the arc chamber. In the investigated range of parameters the influence of small perturbations on the heat flows in the electrodes was found.

Thus, the obtained results are explained by the fact that fluctuations of the electric parameters of the arc contribute to an additional turbulence of the flow and change of conditions of heat and mass transfer between the arc and the plasma gas. Most clearly these effects are observed at low flow rates of gas on the initial section of the channel, i.e. when the flow without perturbation of the current is laminar. With increasing flow rate and with distance from the cathode the effect of electrical disturbances on the arc characteristics is weakened, since it increases the role of disturbances with a gas-dynamic nature.

In the electrical submersible diaphragm pump working gas, interacting with the electric arc, it is heated to high temperatures and expire through the nozzle as a plasma jet. The parameters of the plasma jet depends on the choice of power scheme and design features of the electrical submersible diaphragm pump, sort of working gas and its flow, input power, etc. the studies showed that the nature of the flow of plasma and the size of the plasma jet is significantly influenced by voltage fluctuations

Journal of Physics: Conference Series **669** (2016) 012007 doi:10.1088/1742-6596/669/1/012007

of the power source. For example, at low flow rates of gas smoothing the ripple of the power source has led to a significant reduction in the intensity of background noise and increasing the length of the plasma jet. This is because the smoothing voltage of the power source modifies the pulsation characteristics of the flow and reduces the amplitude of acoustic oscillations of the plasma. When the constancy of all other parameters of the electrical submersible diaphragm pump there is maximum gas flow rate above which the perturbation current is almost no influence on the length of the jet. For example, in our case $G=6\cdot10^{-4}kg\cdot s^{-1}$. With further increase in flow rate disturbed the spatial stabilization of the plasma jet and its length decreases sharply. As can be seen from Fig.2, this value of gas flow corresponds to the transition from laminar to turbulent mode of combustion of the arc. Thus, the state of the plasma jet is directly related to the processes within the arc chamber.

References

[1] Dautov G. Yu., Dzyuba V. L., Karp I. N. The plasma torch with stabilized electric arcs. – Kiev: Naukova Dumka, 1984., 168p.

[2] Zhukov M. F., Smolyakov V. J., B. A. Afukov electric-Arc gas heaters (plasma torches). – M,; Science, 1973. – 232p.

[3] O. I. Yasko, Electric arc in the plasmatron. – Minsk: Science and technology, 1977. – 151p.

[4] M. F. Zhukov, A. S. An'shakov, I. M. Zasypkin, etc. Electric arc generators with inter-electrode inserts. – Novosibirsk: Nauka, 1981. – 221p.

[5] Dautov G, Dautov I, Fayrushin I and Kashapov N 2013 *J. Phys.: Conf. Series* **479** 012001

[6] Dautov G, Kashapov N, Fayrushin I and Dautov I 2013 *J. Phys.: Conf. Series* **479** 012014

[7] Dautov G, Fayrushin I, Kashapov N and Dautov I 2014 *J. Phys.: Conf. Series* **567** 012008

[8] Fayrushin I, Kashapov N and Dautov I 2014 *J. Phys.: Conf. Series* **567** 012009

Journal of Physics: Conference Series **669** (2016) 012008 doi:10.1088/1742-6596/669/1/012008

Equations of spatial hydrodynamic interaction of weakly nonspherical gas bubbles in liquid in an acoustic field

A I Davletshin and T F Khalitova

Institute of Mechanics and Engineering, Kazan Science Center, Russian Academy of Sciences, 2/31 Lobachevsky str., Kazan 420111, Russia

E-mail: anas.davletshin@gmail.com

Abstract. A mathematical model of spatial hydrodynamic interaction of gas bubbles in liquid in an acoustic field taking into account small deformations of their surfaces is proposed. It is a system of ordinary differential equations of the second order in radii of the bubbles, the position vectors of their centers and the amplitudes of deviation of their shape from the spherical one in the form of spherical harmonics. The equations derived are of the first order of accuracy in A / R and of the fourth order in R / D, where R is the characteristic radius of the bubbles, A is the amplitude of characteristic deviation of their surface from the spherical one in the form of spherical harmonics, D is the characteristic distance between bubbles. The derivation of the equations is carried out by the method of spherical functions with the use of the Bernoulli integral, the kinematic and dynamic boundary conditions on the surface of the bubbles. The effects of viscosity and compressibility of the liquid are considered approximately, the gas in the bubbles is assumed homobaric.

1. Introduction

Studying of hydrodynamic interaction of bubbles in liquid is of interest to various applications in chemistry, power, medicine, etc. In literature at theoretical studying of interaction of bubbles analytic-numerical methods are widely applied, in which the governing relations are a system of ordinary differential equations of the second order [1, 2]. And if the spatial interaction is considered, due to complexity of the task, the bubbles are, as a rule, assumed spherical. However as a result of interaction the surfaces of the bubbles can be deformed. At deformation of any bubbles in a group the property of the group can significantly change. In such cases deformation of the bubbles should be considered, otherwise the obtained predictions of their behavior can be incorrect. In the present paper the equations of spatial hydrodynamic interaction of gas bubbles in liquid in an acoustic field taking into account their small deformations are derived.

2. Problem definition

It is supposed that the bubbles are in the antinode of intensive standing wave with the pressure changing with respect to the harmonious law

$$p_\infty = p_0 - \Delta p \sin \omega t ,$$

where t is time, Δp, ω are the amplitude and frequency of oscillations, p_0 is the static pressure of the liquid.

The gas in the bubbles is assumed ideal homobaric with the pressure changing according to the adiabatic law

$$p_k = \left(p_0 + \frac{2\sigma}{R_{0k}}\right)\left(\frac{R_{0k}}{R_k}\right)^{3\kappa},$$

where σ is the coefficient of the surface tension, κ is the isentropic exponent, R_k, R_{0k} are the current and initial radii of the k-th bubble, $k = 1, 2..., K$, K being the number of the bubbles.

The effects of viscosity and compressibility of liquid are assumed small. Therefore they are taken into account by means of corrections to the equations of interaction of the bubbles in the assumption that the liquid is ideal incompressible. The dynamics of the liquid in terms of the velocity potential Φ is described by the equations

$$\nabla^2\Phi = 0, \quad \frac{\partial\Phi}{\partial t} - \mathbf{w}_k \cdot \nabla\Phi + \frac{1}{2}\nabla\Phi\cdot\nabla\Phi + \frac{p - p_\infty}{\rho_0} = 0, \tag{1a, b}$$

where ρ_0 is the density of the liquid, p is the pressure, $\mathbf{w}_k = \dot{x}_k\,\mathbf{i} + \dot{y}_k\,\mathbf{j} + \dot{z}_k\,\mathbf{k}$, x_k, y_k, z_k are the coordinates of the center of the k-th bubble, **i**, **j**, **k** are the unit vectors of the Cartesian coordinate system, the point from above means differentiation with respect to time. The effects of viscosity and compressibility of liquid are assumed small and considered by means of corrections. On the surface of the bubbles the following kinematic and dynamic contact conditions are posed

$$\frac{\partial F_k}{\partial t} + \nabla\Phi\cdot\nabla F_k = 0, \quad p = p_k - \frac{2\sigma}{R_k}. \tag{2a, b}$$

Here $F_k(r_k, \theta_k, \varphi_k, t) = 0$ is the equation of the surface of the k-th bubble in its local spherical coordinate system r_k, θ_k, φ_k with the origin in its center.

3. Equations of interaction of bubbles

The liquid velocity potential Φ is presented in the form of a series

$$\Phi = \sum_{k=1}^{K}\sum_{\gamma=0}^{\infty}\sum_{\varsigma=-\gamma}^{\gamma} B_{\gamma k}^{\varsigma}(t)\, Y_\gamma^\varsigma(\theta_k, \varphi_k)\, r_k^{-\gamma-1}, \tag{3}$$

and the equation of the surface of the k-th bubble is written as

$$F(r_k, \theta_k, \varphi_k, t) = r_k - R_k(t) - \sum_{n=2}^{N}\sum_{m=-n}^{n} a_{nk}^m(t)\, Y_n^m(\theta_k, \varphi_k) = 0, \tag{4}$$

where $Y_\gamma^\varsigma(\theta_k, \varphi_k) = P_\gamma^{|\varsigma|}(\cos\theta_k)e^{i\varsigma\varphi_k}$ are the spherical functions, $P_\gamma^{|\varsigma|}$ are the associated Legendre polynomials of degree γ and order $|\varsigma|$, i is the imaginary unit, $e^{i\varsigma\varphi_k} = \cos\varsigma\varphi_k + i\sin\varsigma\varphi_k$, N is the number of the harmonics used in representation of the surface of the bubbles, a_{nk}^m is the amplitude of deviation of the surface of the k-th bubble from the spherical one ($r_k = R_k$) in the form of the surface harmonic Y_n^m. In the present paper the distortions of the spherical shape of the bubbles, characterized by the parameter $\varepsilon = \max\limits_{n,m,k}\left|\varepsilon_{nk}^m\right|$ ($\varepsilon_{nk}^m = a_{nk}^m / R_k$), are assumed small (powers ε^2 and above can be neglected in comparison with 1). Moreover, it is assumed that the bubbles are not very close to each other so that the quantity δ^5 can also be neglected in comparison with 1. Here $\delta = \max\limits_{k,j}\left[\left(R_k + R_j\right)/d_{kj}\right]$, $k, j = 1, 2..., K$, $k \neq j$, d_{kj} is the distance between the centers of the k-th and j-th bubbles.

The coefficients $B_{\gamma k}^{\varsigma}$ of the potential are found by substituting (3) and (4) in kinematic boundary conditions (2*a*) with allowing for the orthogonality of the spherical functions. Further, substitution in Bernoulli integral (1*b*) (with taking into account dynamic boundary conditions (2*b*)) of potential expressions (3) with the found coefficients $B_{\gamma k}^{\varsigma}$ and a number of rather bulky transformations with allowing for the orthogonality of the spherical functions, give the following system of equations in radii of the bubbles R_k, the position vectors of their centers $\mathbf{p}_k$ and the amplitudes of deviation of their shape from the spherical one a_{nk}^m

$$R_k\ddot{R}_k+\frac{3\dot{R}_k^2}{2}-\frac{\dot{\mathbf{p}}_k^2}{4}-\frac{p_k-p_\infty}{\rho_0}+\psi_{0k}+\Delta_k=\sum_{j=1,j\neq k}^{K}\left[\frac{\dot{B}_{0j}}{d_{kj}}-\frac{R_j^2\,\mathbf{p}_{kj}\cdot\left(R_j\ddot{\mathbf{p}}_j+\dot{R}_j\dot{\mathbf{p}}_k+5\dot{R}_j\dot{\mathbf{p}}_j\right)}{2d_{kj}^3}+\right.$$

$$+\frac{R_j^3\dot{\mathbf{p}}_j\cdot\left(\dot{\mathbf{p}}_k+2\dot{\mathbf{p}}_j\right)}{4d_{kj}^3}-\frac{3R_j^3\mathbf{p}_{kj}\cdot\dot{\mathbf{p}}_j\mathbf{p}_{kj}\cdot\left(\dot{\mathbf{p}}_k+2\dot{\mathbf{p}}_j\right)}{4d_{kj}^5}+\sum_{l=1,l\neq k}^{K}\frac{3B_{0j}B_{0l}\mathbf{p}_{kj}\cdot\mathbf{p}_{kl}}{4d_{kj}^3d_{kl}^3}-$$

$$\left.-\sum_{l=1,l\neq j}^{K}\frac{\left(R_j^3B_{0l}\right)'\mathbf{p}_{jl}\cdot\mathbf{p}_{kj}}{2d_{kj}^3d_{jl}^3}-\frac{3C_{10}^{\gamma 0}\Theta_{211}^{s\varsigma-\gamma,2}\left(R_j^2a_{2j}^sY_{1kj}^{\gamma}\boldsymbol{\delta}_\varsigma\cdot\dot{\mathbf{p}}_j^*\right)'}{4d_{kj}^2}\right],$$

$$\frac{R_k\ddot{\mathbf{p}}_k}{3}+\dot{R}_k\dot{\mathbf{p}}_k+\left(\frac{2\Theta_{211}^{s,\varsigma,-s-\varsigma,0}}{3}\dot{R}_k\varepsilon_{2k}^s-\frac{3\alpha_{211}^{s,\varsigma,-s-\varsigma}}{2}\dot{a}_{2k}^s\right)\boldsymbol{\delta}_\varsigma\cdot\dot{\mathbf{p}}_k^*-\frac{7\alpha_{211}^{s,\varsigma,-s-\varsigma}}{6}R_k\boldsymbol{\delta}_\varsigma\cdot\ddot{\mathbf{p}}_k^*\varepsilon_{2k}^s+$$

$$+\left(\Theta_{n12}^{m,\varsigma,-s,2}\Theta_{211}^{s,\varsigma,-s-\varsigma,0}-3\Theta_{n111}^{m,\varsigma,\gamma,-m-\varsigma-\gamma,0}\right)\frac{\boldsymbol{\delta}_\varsigma\cdot\dot{\mathbf{p}}_k^*\boldsymbol{\delta}_\gamma\cdot\dot{\mathbf{p}}_k^*\varepsilon_{nk}^m}{2}+\boldsymbol{\psi}_{1k}=\sum_{j=1,j\neq k}^{K}\left\{-\frac{\left(R_kB_{0j}\right)'\mathbf{p}_{kj}}{d_{kj}^3}-\right.$$

$$-\frac{\left(R_kR_j^3\dot{\mathbf{p}}_j\right)'-2R_kB_{0j}\dot{\mathbf{p}}_j}{2d_{kj}^3}-\frac{3R_j^3R_k\left(\dot{\mathbf{p}}_j\cdot\dot{\mathbf{p}}_j\mathbf{p}_{kj}+2\mathbf{p}_{kj}\cdot\dot{\mathbf{p}}_j\dot{\mathbf{p}}_j\right)}{2d_{kj}^5}+\frac{3\mathbf{p}_{kj}\cdot\left[\left(R_kR_j^3\dot{\mathbf{p}}_j\right)'-2R_kB_{0j}\dot{\mathbf{p}}_j\right]\mathbf{p}_{kj}}{2d_{kj}^5}+$$

$$+\frac{15R_j^3R_k\left(\mathbf{p}_{kj}\cdot\dot{\mathbf{p}}_j\right)^2\mathbf{p}_{kj}}{2d_{kj}^7}+C_{01}^{0\varsigma}\left[\frac{\Theta_{211}^{s,\varsigma,-s-\varsigma,2}\left(B_{0j}Y_{1kj}^{-\varsigma}a_{2k}^s\right)'}{2d_{kj}^2}+\frac{\Theta_{211}^{s,\varsigma,-s-\varsigma,0}B_{0j}Y_{1kj}^{-\varsigma}\left(\dot{a}_{2k}^s+2\dot{R}_k\varepsilon_{2k}^s\right)}{3d_{kj}^2}\right]+$$

$$\left.+C_{01}^{0\gamma}\left[\left(\Theta_{n12}^{m,\varsigma,-s,2}\Theta_{211}^{s,\gamma,-s-\gamma,0}+\Theta_{n12}^{m,\gamma,-s,2}\Theta_{211}^{s,\varsigma,-s-\varsigma,0}\right)-6\Theta_{n111}^{m,\varsigma,\gamma,-m-\varsigma-\gamma,0}\right]\frac{B_{0j}Y_{1kj}^{-\gamma}\boldsymbol{\delta}_\varsigma\cdot\dot{\mathbf{p}}_k^*\varepsilon_{nk}^m}{2d_{kj}^2}\right\},$$

$$\frac{R_k\ddot{a}_{ck}^{-\sigma}}{c+1}+\frac{3\dot{R}_k\dot{a}_{c,k}^{-\sigma}}{c+1}-\frac{(c-1)\ddot{R}_ka_{c,k}^{-\sigma}}{c+1}-\frac{3\Upsilon_{n1c}^{m,\varsigma,\sigma,2(c-2)/3}}{2(1+c)}R_k\boldsymbol{\delta}_\varsigma\cdot\ddot{\mathbf{p}}_k^*\varepsilon_{nk}^m+\frac{3\Theta_{n1c}^{m,\varsigma,\sigma,2}}{2(1+c)}\boldsymbol{\delta}_\varsigma\cdot\dot{\mathbf{p}}_k^*\dot{a}_{nk}^m+$$

$$+\boldsymbol{\delta}_\varsigma\cdot\dot{\mathbf{p}}_k^*\left(\frac{9\alpha_{11c}^{\varsigma,\gamma,\sigma}}{8}\boldsymbol{\delta}_\varsigma\cdot\dot{\mathbf{p}}_k^*+\frac{9}{4}\left(\frac{\Theta_{n1\zeta}^{m,\varsigma,-\xi,2}\Theta_{\zeta 1c}^{\xi,\varsigma,\sigma,0}}{\zeta+1}-\Theta_{n11c}^{m,\varsigma,\gamma,\sigma,0}\right)\boldsymbol{\delta}_\gamma\cdot\dot{\mathbf{p}}_k^*\varepsilon_{nk}^m+\frac{3\Theta_{n1c}^{m,\varsigma,\sigma,0}}{2(n+1)}\left(\dot{a}_{nk}^m+2\dot{R}_k\varepsilon_{nk}^m\right)\right)=$$

$$=\sum_{j=1,j\neq k}^{K}\left\{\boldsymbol{\delta}_\varsigma\cdot\dot{\mathbf{p}}_k^*\left(\frac{9C_{01}^{0\gamma}\Theta_{11c}^{\varsigma,\gamma,\sigma,0}B_{0j}Y_{1kj}^{-\gamma}}{4d_{kj}^2}-\frac{9R_j^3C_{11}^{\gamma\zeta}Y_{2kj}^{\gamma-\zeta}\Theta_{11c}^{\varsigma,\zeta,\sigma,0}\boldsymbol{\delta}_\gamma\cdot\dot{\mathbf{p}}_j^*}{8d_{kj}^3}+\frac{5C_{02}^{0s}\Theta_{21c}^{s,\varsigma,\sigma,0}B_{0j}R_kY_{2kj}^{-s}}{2d_{kj}^3}\right)+\right.$$

$$+\frac{5C_{02}^{0-\sigma}\delta_c^2\left(B_{0j}R_k^2Y_{2kj}^{\sigma}\right)'}{3d_{kj}^3}-\frac{5C_{12}^{\gamma s}\Theta_{21c}^{s,\varsigma,\sigma,0}R_j^3R_kY_{3kj}^{\gamma-s}\boldsymbol{\delta}_\gamma\cdot\dot{\mathbf{p}}_j^*\boldsymbol{\delta}_\varsigma\cdot\dot{\mathbf{p}}_k^*}{4d_{kj}^4}+\frac{21B_{0j}R_k^2C_{03}^{0\xi}Y_{3kj}^{-\xi}\Theta_{31c}^{\xi,\varsigma,\sigma,0}\boldsymbol{\delta}_\varsigma\cdot\dot{\mathbf{p}}_k^*}{8d_{kj}^4}-$$

$$-\frac{5\delta_c^2\left(C_{12}^{\varsigma-\sigma}\left(R_j^3R_k^2Y_{3kj}^{\sigma+\varsigma}\boldsymbol{\delta}_\varsigma\cdot\dot{\mathbf{p}}_j^*\right)'+6B_{0j}R_k^2C_{02}^{0-\sigma}\dot{d}_{kj}Y_{2kj}^{\sigma}\right)}{6d_{kj}^4}+\frac{7C_{03}^{0-\sigma}\delta_c^3\left(B_{0j}R_k^3Y_{3kj}^{\sigma}\right)'}{4d_{kj}^4}-$$

$$-\frac{9C_{01}^{0\varsigma}C_{01}^{0\gamma}\Theta_{11c}^{\varsigma\gamma\sigma,0}B_{0j}B_{0l}Y_{1kj}^{-\varsigma}Y_{1kl}^{-\gamma}}{8d_{kj}^2d_{kl}^2}+\frac{3C_{01}^{0\varsigma}\Theta_{n1c}^{m,\varsigma,\sigma,2}\left(B_{0j}Y_{1kj}^{-\varsigma}a_{nk}^m\right)'}{2(c+1)d_{kj}^2}+$$

$$+\frac{9C_{01}^{0\gamma}\left(\Theta_{211}^{s,\varsigma,-\zeta,2}\Theta_{11c}^{\gamma,\zeta,\sigma,0}+\Theta_{211}^{s,\gamma,-\zeta,2}\Theta_{11c}^{\varsigma,\zeta,\sigma,0}\right)B_{0j}Y_{1kj}^{-\gamma}\boldsymbol{\delta}_{\varsigma}\cdot\dot{\mathbf{p}}_k^{*}\varepsilon_{2k}^{s}}{8d_{kj}^2}+\frac{3C_{01}^{0\varsigma}\Theta_{n1c}^{m,\varsigma,\sigma,0}B_{0j}Y_{1kj}^{-\varsigma}\left(\dot{a}_{nk}^{m}+2\dot{R}_k\varepsilon_{nk}^{m}\right)}{2(n+1)d_{kj}^2}+$$
$$\left.+\frac{9C_{01}^{0\gamma}\left(\Theta_{n1p}^{m,\varsigma,-q,2}\Theta_{p1c}^{q,\gamma,\sigma,0}+\Theta_{n1p}^{m,\gamma,-q,2}\Theta_{p1c}^{q,\varsigma,\sigma,0}-2(p+1)\Theta_{n11c}^{m,\varsigma,\gamma,\sigma,0}\right)B_{0j}Y_{1kj}^{-\gamma}\varepsilon_{nk}^{m}\boldsymbol{\delta}_{\varsigma}\cdot\dot{\mathbf{p}}_k^{*}}{4(p+1)d_{kj}^2}\right\}.$$

Here the points from above and the primes mean differentiation with respect to time, $k = 1, 2, \ldots, K$ (K is the number of the interacting bubbles), $c = 2, 3, \ldots, N$ (N is the number of harmonics), $s = -c, -c+1, \ldots, c-1, c$, summation is supposed over the repeating indexes: from –1 to 1 over γ, ς, ζ, from –2 to 2 over s, from –3 to 3 over ξ, from 2 to N over n and p, from $-n$ to n over m, from $-p$ to p over q; $B_{0k} = -R_k^2\dot{R}_k$, $\boldsymbol{\delta}_{\varsigma} = \left(\delta_{\varsigma}^{1}, \delta_{\varsigma}^{-1}, \delta_{\varsigma}^{0}\right)$, $\delta_{\varsigma}^{\gamma}$ is the Kronecker delta, $\mathbf{p}_k = \left(x_k, y_k, z_k\right)$, $\mathbf{p}_k^{*} = \left(\frac{x_k - i\,y_k}{2}, \frac{x_k + i\,y_k}{2}, z_k\right)$, x_k, y_k, z_k are the coordinates of the center of the k-th bubble in the system of the global Cartesian coordinates, $\mathbf{p}_{kj} = \mathbf{p}_k - \mathbf{p}_j$, $d_{kj} = |\mathbf{p}_{kj}|$ is the distance between the centers of the k-th and j-th bubbles, ψ_{0k}, $\boldsymbol{\psi}_{1k}$, Δ_k are the corrections for viscosity and compressibility of the liquid [2]. Numbers $Y_{\gamma kj}^{\varsigma}$, $C_{\gamma\varsigma}^{\gamma'\varsigma'}$, $\Theta_{n_1n_2n_3}^{m_1,m_2,m_3,k}$, $\Upsilon_{n_1n_2n_3}^{m_1,m_2,m_3,k}$ and $\alpha_{n_1n_2n_3}^{m_1m_2m_3}$ are defined as follows

$$Y_{\gamma kj}^{\varsigma} = Y_{\gamma}^{\varsigma}\left(\theta_{kj}, \varphi_{kj}\right),\ C_{\gamma\varsigma}^{\gamma'\varsigma'} = (-1)^{\gamma+(|\gamma'-\varsigma'|-|\gamma'|-|\varsigma'|)/2}\left(\gamma+\varsigma-|\gamma'-\varsigma'|\right)!/\left[\left(\gamma-|\gamma'|\right)!\left(\varsigma+|\varsigma'|\right)!\right],$$

$$\Theta_{n_1n_2n_3}^{m_1,m_2,m_3,k} = \frac{2k - n_1(n_1+1) - n_2(n_2+1) + n_3(n_3+1)}{2}\alpha_{n_1n_2n_3}^{m_1m_2m_3},$$

$$\Upsilon_{n_1n_2n_3}^{m_1,m_2,m_3,k} = \frac{2k + n_1(n_1+1) + n_2(n_2+1) - n_3(n_3+1)}{2}\alpha_{n_1n_2n_3}^{m_1m_2m_3},$$

$$\alpha_{n_1n_2n_3}^{m_1m_2m_3} = \frac{2n_3+1}{4\pi}\frac{\left(n_3-|m_3|\right)!}{\left(n_3+|m_3|\right)!}\int_0^{2\pi}\int_0^{\pi}\sin\theta Y_{n_1}^{m_1}(\theta,\varphi)Y_{n_2}^{m_2}(\theta,\varphi)Y_{n_3}^{m_3}(\theta,\varphi)\,d\theta\, d\varphi\ .$$

4. Conclusion

The system of ordinary differential equations of the second order in radii of bubbles, the position vectors of their centers and the amplitudes of deviation of their shape from the spherical one, describing the spatial hydrodynamic interaction of weakly nonspherical gas bubbles in an acoustic field, is derived. The derivation of the equations is carried out by the method of spherical functions with the use of Bernoulli integral, the kinematic and dynamic boundary conditions on the surfaces of the bubbles. The effects of viscosity and compressibility of the liquid are taken into account approximately, the gas in the bubbles is assumed homobaric.

Acknowledgments
This work was performed within the Program of Basic Research of OEM RAS No12.

References

[1] Doinikov A A 2004 *JASA* **116** 821
[2] Aganin A A and Davletshin A I 2014 *Waves and vortices in complex media: 5th international scientific school of young scientists. Collected materials of school* (Moscow: MAKS Press) 105

Features of hydrogen trapping and desorption during deposition of yttrium coating on zirconium in a gas discharge

A E Evsin, L B Begrambekov and S S Dovganyuk

National Research Nuclear University MEPhI, 31 Kashirskoe shosse, Moscow, 115409, Russian Federation

E-mail: evsin@plasma.mephi.ru

Abstract. Transport of hydrogen isotopes during the various regimes of deposition of yttrium coating on zirconium in argon plasma with addition of deuterium is studied. The influence of oxygen contamination in plasma-generating gas on the processes of trapping and desorption of hydrogen isotopes is also investigated. It is shown that deposition of yttrium coating on zirconium in $Ar+5\%D_2$ plasma enhances both hydrogen desorption from zirconium and deuterium trapping into zirconium in comparison to those under plasma exposure without deposition. Yttrium deposition in $Ar+25\%O_2+5\%D_2$ plasma, conversely, mitigates both hydrogen desorption and deuterium trapping. Hydrogen desorption from zirconium increases with the increase of energy of ions, bombarding the sample during deposition of the coating in oxygen-free plasma, but it, on the contrary, decreases in oxygen-containing plasma.

1. Introduction

Protective coatings preventing hydrogen penetration into metal are widely used in many applications. Particularly, the coatings, protecting zirconium components of active core of light-water reactors from oxygen and hydrogen corrosion, are actively developed. Encouraging results of reactor-simulation tests of yttrium-based protective coatings deposited on zirconium in a gas discharge [1] demonstrate the prospects of development of these coatings. An important step of protective coatings creation is investigation of hydrogen behavior in multicomponent system consisting of materials of the coating and the substrate. The understanding of these regularities is also needed at the stage of deposition of the coating in a gas discharge, when the sample undergoes plasma irradiation. Features of trapping and desorption of hydrogen isotopes during the various regimes of yttrium deposition on zirconium in a gas discharge were investigated by thermal desorption spectroscopy (TDS) in a present work.

2. Experimental

Coatings were deposited in experimental device for thin films deposition in a gas discharge with a hot tungsten cathode. The experiments were carried out in two plasma-generating gas compositions – $95\%Ar+5\%D_2$ and $70\%Ar+25\%O_2+5\%D_2$. Samples of zirconium alloy E110 (Zr–1%Nb) in the form of tube sections with a size of ~7×7×1 mm preliminarily rinsed in ultrasonic ethanol bath were used as substrates. According to the results of previous experiments the substrate should be irradiated by argon ions before the deposition of the coating to demonstrate the proper coating adhesion.

There were several types of the investigated coatings: yttrium coating and zirconium coating deposited on zirconium separately and two-component layer consisting of zirconium and yttrium as well. Coatings were formed by yttrium or zirconium atoms sputtered by ions of plasma from targets made of corresponding metals. Joint deposition of zirconium and yttrium was performed by simultaneous gradual decrease of zirconium deposition rate and increase of yttrium deposition rate. Separate deposition of zirconium and yttrium coatings was carried out at floating potential (≈12 V) on the substrate. Two-component layers were deposited at two values of potential on the substrate: floating potential and 200 V. Deposition rate of all of the coatings in oxygen-free plasma at floating potential on the substrate was ≈800 nm per hour.

Pressure of plasma-generating gas in all the experiments was equal to 1.3×10^{-1} Pa, residual gas pressure did not exceed 2.6×10^{-3} Pa, the experiments lasted 80 min.

Trapping and desorption of hydrogen isotopes under deposition of coatings were investigated via thermal desorption spectroscopy in MIKMA device [2]. The results were compared with those of plasma exposure of the samples under the same conditions, but without deposition of any coating. Residual gas pressure in TDS chamber did not exceed 1.3×10^{-5} Pa. Samples were heated up linearly at a rate of 5 K/s. Molecules of H_2, HD, D_2, H_2O, HDO, D_2O were measured. Desorption of other molecules, containing hydrogen isotopes, was negligible.

3. Separate deposition of zirconium and yttrium at floating potential on the substrate

The amount of hydrogen atoms in the samples after preliminary ion bombardment before deposition was determined as $\approx6\times10^{18}$ cm^{-2}. Figure 1 illustrates that exposure of zirconium to Ar+5%D_2 plasma as well as to Ar+25%O_2+5%D_2 plasma at floating potential causes desorption of ≈60% of hydrogen atoms from zirconium and trapping of a much lesser amount of deuterium atoms. In our previous work [3] it was shown that ion irradiation of a metal sample with oxidized surface initiates hydrogen exchange between the sample and the gas ambience of the experimental stand. The mechanism of this process is based on surface reactions occurring between hydrogen-containing and oxygen-containing particles of gaseous phase, from one side, and those of oxidized surface layer, from the other side.

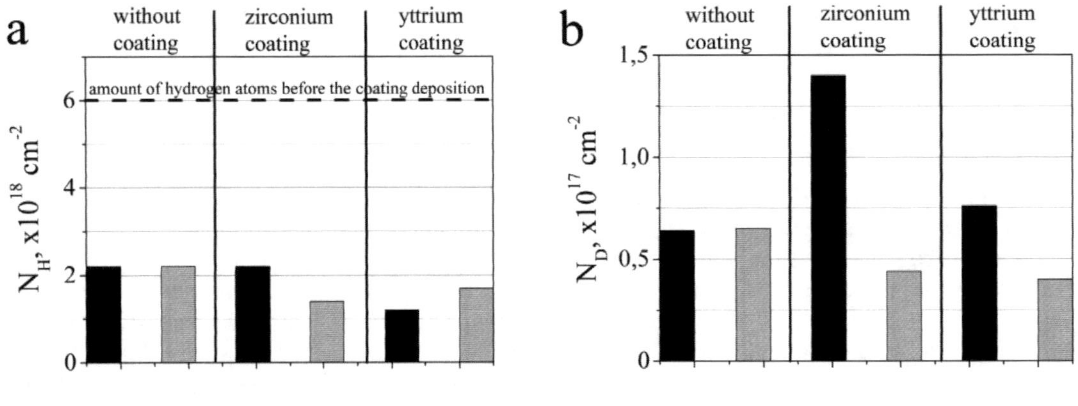

Figure 1. The amount of hydrogen (a) and deuterium (b) atoms in zirconium samples after exposure to plasma without deposition of any coating and after deposition of zirconium and yttrium (black – Ar+5%D_2 plasma, gray - Ar+25%O_2+5%D_2 plasma).

Deposition of yttrium coating in Ar+5%D_2 plasma enlarges hydrogen desorption by more than 25% in comparison to that under exposure to plasma without deposition. Besides, deuterium trapping increases by ≈17%. Hydrogen desorption under deposition of zirconium coating in Ar+5%D_2 plasma is almost the same as that under plasma exposure. Deuterium trapping under zirconium deposition increases by more than 2 times.

Deposition of yttrium coating in Ar+25%O_2+5%D_2 plasma retards hydrogen desorption in comparison to that under the similar deposition in Ar+5%D_2 plasma. Deposition of zirconium coating in oxygen-containing plasma, conversely, leads to the increase of hydrogen desorption. Deuterium trapping under deposition of any of the investigated coatings in Ar+25%O_2+5%D_2 plasma is 2-3 times less than that under its deposition in oxygen-free plasma.

4. Joint deposition of zirconium and yttrium at floating potential on the substrate

Hydrogen exchange under deposition of two-component layer, containing zirconium and yttrium, at floating potential on the substrate is similar to that observed during the deposition of yttrium coating (figure 2). Both hydrogen desorption and deuterium trapping decreases when oxygen content in plasma-generating gas increases. It should be also mentioned that the values of hydrogen desorption from zirconium under deposition of two-component layer in Ar+5%D_2 plasma are close to those under deposition of the components separately. At the same time, deuterium trapping under the joint deposition of zirconium and yttrium in oxygen-free plasma increases many times over the corresponding values obtained after deposition of zirconium or yttrium singly.

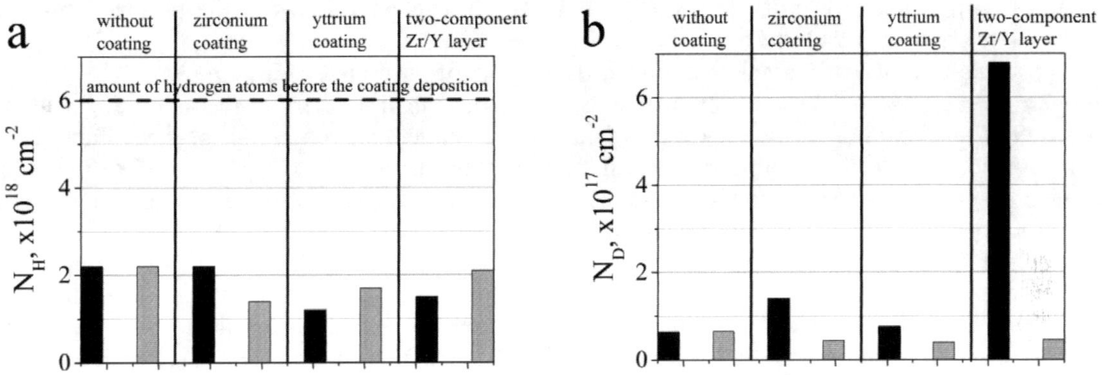

Figure 2. The amount of hydrogen (a) and deuterium (b) atoms in zirconium samples after exposure to plasma without deposition of any coating and after deposition of zirconium, yttrium and two-component layer Zr/Y (black – Ar+5%D_2 plasma, gray - Ar+25%O_2+5%D_2 plasma).

5. Joint deposition of zirconium and yttrium at 200 V on the substrate

Dependence of hydrogen desorption from zirconium under irradiation by ions of ≈200 eV (without deposition of any coating) on oxygen content in plasma differs from that observed when zirconium was exposed to plasma at floating potential. Addition of oxygen to plasma-generating gas leads to sufficient increase of hydrogen desorption from the sample (figure 3).

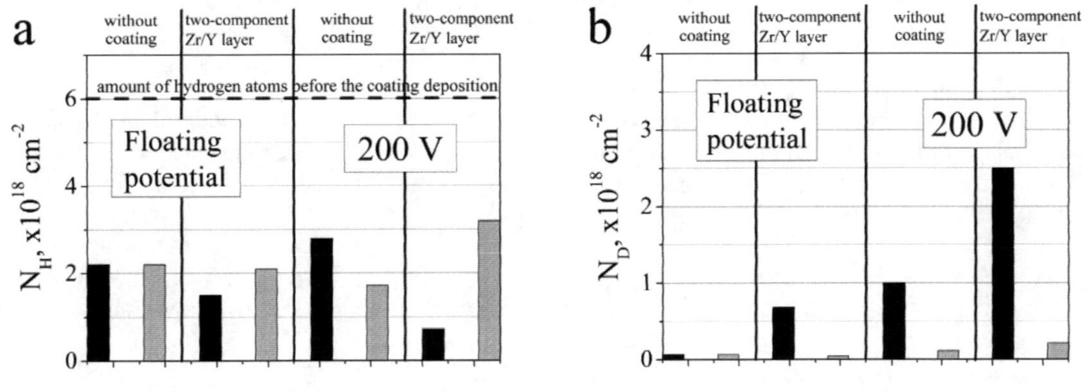

Figure 3. The amount of hydrogen (a) and deuterium (b) atoms in zirconium samples after exposure to plasma without deposition of any coating and after deposition of two-component layer Zr/Y (black – Ar+5%D_2 plasma, gray - Ar+25%O_2+5%D_2 plasma).

At the same time, regularities of hydrogen desorption under deposition of two-component layer at biasing potential of 200 V on the substrate are similar to that observed during the deposition of this coating at floating potential. Namely, joint deposition of zirconium and yttrium in Ar+5%D_2 plasma causes the increase of both hydrogen desorption from the sample and deuterium trapping into the sample in comparison to those under exposure to plasma without deposition of any coating. Deposition of two-component layer in oxygen-containing plasma, conversely, decreases both hydrogen desorption and deuterium trapping. Besides, when the biasing potential on the sample increases from floating potential to 200 V, discrepancy in hydrogen desorption under deposition of the coating in Ar+5%D_2 plasma and in Ar+25%O_2+5%D_2 plasma reveals to a greater extent.

6. Conclusion

The experiments have shown that deposition of yttrium coating on zirconium sample in a gas discharge significantly affects the regularities of hydrogen exchange between the plasma irradiated sample and the gaseous ambience of the experimental stand.

In particular, it was observed that under deposition of yttrium coating in Ar+5%D_2 plasma hydrogen atoms desorb from zirconium more intensively than under plasma exposure without deposition of any coating. At the same time, deposition of zirconium coating at the same parameters does not affect the rate of this process. Deposition of yttrium in Ar+25%O_2+5%D_2 plasma mitigates hydrogen desorption in comparison to that under plasma exposure without deposition of the coating, whereas zirconium deposition under the same conditions, conversely, intensifies it.

Deuterium trapping under deposition of any of the investigated coatings in Ar+5%D_2 plasma is 2-3 times larger than that under exposure to plasma without deposition. Besides, under the joint deposition of zirconium and yttrium deuterium atoms trap at a significantly greater extent than under deposition of zirconium or yttrium separately. Under deposition of any of the investigated coatings in oxygen-containing plasma deuterium trapping decreases, apparently, due to enhancement of deuterium desorption from the surface as a part of water molecules.

The increase of energy of ions irradiating the sample enlarges the difference in the amounts of hydrogen atoms desorbing during the joint deposition of zirconium and yttrium in oxygen-containing plasma and in oxygen-free plasma.

One could propose that the main reasons of alteration of hydrogen behavior during yttrium deposition (in comparison to that during zirconium deposition or plasma exposure without deposition) are different values of enthalpy of hydrogen solution in these metals (ΔH_s=-0.66 eV/at.H for Zr and ΔH_s=-0.85 eV/at.H for Y [4]) as well as different values of standard Gibbs energy of formation of their oxides (ΔG=-1042.8 kJ/mol for ZrO_2 and ΔG=-1816.6 kJ/mol for Y_2O_3 [5]). These parameters are supposed to determine thermodynamic conditions of hydrogen transport in systems "zirconium-yttrium" and "zirconia-yttria" as well as surface reactions, regulating hydrogen exchange [3].

Acknowledgments

This work was supported by Competitiveness Growth Program of the Federal Autonomous Educational Institution of Higher Professional Education National Research Nuclear University MEPhI and by the Ministry of Education and Science of the Russian Federation (agreement contract №14.575.21.0049 of 27.06.14).

References

[1] Begrambekov L B, Gordeev A A, Evsin A E, Ivanova S V, Kaplevsky A S, and Sadovskiy Ya A 2015 *Phys. Atom. Nucl.* **78** (14) at press (DOI: 10.1134/S1063778815140033).
[2] Airapetov A, Begrambekov L, Brémond S, Douai D, Kuzmin A, Sadovskiy Ya, Shigin P, Vergasov S 2011 *J. Nucl. Mater.* **415** pp S1042–45.
[3] Begrambekov L, Dvoychenkova O, Evsin A, Kaplevsky A, Sadovskiy Ya, Schitov N, Vergasov S and Yurkov D 2014 *J. Phys. Conf. Ser.* **567** 012003.
[4] Fukai Y 2005 *The Metal-Hydrogen System: Basic Bulk Properties* (Berlin: Springer-Verlag Berlin Heidelberg) p 16.
[5] Lide D R 2005 *CRC Handbook of Chemistry and Physics* (Boca Raton: CRC Press) pp 5-5–5-25.

Journal of Physics: Conference Series **669** (2016) 012010 doi:10.1088/1742-6596/669/1/012010

Transverse acoustic oscillations in cylindrical chambers with the radial temperature gradient

S A Fadeev, A I Saifutdinov and N F Kashapov
Kazan Federal University, 18 Kremlyovskaya str., Kazan 420008, Russian Federation

E-mail: sergey_fadeev@kpfu.ru

Abstract. In this article the transverse acoustic oscillations in cylindrical chambers with a radial temperature gradient are considered. Expression for the velocity potential is calculated. This expression describes the free transverse oscillations in cylindrical chamber with radial temperature gradient.

1. Introduction

The interest in acoustic oscillations of the gas is maintained because of the wide spread them in various fields of science and technology. Many technical devices are some bounded by solid walls, volume, that is are these or other resonators. For example, the basic elements of gas turbines, steam generators, aircraft engines, gas lasers, plasma generators, and others. Therefore, the study of acoustic oscillations in the resonant modes of operation such devices is a major challenge. When using all these devices, oscillatory processes occur in a non-uniform temperature field.

Longitudinal acoustic oscillations with axial temperature gradient in a large number of works are studied [1, 2]. In this distribution of transverse of acoustic oscillations in the tube with a radial temperature gradient has not been considered. An example of a device with a radial temperature gradient can serve as generators of low-temperature plasma – plasma generators, high-power gas lasers, spectrum analyzers, and others [3-5]. Various methods of controlling the glow discharge parameters have been considered in [6-11]. In particular, excitation of transverse of acoustic oscillations in the discharge chamber can intensified heat transfer in the radial direction compared with the longitudinal acoustic oscillations [12]. To solve the self-consistent problem on the influence of the transverse acoustic oscillations on the plasma glow discharge is necessary to consider the reverse effect of the discharge on the characteristics of transverse acoustic waves. In this paper we consider the influence of radial temperature gradient on the parameters of the transverse acoustic oscillations in the chambers with a cylindrical geometry.

2. Transverse acoustic oscillations in cylindrical chamber with the radial temperature gradient

For the description of acoustic oscillations in the tube with a radial temperature gradient write the equations of motion and continuity in the cylindrical coordinate system.

$$\frac{\partial u_r}{\partial t} + u_r \frac{\partial u_r}{\partial r} + \frac{u_\theta}{r}\frac{\partial u_r}{\partial \theta} + u\frac{\partial u_r}{\partial x} - \frac{u_\theta^2}{r} = -\frac{1}{\rho}\frac{\partial p}{\partial r}, \tag{1}$$

$$\frac{\partial u_\theta}{\partial t} + u_r \frac{\partial u_\theta}{\partial r} + \frac{u_\theta}{r}\frac{\partial u_\theta}{\partial \theta} + u\frac{\partial u_\theta}{\partial x} - \frac{u_r u_\theta}{r} = -\frac{1}{\rho}\frac{\partial p}{r\partial \theta}, \tag{2}$$

$$\frac{\partial u_x}{\partial t}+u_r\frac{\partial u_x}{\partial r}+\frac{u_\theta}{r}\frac{\partial u_x}{\partial \theta}+u\frac{\partial u_x}{\partial x}=-\frac{1}{\rho}\frac{\partial p}{\partial x}, \tag{3}$$

$$\frac{\partial \rho}{\partial t}+\frac{\partial(\rho u_x)}{\partial x}+\frac{1}{r}\frac{\partial(r\rho u_r)}{\partial r}+\frac{1}{r}\frac{\partial(\rho u_\theta)}{\partial \theta}=0, \tag{4}$$

where x, r, θ – longitudinal, radial and angular coordinates, respectively, u_x, u_r, u_θ – the speed of the projection of the gas corresponding to the longitudinal, radial and angular coordinates, ρ – density, p – pressure.

Represent the flow parameters as the stationary and pulsating parts

$$\begin{gathered} u_r=\bar{u}_r+u'_r,\\ u_\theta=\bar{u}_\theta+u'_\theta,\\ u_x=\bar{u}_x+u'_x,\\ \rho=\rho_0+\rho',\\ p=p_0+p'. \end{gathered} \tag{5}$$

Due to the radial temperature gradient and the constant pressure in the tube $\rho_0=\rho_0(r)$. From (1-4) and (5) we obtain a system of equations, which is linearized by taking into account the following equation $\bar{u}_r=\bar{u}_\theta=0$

$$\frac{\partial u'_r}{\partial t}+\bar{u}_x\frac{\partial u'_r}{\partial x}=-\frac{1}{\rho_0}\frac{\partial p'}{\partial r}, \tag{7}$$

$$\frac{\partial u'_\theta}{\partial t}+\bar{u}_x\frac{\partial u'_\theta}{\partial x}=-\frac{1}{\rho_0 r}\frac{\partial p'}{\partial \theta}, \tag{8}$$

$$\frac{\partial u'_x}{\partial t}+\bar{u}_x\frac{\partial u'_x}{\partial x}=-\frac{1}{\rho_0}\frac{\partial p'}{\partial x}. \tag{9}$$

$$\frac{\partial \rho'}{\partial t}+\bar{u}_x\frac{\partial \rho'}{\partial x}+u'_r\frac{\partial \rho_0}{\partial r}+\rho_0\left(\frac{\partial u'_x}{\partial x}+\frac{u'_r}{r}+\frac{\partial u'_r}{\partial r}+\frac{1}{r}\frac{\partial u'_\theta}{\partial \theta}\right)=0. \tag{10}$$

Using expressions $\rho_0=\gamma\frac{p_0}{c^2(r)}$, $\rho'=\frac{p'}{c^2(r)}$, $c(r)=\sqrt{\gamma R_g T(r)}$ [13], where γ – adiabatic index, c – velocity of sound, R_g – gas constant, T – temperature, rewrite the equation (10)

$$\frac{\partial p'}{\partial t}+\bar{u}_x\frac{\partial p'}{\partial x}+\gamma p_0\left(\frac{\partial u'_x}{\partial x}+\frac{u'_r}{r}+\frac{\partial u'_r}{\partial r}+\frac{1}{r}\frac{\partial u'_\theta}{\partial \theta}-\frac{2u'_r}{c}\frac{\partial c}{\partial r}\right)=0. \tag{11}$$

We express the velocity through the velocity potential [14] by means of relations

$$u'_x=\frac{\partial \varphi}{\partial x},\ u'_r=\frac{\partial \varphi}{\partial r},\ u'_\theta=\frac{1}{r}\frac{\partial \varphi}{\partial \theta}. \tag{12}$$

From the equations (7) and (9) we obtain

$$p'=-\rho_0\left(\partial\varphi/\partial t+\bar{u}_x\partial\varphi/\partial x\right) \tag{13}$$

Substitute (12), (13) into (11) and obtain the wave equation for the velocity potential

$$c^2\left(\frac{\partial^2\varphi}{\partial x^2}+\frac{\partial^2\varphi}{\partial r^2}+\frac{1}{r^2}\frac{\partial^2\varphi}{\partial \theta^2}+\frac{1}{r}\frac{\partial\varphi}{\partial r}-\frac{2}{c}\frac{\partial\varphi}{\partial r}\frac{\partial c}{\partial r}\right)-2\bar{u}_x\frac{\partial^2\varphi}{\partial x\partial t}-\bar{u}_x^2\frac{\partial^2\varphi}{\partial x^2}=\frac{\partial^2\varphi}{\partial t^2} \tag{14}$$

The velocity potential must satisfy the following conditions

$$\left.\frac{\partial \varphi}{\partial r}\right|_{r=r_0}=0,\ \ \varphi(\theta+2\pi)=\varphi(\theta), \tag{15}$$

where r_0 – the radius of the tube.

Suppose that the average flow in the tube is missing $\bar{u}_x=0,$. Then (14) takes the form

$$\frac{\partial^2\varphi}{\partial x^2}+\frac{\partial^2\varphi}{\partial r^2}+\frac{1}{r^2}\frac{\partial^2\varphi}{\partial\theta^2}+\frac{1}{r}\frac{\partial\varphi}{\partial r}-\frac{2}{c}\frac{\partial c}{\partial r}\frac{\partial\varphi}{\partial r}=\frac{1}{c^2}\frac{\partial^2\varphi}{\partial t^2}. \tag{16}$$

Consider the mode of harmonic oscillations. In this case, the velocity potential may be represented as $\varphi=\bar{\varphi}(r,\theta,x)e^{i\omega t}$, where ω – angular frequency. Velocity potential due the boundary condition (15) is a periodic function of angular coordinate θ. Therefore, it can be expanded in a trigonometric Fourier series. Then we obtain

$$\varphi=\sum_0^\infty\varphi_m(r,x)\cos m\theta e^{i\omega t}, \tag{17}$$

where $m=1,2,\ldots$.

Substituting (17) into (16) obtain

$$\sum_0^\infty\left(\frac{\partial^2\varphi_m}{\partial x^2}+\frac{\partial^2\varphi_m}{\partial r^2}-\frac{m^2}{r^2}\varphi_m+\frac{1}{r}\frac{\partial\varphi_m}{\partial r}-\frac{2}{c}\frac{\partial c}{\partial r}\frac{\partial\varphi_m}{\partial r}+\frac{\omega^2}{c^2}\varphi_m\right)\cos m\theta=0. \tag{18}$$

To solve the equation (19) using Fourier method represent $\varphi_m=R_m(r)X_m(x)$. This results in two equations

$$\frac{\partial^2X_m}{\partial x^2}+\lambda X_m=0, \tag{20}$$

$$\frac{\partial^2R_m}{\partial r^2}+\left(\frac{1}{r}-\frac{2}{c}\frac{\partial c}{\partial r}\right)\frac{\partial R_m}{\partial r}+\left(\frac{\omega^2}{c^2}-\frac{m^2}{r^2}-\lambda\right)R_m=0, \tag{21}$$

Where λ – constant.

The general solution of the equation (20) is

$$X_m=A\cos\left(\sqrt{\lambda}x\right)+B\sin\left(\sqrt{\lambda}x\right) \tag{22}$$

Suppose that the tube is closed at both ends. Then boundary conditions are

$$\left.\frac{\partial X_m}{\partial x}\right|_{x=0}=\left.\frac{\partial X_m}{\partial x}\right|_{x=L}=0, \tag{23}$$

where L – length of the tube.

From (23) and (22) we obtain $B=0$ and $\lambda=\left(\frac{\pi n}{L}\right)^2$. The solution of equation (20) will take the final form

$$X_m=A\cos\left(\frac{\pi n}{L}x\right),. \tag{24}$$

where $n=0,1,2,\ldots$.

It is known that in the cylindrical discharge chambers of gas lasers, the temperature distribution along the radius is parabolic [6]. Then, since c □ $\sqrt{T}$, the velocity of sound can be represented as

$$c=\sqrt{b-ar^2}, \tag{25}$$

where a and b – constants.

Substituting (25) into (21) consider only the transverse oscillations. For this set $n=0$, $\lambda=0$ and obtain

$$\frac{\partial^2 R_m}{\partial r^2}+\left(\frac{1}{r}+\frac{4ar}{b-ar^2}\right)\frac{\partial R_m}{\partial r}+\left(\frac{\omega^2}{b-ar^2}-\frac{m^2}{r^2}\right)R_m=0. \quad (26)$$

The solution of equation (26) can be written as

$$R_m=C_1r^mF(\alpha_1,\beta_1;\chi_1;\delta)+C_2r^{(-m)}F(\alpha_2,\beta_2;\chi_2;\delta), \quad (27)$$

where C_1, C_2 – constants, F – hypergeometric function,

$$\alpha_1=\frac{1}{2}\left(\frac{-2a+\sqrt{a(4a+\omega^2+m^2a)}+ma}{a}\right),\qquad \beta_1=\frac{1}{2}\left(\frac{2a+\sqrt{a(4a+\omega^2+m^2a)}-ma}{a}\right),$$

$$\chi_1=1+m,\qquad \delta=\frac{r^2a}{b},\qquad \alpha_2=\frac{1}{2}\left(\frac{-2a+\sqrt{a(4a+\omega^2+m^2a)}-ma}{a}\right),$$

$$\beta_2=\frac{1}{2}\left(\frac{2a+\sqrt{a(4a+\omega^2+m^2a)}+ma}{a}\right),\quad \chi_2=1-m.$$

Second term on the right side of (27) tends to infinity at $r=0$. From the condition of finiteness, it follows that $C_2=0$ and equation (27) takes the form

$$R_m=C_1r^mF(\alpha_1,\beta_1;\chi_1;\delta). \quad (28)$$

Then solution of the wave equation (16), which takes into account only the transverse oscillations is

$$\varphi=Cr^mF(\alpha_1,\beta_1;\chi_1;\delta)\cos m\theta e^{i\omega t}, \quad (29)$$

where $C=C_1A$.

3. Conclusions

Thus, the transverse acoustic oscillations in cylindrical chamber with a radial temperature gradient describes by the velocity potential, which defined by (29). Expression (29) can be used to determine the natural frequencies of transverse oscillations in various devices, operating environment which is the discharge plasma. In particular, in high-power gas lasers with diffuse cooling.

Acknowledgments
The work is performed according to the Russian Government Program of Competitive Growth of Kazan Federal University. The reported study was funded by RFBR, according to the research project No. 16-38-60187 mol_a_dk.

References

[1] Peat K S 1997 *Journal of Sound and Vibration* **203** 855–866
[2] Larionov V M, Nazarenko T I and Filipov S E 2004 *Izvestiya Vysshikh Uchebnykh Zavedenij. Aviatsionnaya Tekhnika* (**1**) 32–34
[3] Raizer Y P 1992 *Gas Discharge Physics* (Moscow, Science) p 536
[4] Kudryavtsev A A, Smirnov A S and Tsendin L D 2010 *Physics of Glow Discharges* (St. Petersburg, Lan) p 512
[5] Israfilov Z K and Kashapov N F 1991 *Journal of Engineering Physics* **60** 364–368
[6] Saifutdinov A I, Timerkaev B A, and Zalyaliev B R, 2014, *J. Phys.: Conf. Ser.* **567** 012031
[7] Aramyan A R and Galechyan G A 2007 *Physics-Uspekhi* **50** 1147–69

[8] Kashapov N F, Saifutdinov A I and Fadeev S A 2014 *J. Phys.: Conf. Series* **567** 012004
[9] Saifutdinov A I, Fadeev S A, Saifutdinova A A and Kashapov N F 2015 *JETP Lett.* **102** 637–642
[10] Fadeev S A, Kashapov N F and Larionov V M 2014 *IOP Conf. Ser.: Mater. Sci. Eng.* **69** 012006
[11] Gubaidullin D A, Zaripov R G, Tkachenko L A and Shaidullin L R 2015 *J. of Engin. Physics and Thermophysics* **88** 843–847
[12] Fadeev S A and Kashapov N F 2013 *J. Phys.: Conf. Series* **479** 012009
[13] Landau L D and Lifshits E M 1988 *Hydrodynamics* vol 6, ed Y G Rudoi (Moscow, Science) p 736
[14] Skudrzyk E 1976 *The foundations of acoustics* vol 2, ed L M Liamsheva (Moscow, World) p 544

About the equilibrium speed of sound in a liquid with gas-vapor bubbles

D A Gubaidullin, D D Gubaidullina, Yu V Fedorov

Institute of Mechanics and Engineering, Kazan Science Center, Russian Academy of Sciences, 2/31 Lobachevsky str., Kazan 420111, Russia
Kazan (Volga Region) Federal University, 18 Kremlyovskaya str., Kazan 420008, Russia

E-mail: kopperfildd@yandex.ru

Abstract. The general expression of an equilibrium velocity of a sound in vapor-gas-liquid mixtures is presented and influence of concentration of vapor and a volume content of bubbles on the received expression is analyzed. In special cases, for gas-liquid and vapor-liquid mixtures expressions of an equilibrium velocity are presented and the satisfactory consent of the received expressions with known experimental data is discovered.

1. Introduction

It is known that the presence of vapor or gas bubbles in the liquid significantly changes its acoustic properties. There is a considerable amount of works on this problem. Propagation of small disturbances in a liquid with bubbles of vapor or gas in a particular formulation is theoretically investigated in works [1] - [5], with the gas-vapor bubbles – in works [6] - [9]. Some of them can be found in [10], [11].

The results of experiments on the study of the propagation of low-frequency pressure waves in the vapor-liquid flow moving through the close-packed layer of solid spherical particles are presented in [12], [13]. The experimental results allowed to determine the characteristic parameters and conditions under which the speed of propagation of the pressure waves coincides with the thermodynamic equilibrium speed of sound in the liquid-vapor mixture. Thus, it has been found experimentally that the rate of low-frequency disturbances in a vapor-water environment may be a few meters per second, and takes a value close to the sound velocity of Landau. When the carrier phase is a gas or vapor, and the disperse - droplets or particles expressions of equilibrium sound of velocity are obtained in [14], [15].

In this paper we analyzed the expression of the equilibrium speed received by us in [9].

2. The analytical expression.

The expression of the equilibrium speed of sound C_e for the vapor-gas-liquid mixture, resulting in [9], is written as follows

$$C_e^2 = \frac{p_0}{\alpha_{10}\rho_{10}^{\circ}} \frac{\xi_4}{\xi_1 + \xi_2 - \xi_3} \quad (1)$$

$$\xi_1 = k_{G0} m_1 \alpha_{20} \left(ET_0 + GR_1\right), \xi_2 = (1 - G)\frac{m l_0 \alpha_{20}}{c_{p1}}$$

$$\xi_3 = \frac{k_{G0}\alpha_{20}p_0 m}{\rho^\circ_{20}c_{p1}}\left(\frac{R_1}{T_0}+E\right), \xi_4 = \frac{ml_0}{c_{p1}} + m_1 k_{G0} E T_0$$

$$E = \frac{R_{V0}R_{G0}}{R_{20}^2}, G = k_{V0}\frac{R_{V0}}{R_{20}}, R_{20} = k_{V0}R_{V0} + k_{G0}R_{G0}$$

$$m_1 = 1 + m\frac{c_{p2}}{c_{p1}}, R_1 = \left(\Delta R + \frac{1-m^\circ}{k_{G0}}\right)T_0, m = \frac{\alpha_{20}}{\alpha_{10}}m^\circ, m^\circ = \frac{\rho^\circ_{20}}{\rho^\circ_{10}}$$

$$\Delta R = \frac{R_{V0} - R_{G0}}{R_{20}}, \rho^\circ_{20} = k_{V0}\rho^\circ_{V0} + k_{G0}\rho^\circ_{G0}$$

Here ρ°_{V0}, ρ°_{G0} – the density of vapor and gas, R_{V0}, R_{G0} – vapor and gas constants, k_{V0}, k_{G0} – mass concentration of vapor and gas, the index 1 refers to the carrier phase, index 2 – to disperse, α – volume content, T_0 – temperature, p_0 – pressure, c_p – specific heat, l_0 – latent heat of vaporization.

Expression of equilibrium speed of sound C_e for the vapor-liquid mixture is obtained from the formula (1) in the limit $k_{V0} \to 1$, and is written as follows:

$$C_e^2 = \frac{p_0 l_0 m \rho^\circ_{V0}}{(mp_0 - c_{p1}m_1T_0\rho^\circ_{V0})(m^\circ - 1)\alpha_{20}\alpha_{10}\rho^\circ_{10}} \quad (2)$$

$$m_1 = 1 + m\frac{c_{p2}}{c_{p1}}, m = \frac{\alpha_{20}}{\alpha_{10}}m^\circ, m^\circ = \frac{\rho^\circ_{V0}}{\rho^\circ_{10}}, \alpha_{10} + \alpha_{20} = 1$$

In the case where the liquid contains only the gas bubbles $k_{V0} \to 0$, the equilibrium sound velocity becomes by Gubaidullin and Fedorov [5], Shagapov and Sarapulova [16]

$$C_e^2 = \frac{p_0}{\alpha_{10}\alpha_{20}\rho^\circ_{10}} \quad (3)$$

3. The results of calculations.

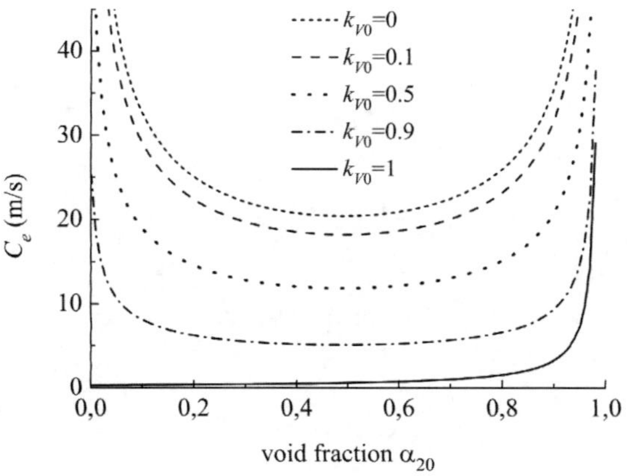

Figure 1. The dependences of the equilibrium speed of sound on the volume of gas or vapor concentration at different values of the initial concentration of vapor.

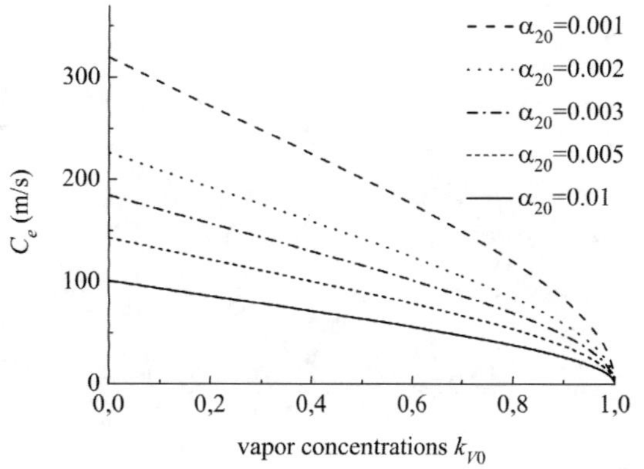

Figure 2. Dependences of the equilibrium speed of sound on the initial concentration of vapor at various volume content of bubbles.

Fig. 1 shows the dependences of the equilibrium speed of sound on the volume content for the mixture of water with vapor-air bubbles at different values of the initial concentration of water vapor. Pressure of mixture $p_0 = 0.1$ MPa. The bottom curve ($k_{V0} = 1$) is built according to the formula (2) and corresponds to a vapor bubble, the upper curve ($k_{V0} = 0$) according to the formula (3) and corresponds to the air bubbles, the other curves by the formula (1) and correspond to vapor-air bubbles.

It is evident that a small amount of gas in the gas-vapor bubbles affects the character of the curve of the equilibrium speed. If for the gas or gas-vapor bubbles curves of equilibrium speed are non-monotonic, then for pure vapor bubbles we observe a monotonically increasing curve.

The following Fig. 2 shows the results of a comparison of the dependency of equilibrium sound speed on the initial concentration of vapor in the bubbles at various volumetric content. The calculations were performed using the formula (1). For the construction of the theoretical curves, the following approximate dependences of thermodynamic parameters on the concentration of vapor were used:

$$T_0 = -68.75k_{V0}^2 + 123.75k_{V0} + 315.3125, l_0 = 10^6(0.875k_{V0}^2 - 1.2975k_{V0} + 2.721)$$

$$c_{pV} = 1000(-0.1281k_{V0}^2 + 0.2719k_{V0} + 1.8861), c_{pG} = 1010$$

$$c_{p1} = 1000(-0.0219k_{V0}^2 + 0.0606k_{V0} + 4.1772), c_{p2} = k_{V0}c_{pV} + k_{G0}c_{pG}$$

$$\rho_1^\circ = 37.5k_{V0}^2 - 70k_{V0} + 992.625, \rho_G^\circ = 0.0781k_{V0}^2 - 0.4469k_{V0} + 0.9439$$

$$\rho_V^\circ = -0.2969k_{V0}^2 + 0.8531k_{V0} + 0.0177, \rho_2^\circ = k_{V0}\rho_V^\circ + k_{G0}\rho_G^\circ, R_{V0} = 460, R_{G0} = 287$$

In particular, we see that the volume content of bubbles $\alpha_{20} = 0.01$ and vapor concentration $k_{V0} = 0$ (air bubbles in the water), the equilibrium speed of sound is about 100 m/s. This fact is also theoretically shown, for example, by Commander and Prosperetti [1], Temkin [17] and experimentally by Silberman [18], Karplus [19]. Further, with an increase of the initial concentration of vapor in bubbles at different volumetric contents equilibrium speed decreases and, if the vapor concentration is equal 1, reaches practically the same minimum value.

4. Comparison with experiment.

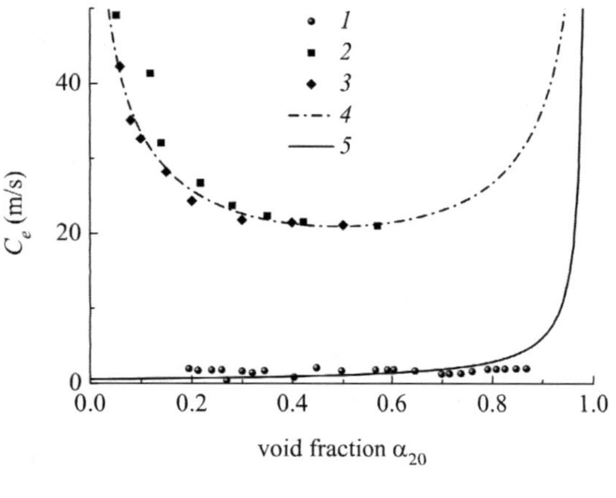

Figure 3. Comparison of the dependence of equilibrium speed of sound on the volume gas or vapor content with the experimental data.

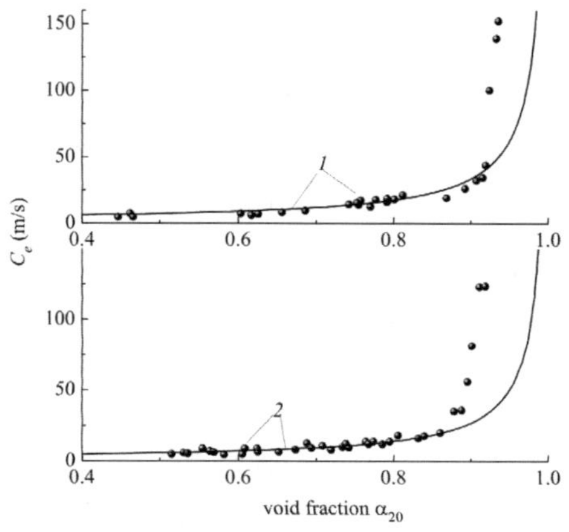

Figure 4. Comparison of the dependences of the equilibrium speed of sound on the volume vapor content with the experimental data.

Fig. 3 shows a comparison of dependences of equilibrium speed of sound on the volume gas or vapor content with the experimental data: *1*, *2* – [12], *3* – [19]. The curve 4 is constructed for an air-water mixture under the pressure $p_0 = 0.1$ MPa by formula (3), the curve 5 – for vapor-water mixture under the pressure $p_0 = 0.2$ MPa by the formula (2). The characteristic frequency of the input disturbance in the experiment was 1.5 - 2.5 Hz. We see that the curve of the equilibrium velocity for the gas-liquid mixture differs significantly from the curve for the vapor-liquid mixture, what is confirmed by experimental data. The authors of the experiment [12] found that when the volume content of water vapor increases velocity of the low-frequency disturbances almost doesn't differ and tends to the speed of sound Landau. The shown theoretical curve 5 is increasing, but, in general, agrees satisfactorily with the experimental data.

The following Fig. 4 shows a comparison of the dependence of the equilibrium speed of sound on the volume vapor content with experimental data [20]. The curve *1* is built for R404A refrigerant at saturation temperature of 293 K, the curve *2* – for R134A refrigerant at saturation temperature 303 K. The calculations were performed using the formula (2). The characteristic frequency of the input disturbance in the experiment was 0.2 – 5 Hz. For these mixtures taking into account phase transformations in the field of moderate vapor content the speed of sound also takes quite low values, and that is about 5 m/s at a volume content $\alpha_{20} = 0.4$. With increasing of volume vapor content equilibrium speed increases and with approximation of the volume content to unit tends to the speed of sound in a clean vapor. The theoretical curves are in good agreement with the experimental data for the vapor content to 85-90%. At high vapor content experimental values of the equilibrium velocity accept higher values compared with the theoretical.

5. Conclusions

1. When the concentration of vapor in bubbles increases the equilibrium speed of sound decreases.
2. The presence of gas in the bubbles leads to a nonmonotonic dependence of the equilibrium speed of sound on a volume content.
3. Satisfactory agreement of obtained expression with the experimental data was disclosed.

Acknowledgment
This work was funded by the subsidy of the Russian Government to support the Program of Competitive Growth of Kazan Federal University among Worlds Leading Academic Centers.

References

[1] Commander K W and Prosperetti A 1989 *J. Acoust. Soc. Am.* **85** 732
[2] Ardron K H and Duffey R B 1978 *Int. J. Multiphase Flow* **4** 303
[3] Nigmatulin R I, Gubaidullin D A and Fedorov Yu V 2013 *Doklady Physics* **58** 261
[4] Gubaidullin D A, Gubaidullina D D and Fedorov Yu V 2013 *Fluid Dynamics* **48** 781
[5] Gubaidullin D A and Fedorov Yu V 2013 *Journal of Applied Mathematics and Mechanics* **77** 532
[6] Gubaidullin D A and Nikiforov A A 2010 *High Temperature* **48** 170
[7] Gubaidullin D A, Nikiforov A A and Gafiyatov R N 2012 *High Temperature* **50** 250
[8] Gubaidullin D A, Nikiforov A A and Gafiyatov R N 2015 *High Temperature* **53** 240
[9] Gubaidullin D A and Fedorov Yu V 2015 *Fluid Dynamics* **50** 61
[10] Varaksin A Yu 2013 *High Temperature* **51** 377
[11] Varaksin A Yu 2015 *High Temperature* **53** 423
[12] Pokusaev B G, Tairov E A and Vasilyev S A 2010 *Acoustical Physics* **56** 306
[13] Pokusaev B G, Tairov E A, Safarov A S and Nekrasov D A 2014 *Thermal Science* **18** 591
[14] Gubaidullin D A, Teregulova E A and Gubaidullina D D 2015 *High Temperature* **53** 714
[15] Gubaidullin D A and Fedorov Yu V 2012 *Fluid Dynamics* **47** 593
[16] Shagapov V Sh and Sarapulova V V 2015 *Acoustical Physics* **61** 37
[17] Temkin S 1992 *Phys. Fluids A.* **4** 2399
[18] Silberman E 1957 *J. Acoust. Soc. Am.* **29** 925
[19] Karplus H B 1961 *Research and Development/Report ARF-4132-12. Atomic Energy Commission.*
[20] Kuczynski W 2013 *Int. J. Heat and Fluid Flow* **40** 135

Deposition of polymer powder particles on the product surface at jet spraying

V L Fedyaev[1,2], E R Galimov[2], I R Gimranov[3], M S Takhaviev[3], I V Morenko[1] and A R Siraev[1]

[1] Institute of Mechanics and Engineering, Kazan Science Center, Russian Academy of Sciences, Kazan, Russian Federation
[2] Kazan National Research Technical University named after A N Tupolev, Kazan, Russian Federation
[3] Gazprom Transgaz Kazan, Russian Federation

E-mail: morenko@imm.knc.ru

Abstract. Gas-thermodynamic processes taking place during spraying of polymer powder materials are considered. Numerical and analytical models describing these processes are proposed. For the case of circular cylinder rotating with a constant angular velocity results of calculations and engineering formulas for estimating the thickness of powder particles layer deposited on the processed surface are present.

1. Introduction

Widespread application of polymeric powder materials for production of coatings on the surface of parts, products, constructions is caused by, on the one hand, the development of chemistry, production of a large variety of synthetic materials that possess valuable properties and performance characteristics, on the other hand, the increasing requirements to coating characteristics [1-3]. In this work among a great number of ways and methods to deposit the polymer powder coatings the jet methods such as flameless gas-thermal spraying are distinguished. When developing the appropriate equipment and technologies we can point out two main problems: to provide proper performance, required quality of the coatings and to minimize the materials loss and energy consumption. One of the effective ways to solve these problems is mathematical modeling of gas thermodynamic processes that take place during the spraying as well as processes of coating formation on the surface of processed bodies.

For the spraying processes of polymer powder on the surface of parts, products and constructions it is characteristic that the jet flowing out of the nozzle, the spraying gun header or the other device can be flat, axisymmetric, circular, conical and etc. Apart from the carrying medium (air fuel combustion products) it contains particles of polymer powder that can be solid, fritted or completely melted. In the presence of temperature drops in the direction of medium motion the heat exchange occurs between powder particles and carrying medium, jet and ambient air.

Particularly complex gas-thermodynamic processes are observed when the jet flows over the processed body. It is very difficult to take fully into account the named and other peculiarities of the considered processes at mathematical modeling, the corresponding problems can be solved numerically.

2. Calculation of the flow past of the cylinder

Below it is assumed that the body being processed is a lengthy circular cylinder of radius a, rotating counter-clockwise at a constant angular velocity ω, the carrying medium – air, lateral dimensions, volume concentration of polymer powder particles are small. For the mathematical description of a medium flow and particles movement we use the averaged Reynolds equations with closing $k-\varepsilon$ turbulence model and Lagrangian continual approach, respectively. Velocity profile, temperature, turbulence intensity and scale of turbulence are specified at the inlet of computational domain in the jet zone. The symmetry conditions are set on the upper and lower boundaries of the computational domain. At the outlet we set boundary conditions that signify the equalization of hydrodynamic and thermal characteristics of the flow.

The no-slip condition are realized at the boundary of the streamlined body, temperature is set constant.

Figure 1 shows the calculation results of the streamlines using triangular mesh and finite volume method at a relative rotational speed of the cylinder surface equal to 0.25.

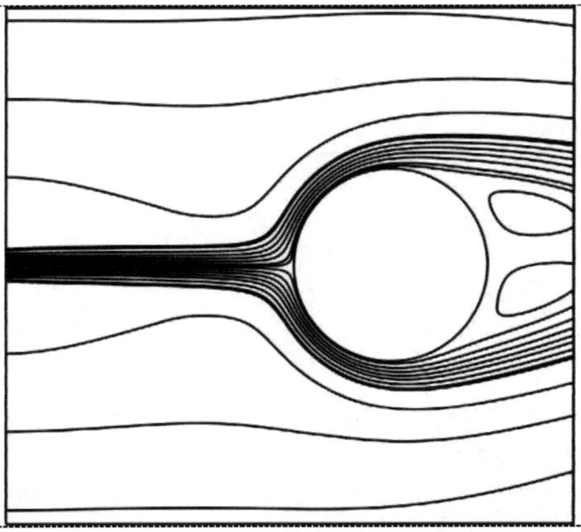

Figure 1. Streamlines.

It can be seen that when the jet flows over the body above and below it air is attracted to the jet, there is an effect of ejection which consists in the fact that at high speeds the pressure is reduced as a result the environment flows into this area. In addition cylinder rotation causes a contraction of the oncoming flow in the upper part of the flow region near the front point of the body and some shift upwards of the vortex region behind the body. It should be noted that unlike the case when rotating cylinder is streamlined by an unlimited flow [4] where Karman Street is formed behind the body, in this case such structures are not observed.

3. Calculation of the thickness of deposited powder layer

In order to obtain engineering evaluations it is additionally assumed that the lateral dimensions of a streamlined circular cylinder are commensurable with the cross section dimensions of flowing over plane jet near its front point. A model of turbulent jet-source is used to describe the jet flow. The fact that when jet flows over an obstacle we can identify the region of a free jet, the zone of flow turning and trickles moving along the surface of the streamlined body [5] is taken into account. Taking into consideration the above and the fact that thickness of boundary layers on the front part of the body surface is small, the movement of air in them has no substantial effect on the dynamics of powder particles deposition in the zone of flow turning we identify the phantom cross section S_0 after which the carrying medium flowing over the body is considered an ideal incompressible fluid. It is necessary to understand that when gas suspension cross flows a circular cylinder including the rotating one the particles are deposited mainly on the front surface of the body [4], close to which the effect of fluid viscosity on the flow is insignificant.

When considering the motion of the powder particles the following assumptions are made: particles have a spherical shape, identical dimensions, the particle radius is $r_p = 8 \cdot 10^{-5}$ m, the density of powder material ρ_p=900 kg/m^3, particles do not rotate around its own axis, their velocity at the nozzle outlet is the same $u_{pc} = 0.8$ m/s, powder mass flow rate through the slotted nozzle of the spraying gun with the size 0.016x0.016 m $G_p = 0.003$ kg/s, the distance between the particles is about $5 \cdot 10^{-4}$ m, volume concentration $1.5 \cdot 10^{-2}$, mass concentration – 0.92. Because of small particle size, volume concentration it is assumed that when moving in the air they do not break up and do not stick together in the jet from the nozzle section to the cross section S_0 particles velocity is close to the velocity of flow particles, gas suspension is a homogeneous medium. Rectangular Cartesian coordinate system xOy is introduced its beginning is at the center of a circular cylinder and axis Ox is directed along the symmetry axis of the jet along the flow. Based on solutions for the main portion of a submerged turbulent jet [5] cross flow of a circular cylinder by an ideal fluid [6] taking a number of assumptions in cross-section S_0 that is located at a distance $|x_s|$ from the center of the rotating cylinder we obtain for the longitudinal particle velocity u_{ps} a dependence

$$u_{ps} = 3.8u_{pc}\left(\frac{\bar{l}_0}{\bar{x}_{s\Pi}}\right)^{1/2}\left(1-9.7\left|\frac{\bar{y}_{ps}}{\bar{x}_{s\Pi}}\right|^{3/2}\right)^2 \tag{1}$$

Here $\bar{l}_0$, $\bar{x}_{s\Pi}$ – dimensionless divided by/ attributed to cylinder radius half the width of the nozzle section, remoteness of cross section S_0 from the pole of the jet source, $a = 0.05$ m, $\bar{l}_0$=0.16, $\bar{x}_{s\Pi}$=5.0, $\bar{x}_s$=2.2. The ordinate of the source from which the particles that hit the control point on the cylinder surface K with dimensionless coordinates $\bar{x}_{pK}$, $\bar{y}_{pK}$ are emitted will be:

$$\bar{y}_{ps} = \bar{y}_{pK}\left[1-0.5\frac{v_0\bar{x}_{pK}}{a}\left(\frac{1}{R_\mu}-\tilde{\tau}_k\right)\right]+\frac{\omega\bar{x}_{pK}}{R_\mu}, \tag{2}$$

where $v_0 = 4.55u_{pc}\bar{l}_0/\bar{x}_{s\Pi}$ – velocity of the oncoming flow of an ideal fluid; $R_\mu = k_c r_p \mu/m_p$ – parameter characterizing the effect of the aerodynamic drag force on the particles, k_c=19÷25 – coefficient of proportionality, μ – gas viscosity, m_p – particle mass, $\tilde{\tau}_k = (|x_s| - 0.5a)/v_0$.

In the case of a fixed cylinder $(\omega = 0)$ separating on its surface closely located control points K_i and K_{i+1} using (2) we find the distance $\Delta_{pi} = \bar{y}_{ps(i+1)} - \bar{y}_{psi}$ in cross section S_0 and using (1) – the velocities $\bar{u}_{psi}$, $\bar{u}_{ps(i+1)}$, mean velocity $u_i = 0.5(\bar{u}_{psi} + \bar{u}_{ps(i+1)})$ on the interval Δ_{pi}.

Having these data for evaluating the thickness of the powder layer sprayed over a period of time τ_H on the portion of a cylinder surface with a width l_i between these control points we obtain a dependency:

$$h_i = k_\Pi \varepsilon_{pi} u_i \tau_H \left(\rho_p \Delta_{pi}/(\rho_i l_i)\right), \tag{3}$$

where k_{Π} – correction coefficient by which you can take into account the movement of the powder particles along the body surface, their blowing away; ρ_i – medium density of powder layer depending on the rate of fall of particles onto the processed surface, their size, degree of fritting and other factors; ε_{pi} – mean volume concentration of particles of gas suspension flowing through the portion Δ_{pi} of the cross section S_0.

If the cylinder rotates $(\omega \neq 0)$ then at inversion of motion the "spot" of deposited particles will move across the cylinder surface in the clockwise direction at a velocity $v_{\Pi} = a\omega$. In this case we can identify three characteristic stages of spraying: initial stage starting from the beginning of particles deposition $\tau = 0$ to $\tau_{\Pi} = l_{\Pi}/v_{\Pi}$ when the spot will cover a distance equal to its width l_{Π}; regular stage during which the thickness of the sprayed powder layer remains constant and final stage over a period of time τ_k before the end of the spraying.

Assuming in (3) $\Delta_{pi}/l_i \approx 2\delta_{ps}/l_{\Pi} = k_{s\Pi}$ (δ_{ps} – half of the width of particles layer in the cross section S_0, which are deposited on the cylinder surface), $\rho_i = \rho_H$, to evaluate the thickness of layer of deposited particles at the initial stage we obtain the relation:

$$h_0 = h_0(l) = \frac{k}{v_{\Pi}} \int_0^l \varepsilon_p(\eta) u_p(\eta) d\eta \quad (0 \le l \le l_{\Pi}). \tag{4}$$

Here $k = k_{\Pi} \cdot k_{s\Pi} \rho_p / \rho_H$, ρ_H – mean material density of the sprayed powder layer, l – length of the circular arc, $\eta = l - \tau v_{\Pi}$ – longitudinal coordinate moving with the sprayed "spot".

When you move to the regular stage ($\tau = \tau_{\Pi}$, $l = l_{\Pi}$) thickness of powder layer h_p is equal to $h_0(l_{\Pi})$. In particular if in cross section S_0 the concentration of powder particles, their longitudinal velocity varies insignificantly then

$$h_p \approx k \varepsilon_p u_p l_{\Pi} / (a\omega), \tag{5}$$

where ε_p, u_p – mean at a section $2\delta_{ps}$ volume concentration and longitudinal velocity of particles.

In contrast to (4) at the final stage of spraying with increasing l the thickness of deposited powder layer h_k decreases according to formula

$$h_k = h_k(l) = L h_p - \frac{k}{v_{\Pi}} \int_0^l \varepsilon_p(\eta) u_p(\eta) d\eta. \tag{6}$$

Here L – number of complete revolutions of the "spot" along the cylinder surface with overlapping the initial section. When thickness of the sprayed layers is small the time T_{Π}, during which such a revolution is performed will be $T_{\Pi} = \tau_{\Pi} + 2\pi/\omega$. If τ_H – total spraying time ($\tau_H > LT_{\Pi}$), then over a period of time $\tau_k = \tau_H - LT_{\Pi}$ the "spot" moves relatively to the cylinder surface on a distance $l_k = v_{\Pi} \cdot \tau_k$. Accordingly, in (6) parameter l will change from 0 to l_k.

4. Conclusions

In conclusion we note that the obtained relationships to calculate the layer thickness of the sprayed polymer powder as in the case of fixed circular cylinder and in the case of rotating circular cylinder contain all main technological parameters of the considered process and allow evaluating their influence and choose rational regimes of spraying.

References

[1] Yakovlev A D 1987 *Powder paints* (Leningrad: Khimiya) 216
[2] Genel' S V, Belyy V A, Bulgakov V Ya and Gekhtman G A 1968 *Application of polymer materials as coatings* (Moskow: Khimiya) 238
[3] Galimov E R, Zverev E V, Tukbaev E E, Galimova N Ya, Kuryntsev S V and Mukhin A M 2012 *Polymer powder coatings for special purposes* (Kazan: Ofset Servis)
[4] Morenko I V and Fedyaev V L 2014 *Journal of Engineering Physics and Thermophysics* **87**(3) 566–572
[5] Abramovich G N, Girshovich T A, Krasheninnikov S Yu, Sekundov A N and Smirnova I P 1984 *Theory of turbulent jets* (Moskow: Nauka)
[6] Kochin N E, Kibel' I A and Roze N V 1963 *Theoretical hydromechanics* (Moskow: Fizmatgiz)

Journal of Physics: Conference Series **669** (2016) 012013 doi:10.1088/1742-6596/669/1/012013

Heat transfer of gas suspension jet flow on the product surface when applying polymer powder coatings

V L Fedyaev[1,2], E R Galimov[2], I R Gimranov[3], M S Takhaviev[3], I V Morenko[1] and L R Fazlyev[3]

[1] Institute of Mechanics and Engineering, Kazan Science Center, Russian Academy of Sciences, Kazan, Russian Federation
[2] Kazan National Research Technical University named after A N Tupolev, Kazan, Russian Federation
[3] Gazprom Transgaz Kazan, Russian Federation

E-mail: morenko@imm.knc.ru

Abstract. Thermal processes occurring at jet spraying of polymer powder materials on the surface of a rotating cylinder are considered. The initial problem is solved numerically using the finite volume method and under certain assumptions analytically. Relationships to calculate the temperature, its rate of change and other indices in the sprayed layer and processed body are presented.

1. Introduction

Research of heat exchange processes at interaction of laminar and turbulent jet flows with streamlined bodies represents both theoretical and practical interest. Turbulent jets are widely used in metallurgy when drying various materials, products, in heating and ventilation systems, at heat treatment, materials cutting and thermal drilling. At the same time, cold gas jets are used for cooling the heated elements of constructions, machines, apparatuses, instruments and in many other cases [1, 2].

When spraying powder on the surface of parts, products, constructions as well as at their preheating, further thermal treatment of sprayed powder layer along with the processes of aerodynamics the heat exchange of working media with the powder layer and base material is of great interest [3-5]. It is necessary that the material of polymer powder layer deposited on the body surface was heated to a temperature that provides its spreading over the surface, reduces the porosity to a minimum and the acceptable adhesion. On the other hand, excessive heating of the powder material may lead to blowing off of the coating by the gas jet, its run-off under action of gravity, the degradation. It should be noted that for most of the used polymer powder materials allowable range of temperature change is very small.

The main factors affecting the dynamics of material heating of the sprayed powder layer are thermal and physical characteristics of the carrier gaseous medium: density ρ, viscosity μ, coefficient of thermal conductivity λ, specific heat at constant pressure c, temperature at the nozzle outlet T_0, volume flow rate G_0, respectively, mean longitudinal velocity u_0. An important role is also played by thermal and physical parameters of powder material its density ρ_p, coefficient of thermal conductivity λ_p, specific heat at constant pressure c_p, mean temperature at the nozzle outlet T_{pc}, velocity u_{pc}, volume concentration ε_{pc}, characteristics size of powder particles d_p, thickness of powder layer on the body surface h_H, mean thermal and physical characteristics of the medium of this layer: density ρ_H, coefficient of thermal conductivity λ_H, specific heat c_H.

Of course, sizes and shape of a processed body, its movement relative to the nozzle or vice versa, movement of the nozzle relative to the body, their relative position and thermal and physical properties of the material of processed body have a noticeable effect on the heat exchange: density ρ_m, coefficient of thermal conductivity λ_m, specific heat c_m, initial temperature T_{m0}, characterizing the heating of material before spraying, including the initial temperature of the treated body surface T_{w0}.

In all these cases, heat transfer at the contact surface of gas suspension and streamlined body is forced convection its intensity is characterized by the Nusselt number including the complex heat exchange: convection in conjunction with thermal radiation when effective coefficient of convective heat exchange is used for an engineering evaluation of heat transfer [6]. For jet spraying of polymer powders the jets are mainly flat or axisymmetric, processed bodies are of various shapes and sizes and the situation when lateral dimensions of the body and width of the jet flow are close is possible.

Taking into account the above mentioned we consider the heat exchange of fixed or rotating circular cylinder at jet spraying of polymer powder.

2. Governing equations

We assume that the nozzle of a spraying device is slotted and the jet of gas suspension is flat. Considering that for jet spraying of polymer powder the most intense heat transfer is observed in the vicinity of the front point of a cross flowed cylinder we confine ourselves to considering of heat exchange processes at this section. Let's introduce the local fixed coordinate system $\xi O \eta$ its center O is located at the position of the cylinder front point, axis $O\xi$ is directed along the contour of the cylinder surface, axis $O\eta$ – along the outer normal to the surface along the jet symmetry axis.

Thickness of the sprayed powder layer $h_H = h_H(\tau, \xi)$ depends on time τ ($0 \le \tau \le \tau_H$, τ_H – spraying time) and longitudinal coordinate ξ. Value h_m is assumed constant. It can be wall thickness of a tube or if the processed cylinder is massive thickness of the heated layer. Further in the named coordinate system we write the heat transfer equations in both layers in the absence of the source terms taking into account the convective components caused by the rotation of a cylinder together with a sprayed layer.

At the initial time $\tau = 0$ it is assumed that on the surface of the considered body section there is a very thin powder layer of constant thickness h_{H0} with mean values ρ_{H0}, λ_{H0}, c_{H0} and temperature T_{H0}. Initial material temperature of the processed body layer $T_{m0} = T_{m0}(\eta)$ varies only along its thickness according to a given law.

On the outer surface of the sprayed layer blown by the jet of gas suspension also being under the influence of infrared radiation we write the Newton's condition:

$$\eta = h_H : \ -\lambda_H \frac{\partial T_H}{\partial \eta} = \alpha \left(T_{H2} - T_g \right). \tag{1}$$

Here $\alpha = \alpha_k + \alpha_r$, $\alpha_k = \alpha_k(\xi)$ – local coefficient of convective heat exchange of carrier gaseous medium with the layer surface [2,7-9], T_g – mass average temperature of this medium at a distance from the surface of a front section of a streamlined body, $\alpha_r = \alpha_r(\xi)$ – component of the effective heat exchange coefficient characterizing the influence of an infrared radiation on a surface, T_{H2} – mean temperature on the outer surface of a layer.

Let's assume that at the boundary between the sprayed layer and the body we have an ideal thermal contact, respectively,

$$\eta = 0: \; T_{H1} = T_{w1}, \; \lambda_H \frac{\partial T_H}{\partial \eta} = \lambda_m \frac{\partial T_m}{\partial \eta}, \tag{2}$$

where T_{H1} – material temperature of the sprayed layer, T_{w1} – material temperature of the body on its surface.

On the back side of the body ($\eta = -h_m$) we assume that

$$T_m = T_{m0} = const. \tag{3}$$

The named boundary conditions should be supplemented by the conditions either on the edges of the selected layers or by periodicity conditions in the circumferential direction.

Specificity of this problem consists of the fact that:

– thermal processes are non-stationary occurring in two thermally coupled layers with different thermal and physical characteristics;

– with time the thickness of sprayed layer changes both because of powder particles coming into the layer and melting of these particles, spreading along the body surface: the first mechanism leads to an increase of h_H , conversely the second leads to a decrease of the thickness because of the decrease of material porosity of the layer;

– at the same time, when porosity of the sprayed layer changes the thermal and physical parameters of its material also change: $\rho_H = \rho_H(\tau)$, $\lambda_H = \lambda_H(\tau)$, $c_H = c_H(\tau)$.

This implies that solution of this problem in a rigorous formulation is very difficult it is reasonable to use numerical methods.

3. Numerical simulation

As an example of the application of numerical methods figure 1 shows the temperature dynamics of polymer powder particles as they fly from the nozzle ($\eta = -0.2$ m) to the surface of rotating counter-clockwise with a relative velocity 0.92 circular cylinder. For the mathematical description of the turbulent air flow the averaged Reynolds equations and closing $k-\varepsilon$ turbulence model are used. Flight simulation of powder particles is carried out using Lagrangian continual approach.

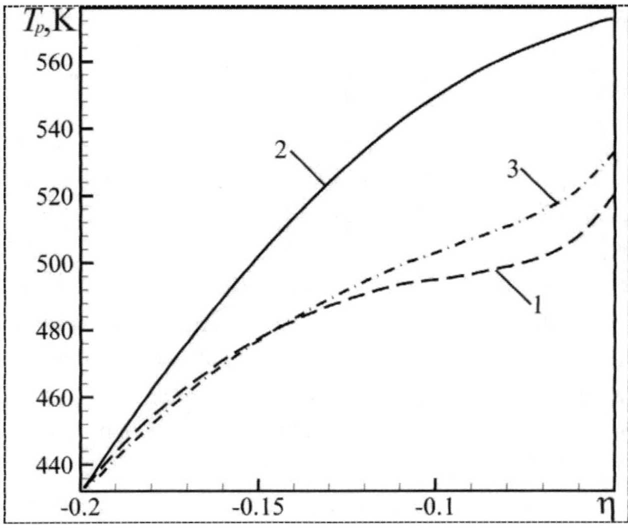

Figure 1. Change of the particle temperature versus its position on the axis η (1 – particle is on the upper boundary of the jet, 2 – in the center of the jet, 3 – particle is on the lower boundary of the jet).

From the results shown in figure 1 it follows that because of the heat exchange with a high temperature gaseous medium of the jet on approaching to the body surface the mean material temperature of powder particles increases notably more intensively increases the particles temperature on the jet axis. The difference in the particles temperature on the upper and lower boundaries of the jet is due to the flow displacement relative to the jet axis at cylinder rotation. Another interesting fact is that behavior of particles temperature on the jet axis and periphery near the cylinder surface differs considerably.

4. Analytical solution

We take additional assumptions to obtain engineering calculating formulas.

1. Assume that the temperature gradient T_H, T_m in the longitudinal direction is small ($\partial T_H/\partial\xi \square 0$, $\partial T_m/\partial\xi \square 0$) because thermal effect of the heated gas jet occurs on a large surface, concentration of this heat source is very small.

2. Assume also that $\partial^2 T_H/\partial\xi^2 \square 0$, $\partial^2 T_m/\partial\xi^2 \square 0$ heat flows in the longitudinal direction change only slightly.

3. Considered nonstationary thermal processes in the sprayed layer, body are, substantially, additional since materials of polymer powder, substrate are preheated. Moreover from the technological point of view these processes represent the greatest interest closer to the end of spraying, i.e. after a certain time. Consequently, analysis can be limited only to the regular heating mode of the system assuming that the heating rate of powder layer and body are constant:

$$w_H = \left(\frac{1}{\rho_H c_H}\right)\left(\frac{\partial(\rho_H c_H T_H)}{\partial\tau}\right) = w_{H0} = const\,; \tag{4}$$

$$w_m = \partial T_m/\partial\tau = w_{m0} = const\,. \tag{5}$$

As a result the initial equations of heat transfer are considerably simplified and take the form:

$$\partial^2 T_H/\partial\eta^2 = \overline{w}_{H0}(\tau),\ \partial^2 T_m/\partial\eta^2 = \overline{w}_{m0}\,. \tag{6}$$

Here $\overline{w}_{H0} = w_{H0}/a_H$, $\overline{w}_{m0} = w_{m0}/a_m$, $a_H = \lambda_H/(\rho_H c_H)$, $a_m = \lambda_m/(\rho_m c_m)$.

Integrating these equations taking into account the boundary conditions (1) and (3) we find:

$$T_H = T_{H1} + \left[\alpha/\lambda_H\left(T_g - T_{H2}\right) - \overline{w}_{H0} h_H\right]\eta + 0.5\overline{w}_{H0}\eta^2\,, \tag{7}$$

$$T_m = T_{w1} + \left[\left(T_{w1} - T_{m0}\right) + 0.5\overline{w}_{m0} h_m^2\right]\eta/h_m + 0.5\overline{w}_{m0}\eta^2\,, \tag{8}$$

wherein in accordance with (2),

$$T_{H1} = T_{w1},\ \alpha\left(T_g - T_{H2}\right) - \lambda_H \overline{w}_{H0} h_H = \left(\lambda_m/h_m\right)\left(T_{w1} - T_{m0} + 0.5\overline{w}_{m0}\eta^2\right). \tag{9}$$

Unfortunately, the relations (9) are not sufficient in order to find parameters T_{H1}, T_{H2}, $\overline{w}_{H0}$, $\overline{w}_{m0}$. Therefore, in a refinement of the problem (1) – (6) we suppose that the material temperature on the outer surface of the powder layer T_{H2} during spraying is close to the mean temperature of powder particles falling on this surface and temperature of gaseous inclusions of carrier medium. The latter can be measured. In any case over a period of time $0 \le \tau \le \tau_H$ the temperature T_{H2} will change only slightly. Secondly as it is customary at regular heating mode we assume that velocity $w_{H0} \cong w_{m0}$, i.e. the heating rate in the sprayed layer and body are close to each other. Thirdly we take into account that the mean heating rate of powder layer

$$w_H = (q_2 - q_1)/(\rho_H c_H h_H),$$

substrate –

$$w_m = (q_1 - q_0)/(\rho_m c_m h_m),$$

where q_0 – heat flux at the back side of the substrate; q_1, q_2 – heat fluxes on the inner and outer surfaces of the sprayed layer, respectively.

Moreover, assuming that the back side of the processed body is insulated (q_0=0) using (8) from the last relation we determine the velocity w_{m0} and from (9) – the temperature T_{H1}:

$$w_{m0} = \left(\frac{\alpha(T_g - T_{H2})}{\rho_m c_m h_m} \right) \Big/ (1 + k_H), \quad (10)$$

$$T_{H1} = T_{m0} + \left[\alpha(T_g - T_{H2}) - \lambda_H \bar{w}_{H0} h_H \right] h_m / \lambda_m - 0.5 \bar{w}_{m0} h_m^2 . \quad (11)$$

Here $w_{m0} = w_{m0} = w_{H0}$, $k_H = (\rho_H c_H h_H)/(\rho_m c_m h_m)$. For small k_H we get:

$$w_{m0} \approx a_m Nu_m (T_g - T_{H2})/h_m^2 , \quad (12)$$

$$T_{H1} \approx T_{m0} + Nu_m (T_g - T_{H2}) - 0.5 w_{m0} h_m^2 / a_m , \quad (13)$$

where $Nu_m = \alpha h_m / \lambda_m$.

We make sure that the temperature on the contact surface of the sprayed layer and the substrate is determined by the temperature T_{m0}, heat input into the layer from the gas suspension , the heating rate of material: the higher is the heating rate the lower is the temperature T_{H1}.

In order to establish an explicit dependence of the thermal indices of the considered system on time τ for the strip of width $\Delta\xi = 1$ we write the heat balance equation:

$$q_2 d\tau = d\left(\rho_H c_H h_H T_H + \rho_m c_m h_m T_m \right).$$

Here T_H, T_m – mean material temperature in the sprayed layer and the substrate, respectively, $q_2 = \alpha(T_g - T_{H2})$. According to (7), (8)

$$T_H = T_{H1} + 0.5 Nu_H (T_g - T_{H2}) - 0.33 \bar{w}_{H0} h_H^2 , \quad T_m = 1.5 T_{H1} - 0.5 T_{m0} + 0.67 \bar{w}_{m0} h_m^2 ,$$

where $Nu_H = \alpha h_H / \lambda_H$.

Assuming that in $T_H(T_{H1}, h_H)$, $T_m(T_{H1})$ variable parameter is the temperature T_{H1} (layer thickness h_H assumed to be known) under the condition that at the initial time ($\tau \to 0$) $T_{H1} = T_{H0}$ we define:

$$T_{H1} = T_{H0} + \chi_H (T_g - T_{H2}) \tau , \quad (14)$$

where $\chi_H = 0.67\alpha / (\rho_m c_m h_m + 0.67 \rho_H c_H h_H)$.

Further, considering the equations (13) – (14) as a system, we find the approximate dependencies for $T_{H1}(\tau)$ and $T_{H2}(\tau)$:

$$T_{H1} \approx T_{m0} + (T_{H0} - T_{m0})/(1 - \tau\chi_H / Nu_m), \quad (15)$$

$$T_{H2} \approx T_g - (T_{H0} - T_{m0})/(Nu_m - \tau\chi_H). \quad (16)$$

Taking into account (16) from (10) it follows:

$$w_{m0} = a_m \left(T_{H0} - T_{m0}\right) / \left(h_m^2 \left(1 + k_H\right)\right) \approx a_m \left(T_{H0} - T_{m0}\right) / h_m^2 .$$

It can be seen that for small k_H thermal and physical parameters of a thin sprayed powder layer has little effect on the heating rate, substrate characteristics, its thickness and temperature diffusivity coefficient of the material are dominating.

Temperature gradient in the sprayed layer is also of interest. Its mean value

$$T_{H\eta} = \left(T_g - T_{m0}\right) / h_H - \left(T_{H0} - T_{m0}\right)\left(1 + Nu_m\right) / \left(h_H \left(Nu_m - \tau\chi_H\right)\right) . \quad (17)$$

In particular, for small $\tau\chi_H / Nu_m$

$$T_{H\eta} \approx \left(T_g - T_{m0}\right) / h_H - \left(T_{H0} - T_{m0}\right)\left(1 + 1 / Nu_m\right) / h_H .$$

5. Conclusions

Despite the relative simplicity of the obtained calculation relationships they fully reflect the influence on the basic parameters of thermal processes occurring at jet spraying of polymer powder materials of all of the above mentioned parameters including the most important mode parameters, such as, gas and powder flow rate, their temperature, nozzle width of a spraying gun, the distance from it to the surface of the processed body, rotation velocity. However, for the most part indirectly through the heat exchange coefficient α and thickness h_H of the powder layer sprayed on the cylinder surface.

References

[1] Abramovich G N, Girshovich T A, Krasheninnikov S Yu, Sekundov AN and Smirnova I P 1984 *Theory of turbulent jets* (Moskow: Nauka) 717

[2] Yudaev B N 1988 *Engineering thermodynamics. Heat transfer* (Moskow: Vysshaya Shkola) 479

[3] Galimov E R, Zverev E V, Tukbaev E E, Galimova N Ya, Kuryntsev S V and Mukhin A M 2012 *Polymer powder coatings for special purposes* (Kazan: Ofset Servis)

[4] Galimov E R, Galimova N Ya and Tukbaev E E 2010 *Problemy energetiki* **4**

[5] Borisov Yu S, Kharlamov Yu A et al. 1983 *Gas-thermal coatings from powder materials* (Kiev: Naukova dumka) 568

[6] Bennett K O and Mayers D E 1966 *Hydrodynamics, heat and mass transfer* (Moskow: Nedra) 727

[7] Kutateladze S S 1990 *Heat transfer and hydrodynamic resistance* (Moskow: Energoatomizdat) 367

[8] Fedyaev V L, Morenko I V, Siraev A R, Galimov E R, Gimranov I R, Fazlyev L R and Takhaviyev M S 2015 *IOP Conf. Series: Materials Science and Engineering* **86** 1–5

[9] Morenko I V and Fedyaev V L 2014 *Journal of Engineering Physics and Thermophysics* **87**(3) 566–572

Journal of Physics: Conference Series **669** (2016) 012014 doi:10.1088/1742-6596/669/1/012014

Improving the efficiency of plasma heat treatment of metals

Az T Gabdrakhmanov, I H Israphilov, A T Galiakbarov, A D Samigullin, Al T Gabdrakhmanov

Department of Energy and Information, Kazan Federal University, Naberezhnye Chelny, Russian Federation

E-mail: ATGabdrahmanov@kpfu.ru

Abstract. This paper proposes an effective way of the plasma hardening the surface layer at the expense combined influence of the plasma jet and a cold air flow. After that influence occurs a distinctive by plasma treatment microstructure with increased microhardness (an increase of 35%) and depth. There is proposed an improved design of the vortex tube for receiving the air flow with a temperature of 20 C to - 120C.

1. Introduction

The plasma hardening of details from a steel and cast iron is one of the most effective and efficient way to increase a resource of work elements of machines and mechanisms, as well as reducing their consumption of materials. Increasing operational properties hardenable parts achieved due to the improvement of technology hardening that, ultimately, to ensure the optimal thermal cycle (heating-cooling) based on the patterns of structural, phase and polymorphic transitions of the hardening material [1, 2].

A significant feature of the surface hardening is to maintain the quality the bulk metal, which is also heats up with the surface layer. In the plasma quenching amount of energy nested in the metal is sufficient to heat the surface, and the mass of the metal remains cold. [3-5]

2. Experimental studies

In this paper proposes an effective way of the plasma hardening the surface layer due to summary influence of the plasma flow and the coolant in the treatment zone.

The air is used as a supporting source of the coolant, which is readily available and cost-effective cooling medium, and its use in the hardening of steel products a laser quenching can be promising trend in improving this technology.

As a result of the conducted investigations was developed an experimental setup - "vortex tube." The invention relates to refrigeration technology in particular, to the installation, using a vortex effect of separating the gas on the hot and cold flows [6, 7]. The flow can be prepared with a temperature of 20 C to - 120C using pneumatic system as a source of compressed air.

The installation of design shows in Figure 1 [8, 9].

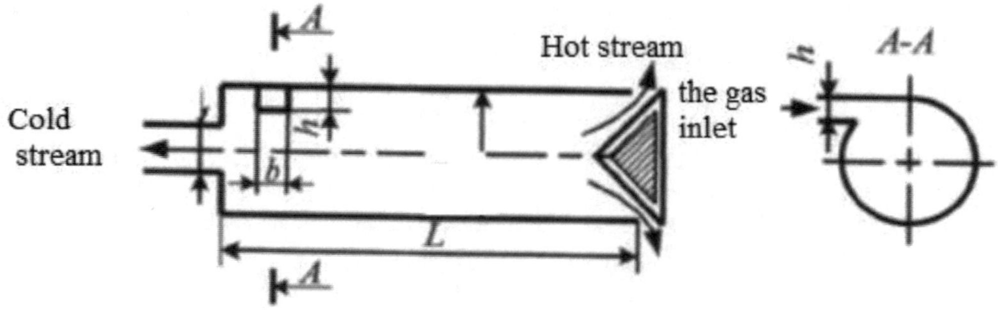

a)

b)

Figure 1. The vortex tube; a) a schematic diagram, b) the appearance

The feature of this device is the vortex zone, having the form of a truncated cone with expansion to the end of the exit stream with a higher temperature. To improve the effectiveness of the "vortex tube", separating chamber provided with finning from outside for cooling the periphery of flow separation.

During the plasma heat treatment on the sample - a steel 45 at the same time supplied cold air flow from the vortex tube with a temperature down to minus 100 ° C. This method of quenching allows to produce a big depth of the hardening and the complex properties of the steel.

3. Results of experimental investigation

The results of experimentally received samples after treatment pulsed plasma generator shows in Fig. 2.

Metallographic researches of the steel 45 in the field of the plasma hardening shows (Fig. 2a), that the area of the thermal impact of the plasma jet has the form of a segment [10]. It is established that, depending on the microstructure and microhardness in the steel 45 the depth formed three layers:

1. The area reflow, consisting of non-equilibrium fine-dispersed structure close to an amorphous, it takes place during quenching from the molten state, has a columnar structure with crystals, extended in the direction of the heat sink. The main structural component is martensite. In this example, the martensite has 6-7 points. This zone microstructurally reveals as weakly etching the layer extremel high hardness and uneven of the treated surface across the width, the maximum depth of the central part is 0,022mm. With the hardness of the surface layer 874HV.

2. The area of the quenching from the solid phase. Directly under the weak etching the layer located area of the plasma influence, which borders with the area complete and incomplete of the hardening with microhardnes, corresponding the hardness of a small spicular martensite and martensite turning into sorbitol of the hardening and sorbitol tempering closer to the main (initial) structure. The thickness of the area is also not uniform and varies 0.45 mm. The depth layer characterized by a strong structural inhomogeneity. In the upper boundary of the layer closer to the surface is observed

martensite and retained austenite. In the lower boundary of the layer closer to the original metal together with martensite is observed elements of the original structure ferrite.

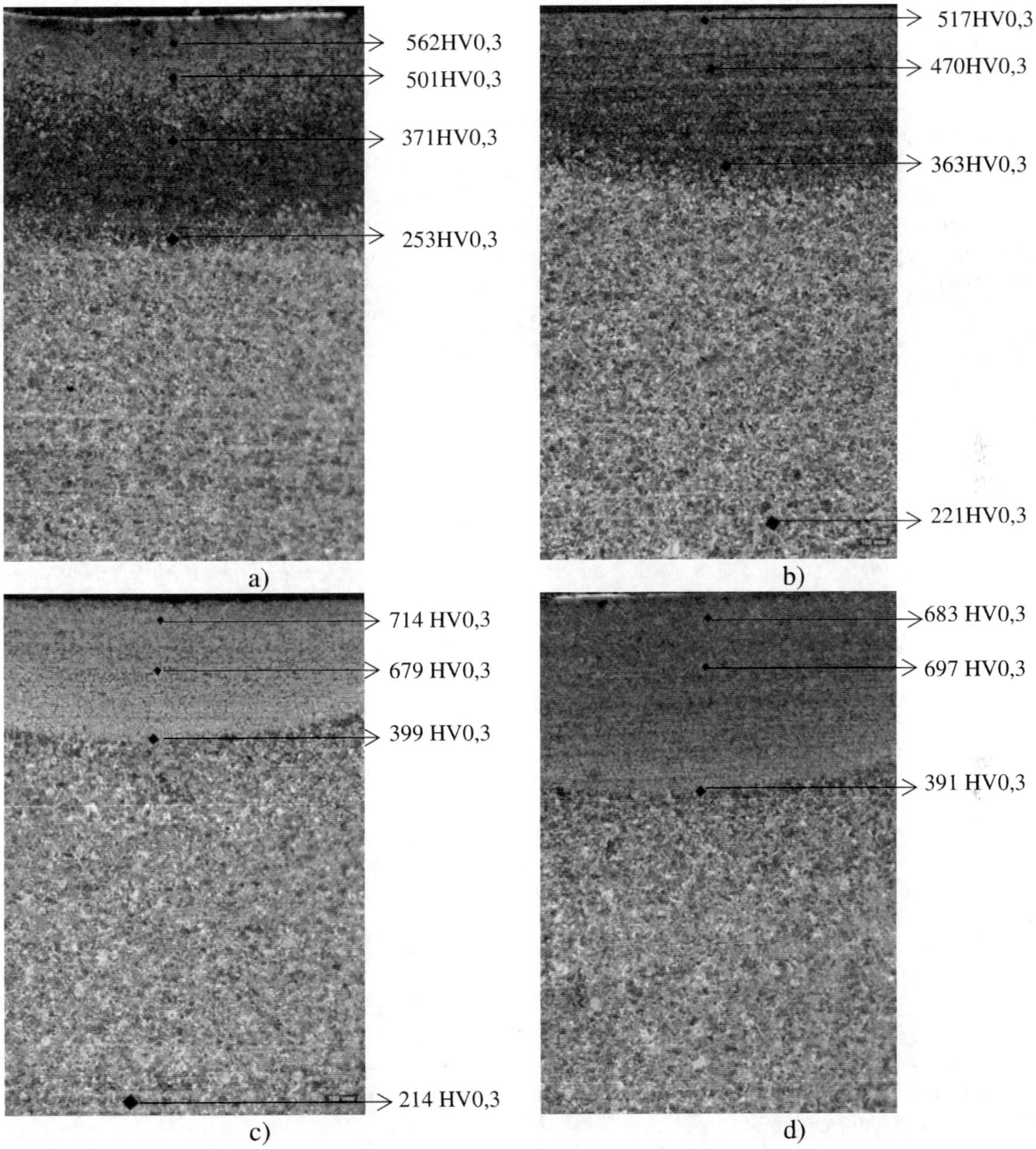

Figure 2. Samples of the steel 45 after plasma treatment × 50 a) without cooling with the reflow of the surface, b) without cooling without the reflow of the surface, c) with cooling with the reflow of the surface, d) with cooling without the reflow of the surface.

3. The transition area. In this zone the metal is heated to a temperature below the Ac1, in which the main structures are structure of the tempering. The zone of the original structure and hardness, representing a fine-dispersed sorbitol tempering. Transitional layer has the most disordered and heterogeneous structure.

Journal of Physics: Conference Series **669** (2016) 012014 doi:10.1088/1742-6596/669/1/012014

The sample in Fig. 2. b) no fusion zone while maintaining the remaining zone and correspondingly less microhardness on the sample surface.

When combined treatment zone thermal effect of the plasma jet also has the shape of a segment, but the third transition area is virtually absent, homogeneous hardened zone and with greater microhardness.

4. The conclusion

After the plasma quenching metals with the cooling zone of the influence by the flow of a cold air increases the quality of treated parts. This technology of the heat treatment metals is the very effective way of directional changes in physical-chemical and mechanical properties of the surface layers, as the combination of these properties of the material determines many operational characteristics products. If the plasma quenching near a cooled object has a factorial or airborne pneumatic system with the surplus resources, there are many applications where compact and inexpensive cold vortex generator called "vortex tube" there is no alternative.

References

[1] Parkin A.A. *A treatment technology of concentrated energy flows*: Textbook. -Samara: SamGTU. 2004-497p.

[2] Zhukov M.F. *The generation of a low-temperature the plasma and the plasma technologies*: Problems and prospects. -Novosibirsk, 2004. p. 246-253.

[3] Israfilov I.H., Saubanov R.R., Rakhimov R.R. *Prospective application of highly concentrated energy for the surface heat treatment products* // Socio-economic and technical systems: research, design, optimization. № 1, 2011, p. 25-30.

[4] Israfilov I.H., Galiakbarov A.T., Gabdrakhmanov Az.T., Nugumanova A.I. *The Improving of the efficiency the plasma processing plant for heat treatment of metals* // Interuniversity scientific book "Designing and research of technical systems", Naberezhnye Chelny, 2008, Issue 12, s.101-105.

[5] Zvezdin V.V., Galiakbarov A.T., Nugumanova A.I., Gabdrakhmanov Az.T., Saubanov R.R. *The investigation of the influence the parameters the pulsed plasma generator on quality process* // Bulletin KSTU. AN Tupolev. Number 2, 2010, s.50-52.

[6] Suslov A.D., Ivanov S.V. *A vortex apparatus*. -M .: Engineering, 1985. -256p.

[7] Merkulov A.P. *A vortex effect and its applications in engineering*. -M .: Engineering, 1969.-185s.

[8] Huzina L.M., Saubanov R.R., Zvezdin V.V., Galiakbarov A.T. *A heat treatment of the metal surfaces with a concentrated source of energy with an impact on the interaction zone of the effect Ranque-Hilsch* // Socio-economic and technical systems: research, design, optimization. Number 1 (61), 2012, p. 27-32.

[9] Huzina L.M., Saubanov R.R., Galiakbarov A.T. *The energy efficient and environmentally generators of the cold* // Socio-economic and technical systems: research, design, optimization. Number 1 (61), 2012, p. 22-26.

[10] Israfilov I.H., Galiakbarov A.T., Gabdrakhmanov Az.T., Nugumanova A.I. *Thermophysical characteristics process of the pulsed plasma treatment steels* // Izvestiya of the Tula State University. Vol. 6. Part 1. - TSU Publishing House, 2013. - S.253-260.

Journal of Physics: Conference Series **669** (2016) 012015 doi:10.1088/1742-6596/669/1/012015

Enhancing of the glow discharge stability in chamber with cathode sections coated by a discontinuous dielectric coating

I G Galeev and T Ya Asadullin

Kazan National Research Technical University named by A.N.Tupolev
420111 Russia Kazan, K.Marx str., 10

galeev.physics@kstu-kai.ru

Abstract. A method of increasing stability of the glow discharge in a gas flow is implemented by reducing the effective current density at the cathode. In this method the current redistributes at the partitioned cathode using the change in the area of the working surface of each section by applying a discontinuous dielectric coating so that it reaches the distribution of the maximum sustainable current discharge across the working surface of each cathode section.

It is known low-temperature plasma of glow discharge is widely used as active media of gas lasers [1]. Flow of gas is effectively used to increase the volumetric energy deposition in a glow discharge both in subsonic and supersonic flow regimes [2,3]. However there is a significant heterogeneity of the distribution of characteristics in the transverse gas flow both along and across the direction of the electric field. It is necessary underline that the electric field in cathode area is much higher than in the main area of the discharge. There is also a significant energy release in the cathode region due to the high cathode potential drop. The discharge current tends to concentrate in the cathode spot with normal current density. Due to the increased electric field strength at the cathode the possibility of development of instability in this discharge region is greater than in the remaining region of the discharge.

This work presents a method of increasing the stability of the glow discharge in the gas flow by reducing the effective current density at the cathode region. According this method the current is redistributed at partitioned cathode by way of changing areas of the working surface each cathode section by applying a discontinuous dielectric coating to achieve the distribution of the maximum sustainable discharge current across the working surface of each cathode section.

Each cathode section is covered by the dielectric discontinuous coating with the ratio of the areas $S_M/S_C < 0,5$, where S_M - the area of the uncovered metal surface, for example in the form of a grid, and S_C - is the total surface area of the cathode section.

As it is known the discharge at the cathode region is compressed in the cathode spot whose dimensions are much smaller than ones of the positive column. Consequently substantial non-uniformity of the discharge is observed in the direction from the cathode to the positive column. In the proposed method the surface area of each cathode section can considerably exceed the working surface area. In this case the effective area through which the current flows on each section of the cathode is much larger than the area through which the current flows to the uncovered metal surface. So the average effective current density on the cathode section is smaller, respectively the dissipation is stretched along the surface of cathode section and the average value it is also much less. The heat flow is distributed on a larger surface section of the cathode to be more efficiently withdraw due to external cooling. As a result the terms of discharge instability worsen, its homogeneity is increased. The range of the ratio of the surface areas $S_M / S_C < 0{,}5$ is defined because when $S_M / S_C > 0{,}5$ experiments have shown insignificant effect.

The method of organization of discharge in the gas flow has been implemented using the chamber depicted in Fig. 1. The anode of the discharge chamber 1 is made solid and flushed with the dielectric wall 2. The partitioned cathode is located between the dielectric plates 3. Cathode sections 4 is separated by a dielectric 5 and connected to the negative pole of the power source through the ballast resistances R. The discharge chamber is connected to gas pumping and cooling of the cathode sections, it is marked by arrows in Fig.1 a). A view of the cathode section from the discharge area is shown in Fig.1 b).

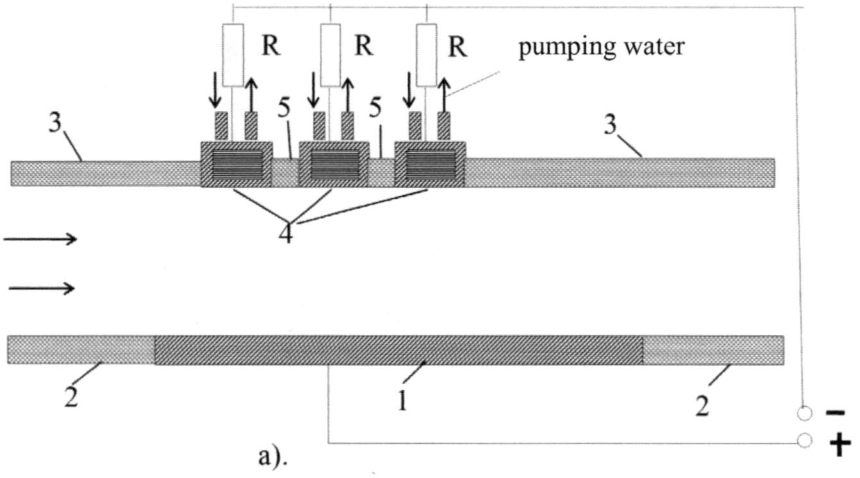

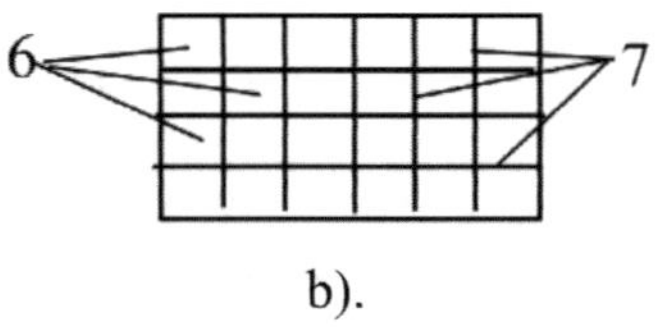

Fig. 1 a). A principal scheme of the discharge chamber. b). View of the cathode section from the discharge area.

The discontinuous dielectric coating is performed on a metal substrate, for instance in the form of squares 6. The remaining open metal surface forms mesh 7 on condition $S_M / S_C < 0{,}5$.

This method is performed in following sequence:

1. to start pumping the gas through discharge chamber, to turn on pumping the cooling water throw the cathode sections 4, to apply voltage to the electrodes 1, 4, to measure the maximum sustainable discharge currents in each cathode section;

2. to register cathode section, which grid 7 is not completely covered by the discharge and to turn off the facility;

3. to recover again above-mentioned cathode section by discontinuous dielectric coating so that the working area of the mesh to reduce;

4. to repeat further the cycle 1-3 to achieve the result that the discharge will completely cover working surfaces of all cathode sections.

Journal of Physics: Conference Series **669** (2016) 012015 doi:10.1088/1742-6596/669/1/012015

Discharge is eventually quasi-distributed along the entire surface of the cathode unit. In this case the average current density and the dissipation along the surface can be reduced more than an order of magnitude.

Experiments on the discharge chamber with the following parameters: the input cross-section of the discharge chamber 5x5 cm^2, gas flow rate of 3.4 g/s, the gas pressure is 38 Torr, pumped gas - air, bare metal surface on each cathode section forming a grid with a cell size of 5x5 mm^2, the area ratio of S_M /S_C was 0.09 (max current per section 0.1 A). Homogeneity and stability of the discharge significantly increased. Volumetric specific energy deposition reached 6.8 W/cm^3. Limit energy deposition is increased more than 2.5 times as compared to the case discharge chamber with uncoated dielectric cathode sections.

References

[1] Raizer Yu P 2009 *Physics of gas discharge* (Moscow: Intellect)

[2] Timerkaev B A 2014 Glow discharge in a transverse supersonic gas flow at low pressures *Therm. Phys. of high temp.* **52** 489 - 93

[3] Saifutdinov A I and Timerkaev B A Drift model of glow discharge with account of nonlocal value of the electric field in the ionization source 2012 *J. Eng. phys*. **85** 1104 - 09

Journal of Physics: Conference Series **669** (2016) 012016 doi:10.1088/1742-6596/669/1/012016

Obtaining fullerene-containing soot during combustion of gaseous hydrocarbons in an external electric field

I G Galeev and T Ya Asadullin

Kazan National Research Technical University named by A.N.Tupolev
420111 Russia Kazan, K.Marx str., 10

galeev.physics@kstu-kai.ru

Abstract. Results of experimental studies are presented which examined the effect of electric field on the soot formation in the combustion of hydrocarbon gases. The peculiarity of these experiments is the possibility of obtaining carbon nanostructures - fullerenes in the combustion of hydrocarbons.

1. Introduction.

Great interest in the study of methods of preparation, structure and properties of nanosized systems is caused by variety and uniqueness of their practical applications. Therefore the increase in the efficiency of production of soot containing carbon nanoparticles - fullerenes is an urgent task. This soot is obtained in the classic method by the use of plasma arc discharge [1]. The discharge is ignited in the helium atmosphere at a pressure about 100 Torr. The arc discharge is used with alternating current I=100 - 200 A, frequency f=60 Hz and voltage U=20-30 V. However this process requires sputtering of graphite electrodes and is relatively ineffective. The authors [2,3] proposed to use the microdischarges for deep processing of oil and producing useful target products. The using of combustion gaseous hydrocarbons with lack of oxygen enables to obtain fullerene-containing soot from relatively cheap raw materials and with high productivity [4].

2. Experimental part.

The experimental stand was assembled allowing changing the relative position of the electrodes, their shape and the distance between them. The experiments were conducted at atmospheric pressure. As a fuel raw material is used a mixture of hydrocarbon gases: propane- 5%, butane - 15%, isobutane - 80%.

3. Results and discussion.

In the first series of experiments the upper electrode was made in the form of a pin (diameter 2 mm) (Fig.1), which served as the anode and the gas-bringing cylindrical copper tube (12mm diameter) acted as the cathode. The burning mode is achieved with maximum smoking at the start of the experimental stand. It is configured by hanging of gas flow. Then voltage is applied to the pin and gas-bringing tube. Typical interelectrode distance was 1.5 cm. The experiments showed that with increasing applied voltage the flame shape changing, the flame starts to go around the pin. And around the pin an air layer is formed. The flame begins oscillating spatially about a vertical axis and the frequency of this low-frequency oscillations enhance with the increase of the applied voltage U. Electrical breakdown was with applied voltages about 3 kV and electric discharge had been occurred between the electrodes. The volumetric fluctuations of glow discharge area was observed in the range of 3 to 3.5 kV. With a further increasing of the applied voltage the discharge area is spatially stabilized and burns smoothly. The formation of soot begins after a period about one minute of operation in this mode. Subsequent experiments have shown that the soot formation occurs more effectively at the cathode and it is best positioned above over the flame.

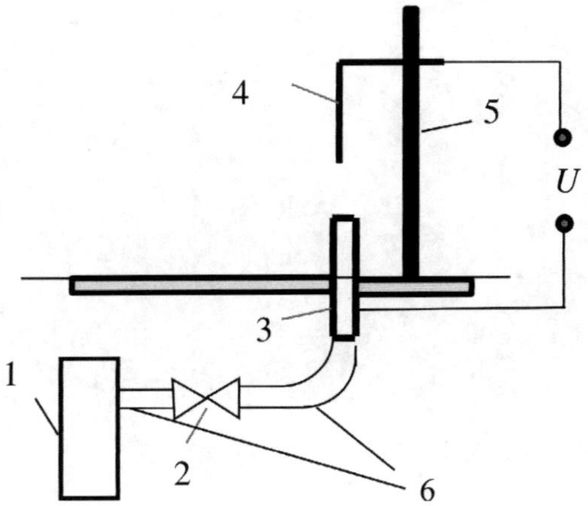

Fig.1 A schematic diagram of the experimental stand. 1 - a gas cylinder, 2 – a valve, 3 – a copper piping for the gas mixture, 4 – the top electrode in the form of a pin (needle), 5 – a dielectric mounting stand, 6 – the inlet hoses.

When the distance between the electrodes is strong reduced then the intensity creation of the soot decreases and the upper electrode is located in the centre of the flame. When electrodes are removed from each other to a distance greater than 2 cm then soot build-up becomes weak, in this case the entire upper electrode is not in flames. The best option for the rate of soot formation was the distance of 1.5 cm between the electrodes.

The father increasing gas flow leads to the immersion of the electrode to flame, which adversely affects soot creation. Too strong reduction of the feed stream led to a lack of food hydrocarbon raw materials, which also negatively affected soot creation. The highest efficiency of the process was observed with average gas stream.

Different variants of upper electrode were considered during the experiments: pin (needle placed vertically), ring, needle (set horizontally). The best variant of the experiment was with the pin positioned vertically.

In the second series of experiments the upper electrode was made in the form of a ring and located in a horizontal plane over the anode tube. There was noticeable stretching region of the flame in the horizontal plane. However the soot formation was less effective.

The electron microscopic photos of images soot samples is obtained for different operation of experimental stand. They show a conglomerate of many interconnected discrete bulk objects. Analysis of photographs with different magnifications showed that they consist of even smaller objects. Fig. 2 shows one of images fullerene-containing soot, wherein the fixed interval of 100 nm shown in the figure. In this Fig. 2 one of the branches of soot is represented which shows the presence of mass objects with dimensions of the order 30 nm and above.

On the basis of obtained data it can be concluded that the electric field of the discharge can have a positive influence on the soot formation, increasing the rate of soot production. It was recorded that the rate of soot formation is significantly affected with both the configuration of the applied electric field and its value and also the presence of discharge currents.

The rate formation of soot was significantly increased in the presence of discharge and the polarity of the applied voltage is very significant. More intensive formation of soot occurred at negative polarity of the upper electrode.

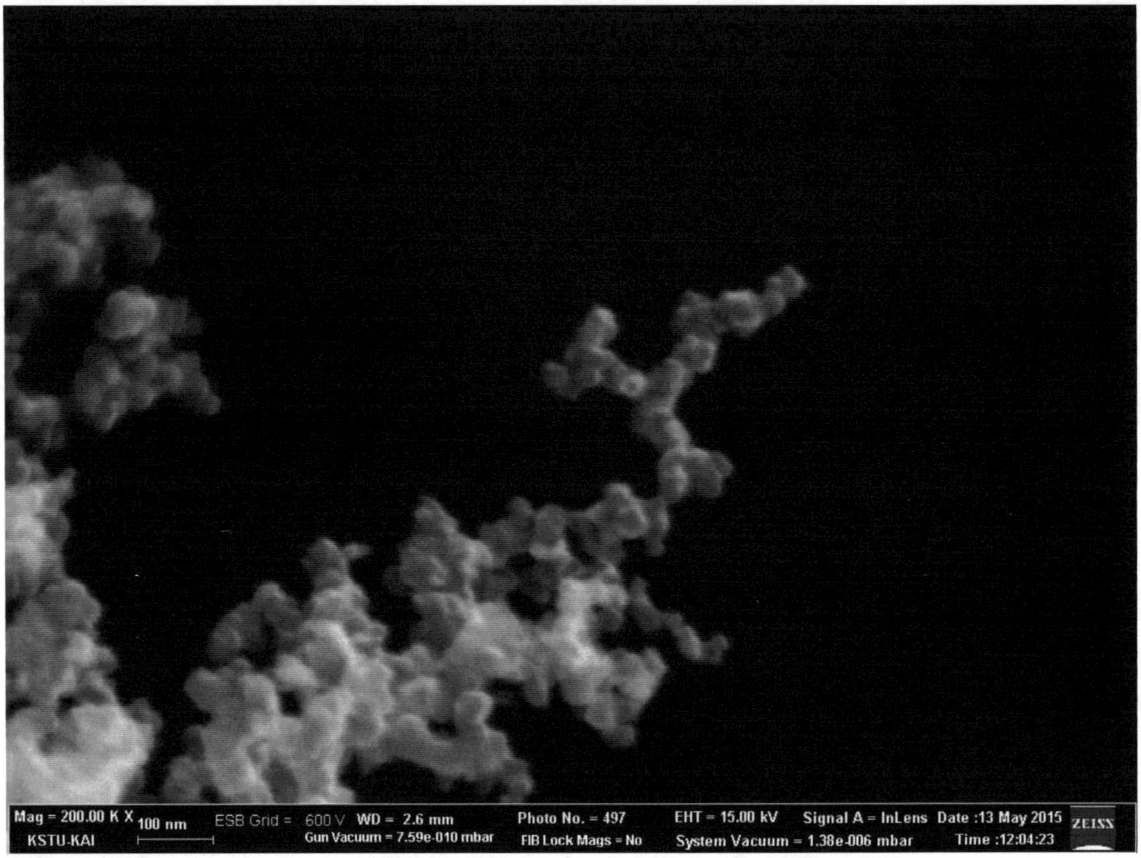

Fig. 2 The electron microscopy photo of a sample of soot. The zoom scale is 100 nm.

References

[1] Smalley R E , Haufler R E Electric arc process for making fullerenes 1993 United States Patent **US 5227038**

[2] Timerkaev B A, Mukhamedyarov H G, Nefedev E S, Fakhrutdinov I M and Yagudin G S 2006 Plasma-chemical *reactor* for deep processing of oil *Russia. Vestnik KSTU. A. N. Tupolev* **№4** 35-38

[3] Timerkaev B A , Ganieva G R, Ziganshin D I, Auhadiev M M Electrical microdischarges in liquids and the prospects for their use in plasma chemistry 2014 *Journal of engineering physics* **87** 677 -81

[4] Mansurov Z A Soot, polycyclic aromatic hydrocarbons, fullerenes, and nanotubes in the combustion of hydrocarbon 2011 *Journal of engineering physics* **84** 116-49

Study of the petroleum schedules thermal cleaning process from asphalt, ressin and paraffin deposits using low-temperature plasma

A D Samigullin, A T Galiakbarov and R T Galiakbarov

Kazan Federal University, 423800, 68/19 Mira avenue, Naberezhnye Chelny, Russia

ADSamigullin@kpfu.ru

Abstract: Petroleum industry uses large amount of pumping and compression pipes. Carrying out the whole range of repair works requires cleaning of the pipe inner surface from deposits which appeared in it during operation [1]. The task of asphalt, resin and paraffin deposits control remains one of the most essential for the branch. The article deals with thermal method and device for asphalt, resin and paraffin deposits removal from pumping and compression pipes inner surface, describes and provides the device application scope for cleaning the pumping and compression pipes inner surface. To deal with borehole equipment and pipe systems waxing problem various deposit prevention and removal methods are used, including mechanical, thermal, chemical, combined and nonconventional methods.

1. Introduction

Thermal method of asphalt, resin and paraffin deposits removal from pumping and compression pipes is based on side-wall melting. This is achieved by warming of the cleaned pipe inner surface by heating with plasma torch from outer surface and high-thermal conductivity of cleaned pipe material. When cleaned pipe inner surface reaches the temperature which exceeds melting temperature of asphalt, resin and paraffin deposits, it leads to softening and melting of the asphalt, resin and paraffin deposits thin side-wall layer adjunct to heated area of the cleaned pipe [2].

2. Device operation

Device for removing the asphalt, resin and paraffin deposits from pumping and compression pipes operates the following way. Circular plasmatron is aligned with cleaned pipe with asphalt, resin and paraffin deposits. The pipe with asphalt, resin and paraffin deposits is mounted to working position at an angle α to the level surface. Angle α to the level surface is adjusted depending on pipe clog level and the length of the cleaned pipe with asphalt, resin and paraffin deposits. Circular plasmatron is activated [3].

Gradual heating of the cleaned pipe starting from the bottom part leads to continuous melting of the side-wall layer of asphalt, resin and paraffin deposits along the cleaned pipe resulting in free vent of gases and melting outputs of asphalt, resin and paraffin deposits without resulting in explosion due to gas development and temperature expansion of side-wall layer of asphalt, resin and paraffin deposits in closed space.

Bulk of asphalt, resin and paraffin deposits in the center of heated area of cleaned pipe does not melt due to low-thermal conductivity of asphalt, resin and paraffin deposits. Circular plasmatron operation rate is adjusted depending on the outside pipe diameter and pipe wall thickness. Calculated data shows that cleaning rate of pipes from asphalt, resin and paraffin deposits for pipe with diameter 70 mm and wall thickness 4 mm is 0.42 m/min i.e. device will clean 6 m pipe within 14 min. It will consume 1.3 kW for the cleaning process. Melted side-wall layer of asphalt, resin and paraffin deposits can function as lubricant at outflow of the melted solid mass of asphalt, resin and paraffin deposits from cleaned pipe under gravity [4].

After outflow of melted solid mass of asphalt, resin and paraffin deposits from the cleaned pipe the heating of the cleaned pipe by several circular plasmatron actuations is continued until all the melted asphalt, resin and paraffin deposits residue flow down to receiving container.

3. Computer modeling

For more complete study of temperature distribution pattern and melting of asphalt, resin and paraffin deposits computer modeling of heating process of pumping and compression pipes with asphalt, resin and paraffin deposits was carried out using Star CCM+ software package [5]. Calculation results are shown in Figure 1 and 2.

Figure 1 shows scalar temperature field of the cleaned pipe in cross section after 140 seconds.

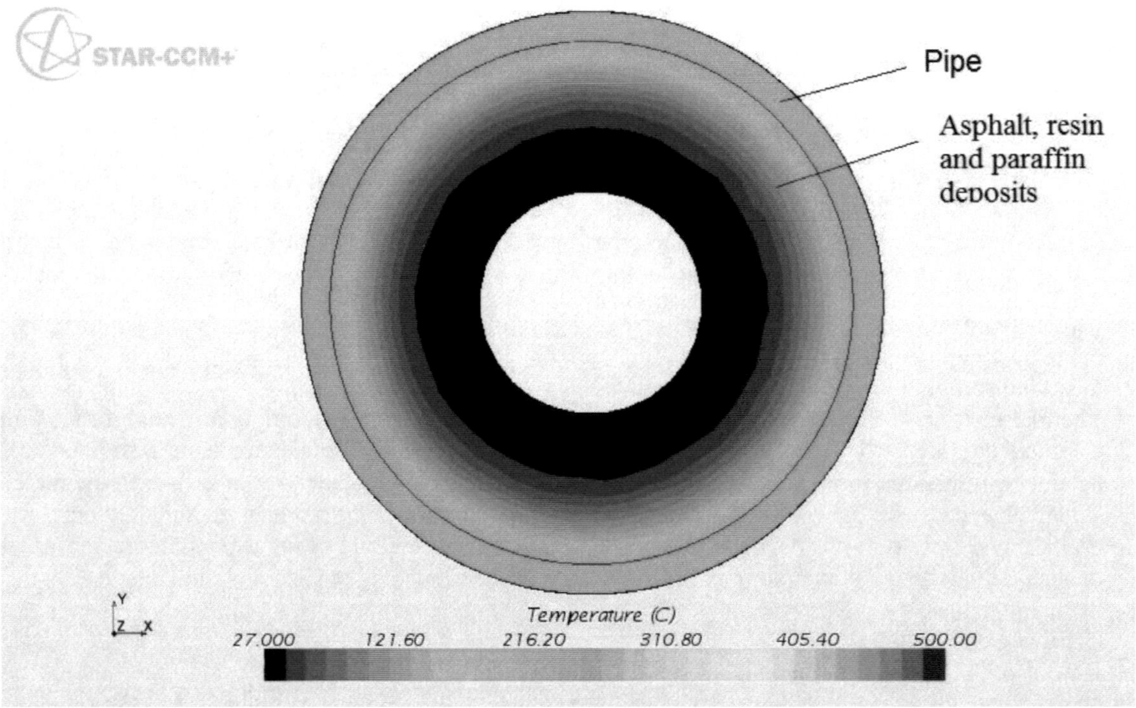

Figure 1. Scalar temperature field in cross section of pipe with asphalt, resin and paraffin deposits.

Figure 2 shows scalar temperature field of the cleaned pipe in longitudinal section with 1 m length after 140 seconds.

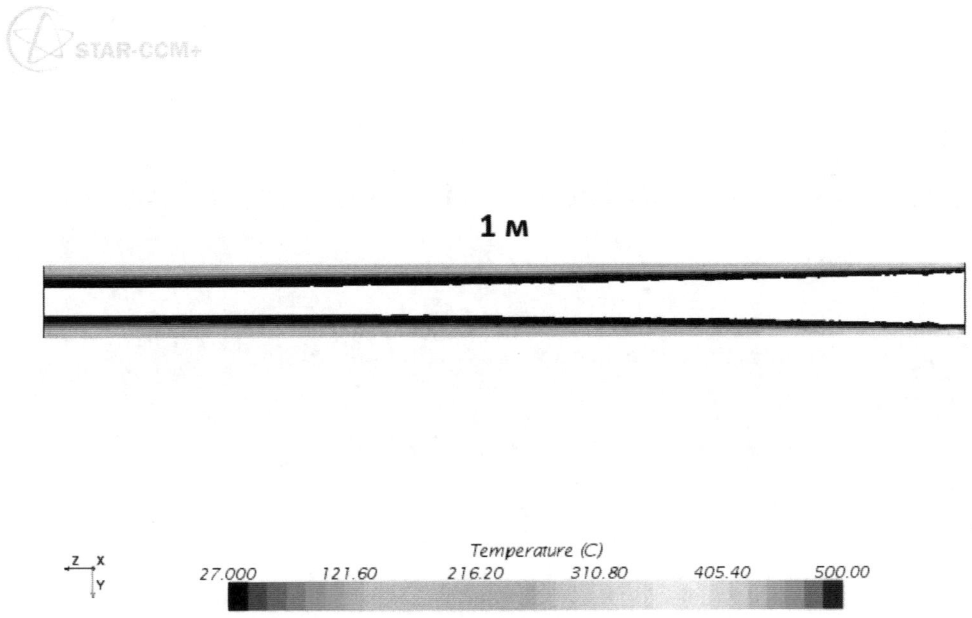

Figure 2. Scalar temperature field in longitudinal section of pipe with asphalt, resin and paraffin deposits.

4. Conclusion

Given thermal cleaning method allows to effectively clean the pumping and compression pipes from asphalt, resin and paraffin deposits. Besides this method is more energy-efficient due to application of low powered plasmatron and low-cost equipment.

References

[1] Ivanova L V, Burov E A and Koshelev V N Asphalt, resin and paraffin deposits in processes of development, transportation and storage // *Online scientific magazine "Oil and Gas Engineering"*. — UGNGU, 2011. — № 1. — Pages 268-284. — ISSN 1813-503X

[2] Russian Federation Patent № 2437726, IPC (2006.01) B08B9/00. Priority from April, 14 2010. Published on December, 27 2011. Patent description

[3] Russian Federation Patent № 2363119, IPC (2006.01) H05H1/26. Priority from July, 03 2006. Published on July, 27 2009

[4] Impulse plasma surface thermohardening of manufacturing engineering parts / Gabdrakhmanov A T, Israphilov I H, Galiakbarov A T, Bashmakov D A and Samigullin A D // *International scientific and technical conference proceedings «Innovative engineering technologies, equipment and materials - 2013» (Interbranch scientific and technical complex «IMTOM-2013») and «Improving competitiveness and energy efficiency of machine builders under conditions of WTO» forum.* Part 2. - Kazan, 2013. - Pages 44-48

[5] Pulse plasma surface thermostrengthening of machine parts /Israfilov, I.H., Galiakbarov, A.T., Bashmakov, D.A., Gabdrakhmanov, A.T., Samigullin, A.D. // *IOP Conf. Ser.: Mater. Sci. Eng.*, Volume 69, Issue 1, 2014, Article number 012037

The effects of bubbles on the structure of upward gas-liquid flow

D A Gubaidulin, B A Snigerev

Institute of Mechanics and Engineering, Kazan science center, Russian Academy of sciences, Russia

E-mail: snigerev@imm.knc.ru

Abstract. The paper presents the results of study of the local structure of turbulent gas-liquid flow in vertical pipe. A mathematical model based on the use of Eulerian description for both phases taking into account the action of different forces of interfacial interaction. Special attention is paid to the development of approaches for the simulation of polydispersed bubbly flows taking into account processes of coagulation and fragmentation. Comparison of simulation results with experimental data showed that the developed approach allows to obtain detailed information about the structures of turbulent gas-liquid flows, the distribution of bubbles by size.

1. Intoduction

Two-phase bubbly turbulent flows are extensively used in various fields of industry such as chemical, food, pharmaceutical, nuclear and thermal power engineering [1-3]. Therefore, the simulation of turbulent bubbly flows is of great practical interest and is the subject of many publications. The complexity of modeling such flows involve a large number of phenomena of different nature, since such flows are characterized by strong mutual influence of the carrier and dispersed phase, accompanied by the processes of heat transfer, phase transitions, coagulation, fragmentation, etc. The structure of two-phase gas-liquid flow significantly depends on flow rate parameters of dispersed phase, the magnitude and direction of velocity of the continuous phase, physical properties of materials for the selected system, the geometry of the modern power plants. Knowledge the information about the structure, the averaged and pulsation characteristics of bubble flows is necessary in the design of modern cost-effective manufacturing equipment.

2. Governing equations

System the Reynolds-averaged Navier-Stokes equations to describe the dynamics of a turbulent mixture of liquid and air bubbles with regard to interfacial interaction has the form

$$\frac{\partial(\rho_k\alpha_k)}{\partial t}+\frac{\partial(\alpha_k\rho_k u_j)}{\partial x_j}=0,\quad \alpha_l+\alpha_g=1,\qquad k=l,g \tag{1}$$

$$\frac{\partial(\alpha_k\rho_k u_{ki})}{\partial t}+\frac{\partial(\alpha_k\rho_k u_{ki}u_{kj})}{\partial x_j}=-\alpha_k\frac{\partial p}{\partial x_i}+\alpha_k\rho_k g_i+F_{ki}+ \qquad (2)$$

$$\frac{\partial}{\partial x_j}\left[(\mu_k+\mu_{kt})\left(\frac{\partial u_{ki}}{\partial x_j}+\frac{\partial u_{kj}}{\partial x_i}-\frac{2}{3}\frac{\partial u_k}{\partial x_k}\delta_{ij}\right)\right]$$

Here x_i - cartesian coordinates,, u_i - components of velocity vector, t - time, p - pressure, ρ_l , ρ_g - density of continuous and dispersed phases respectively, α_l - is the volume fraction of liquid phase in mixture, δ_{ij} - is the Kronecker symbol, μ_l ,μ_g - dynamic coefficient of viscosity of water and air, μ_{kt} - is the coefficient of turbulent viscosity, F_{ki} - the strength of interfacial interaction forces. The turbulent viscosity for the carrier liquid phase are determined using a two-parameter turbulence model, modified for two-phase media [4]. In the equations for the transfer of turbulent kinetic energy and its dissipation introduces additional components to the kinetic energy caused by pulsations of the bubbles. The motion of the dispersed phase is determined by the forces of interfacial interaction. The main forces considered have the following components: Archimedes force, resistance force, joined the force, the rotational force of Magnus, the wall friction force [4]. To describe the distribution of bubble sizes in two-phase flow the equation for conservation of number of particles is written, taking into account the processes of coagulation and fragmentation. To solve the equation of conservation for the number of bubbles we use an approach based on the method of fractions [3]. Range of distribution of particle sizes is divided into a number of fractions with fixed boundaries, there can also be exchange of vesicles between different fractions as a result of coagulation and fragmentation. In this method, the distribution of bubble sizes is approximated by a piecewise uniform distribution, thus, the problem of describing the spectrum of droplet size is reduced to the solution of the equations for the volume concentrations of individual fractions.

3. Results

Verification of this model was carried out by comparing the calculation results with the experimental gas-liquid flow data. We chose the experimental data presented in [5], which has been studied for a polydisperse flow in pipe (pipe diameter D=50.8 mm, height H=3060 mm). Experiments were carried out with the water-air mixture at atmospheric pressure and at a constant temperature. The superficial velocity of liquid at the inlet was $<J_l>$ = 0.49 - 2.01 m/c, which corresponds to a Reynolds number $\mathrm{Re}=\rho_l <J_l> D/\mu_l$=$(0.5-7)\cdot 10^4$. The diameter of the bubbles of dispersed phase ranged d = 0-12 mm, and volumetric gas content expendables in range α_g = 0 - 25%. The superficial velocity of the bubbles $<J_g>$ at the inlet varies from $<J_g>$= 0.02 - 0.5 m/c. Distributions of parameters in all figures are shown in cross-section, which is located at a distance $x_3/D=53$ from the entrance.

On figures 1,2 shows the average axial velocity profiles of the carrier and dispersed phases selected for the following two flow modes: 1- $<J_l>$ = 0.49 m/c, $<J_g>$= 0.19 m/c, α_g = 25.3%; 2 - $<J_l>$= 2.01 m/c, $<J_g>$= 0.47 m/c, α_g= 21.3%. From these figures it is seen that the two-phase flow velocity profiles become more filled with adding high concentration of bubbles in the central part of the tube. The value of the maximum speed is reduced in comparison with single-phase flow. It should be noted that the greatest deformation velocity profiles observed at a small distance from the wall, which is consistent with experimental data [5]. The flat velocity profile in the central part of the tube show a high degree of turbulence of the gas-liquid bubble flow.

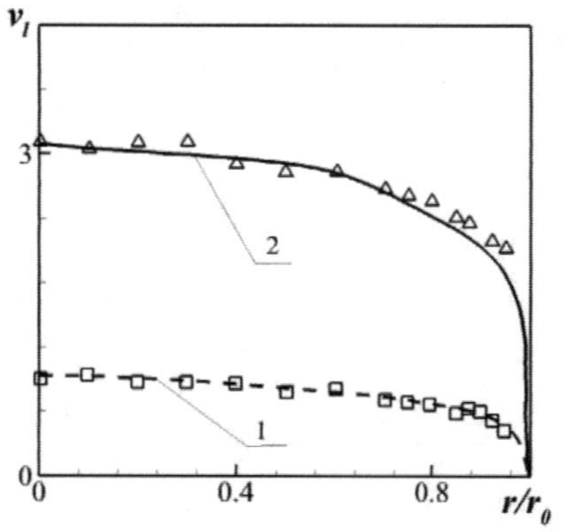

Figure 1. The distribution of the velocity profile of the carrier phase in the cross section of the pipe for different flow regimes:: 1- $< J_l >$ =0.49 m/c; 2-2.01 m/c

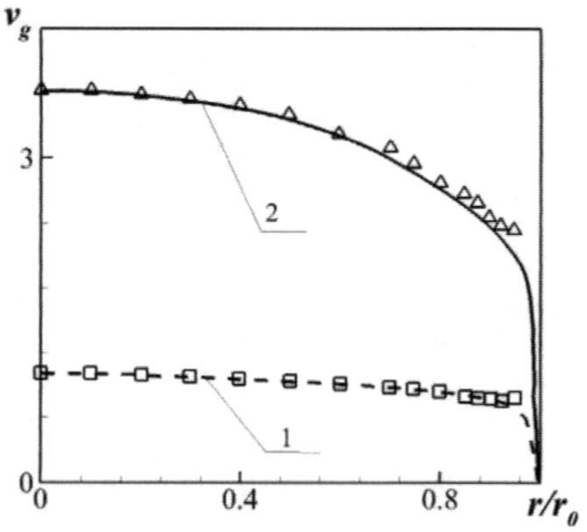

Figure 2. The distribution of the velocity profile of the disperse phase for different flow regimes: $< J_l >$ =0.49 m/c; 2-2.01 m/c

Figure 3 shows the profiles of void fraction along the pipe section for the selected mode. In either mode, the gas is dispersed with significant void fraction α_g , where there are processes of coagulation and fragmentation bubbles in turbulent flow (α_g = 0.23; 0.18). Averaged profiles of experimental data from [5] are shown as symbols (1-□, 2-Δ). For small flow rates the bubble mixture (curve 1) observed peak concentration of bubbles in the boundary region and a shallow uniform concentration distribution of bubbles in the central part of the channel. Figure 4 illustrated the distribution profiles across the pipe the volume-surface average particles diameter d_{32} corresponding to the chosen regime parameters. It can be seen that at low liquid flow rates (curve 1) there observed a uniform distribution of the average bubble diameter in the channel cross section. This indicates that for a given values of gas-liquid flow rate of coagulation resulting in a slight increase the average bubble radius (up to 4mm). However, the speed of the processes of coagulation and fragmentation are small and there is some equilibrium state. The difference of the degree of influence of interfacial forces on bubbles of different sizes leads to the fact that the bubbles in diameter smaller then original size are shifted to the near-wall region. Note that the considered flow regime can be characterized as a flow in which the bulk of the bubbles moves near the wall. Increasing the flow rate to $< J_l >$ = 2.01 m/c significantly changes the structure of the gas-liquid flow stream. The volumetric gas concentration in the central part of the tube reaches a maximum value, which decreases monotonically as it approaches to the channel walls. The growth rate of the liquid leads to the fact that the intensity of collisions of bubbles increases. It grows because of the greater concentration of bubbles per unit volume, leading to smaller distances between the bubbles, comparable to their diameters. Increased involvement of bubbles in the turbulent fluctuations leads to an increase in the intensity of coagulation, as indicated by the large value of the average bubble size in the central part of the channel. However, the decrease in the mean diameter near the walls shows that at this speed the processes of involvement of bubbles in turbulent eddies is sufficient for mass fragmentation of bubbles. The movement of gas-liquid flow at a rate of $< J_l >$ = 2.01 m/c is characterized by the intensification of the processes of coagulation and fragmentation, and the action of the transverse interfacial forces leads to a redistribution of bubbles of various sizes over the cross section of the channel. Due to the action of the interfacial forces is displaced of smaller bubbles to the wall region, while aggregated bubbles accumulate in the center of the channel. The structure of the flow at low flow rates, is characterized by the accumulation of bubbles near the walls, replaced the opposite. At higher speeds, the gas-liquid flow of main mass of bubbles moves in a central region, and near the walls the concentration is negligible.

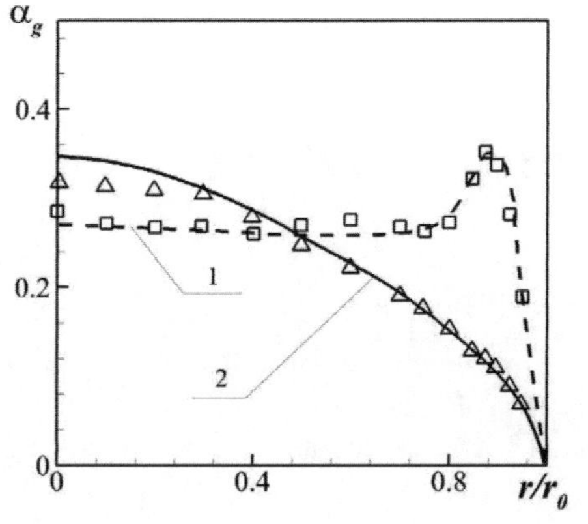

Figure 3. Distribution of the concentration of bubbles in the cross section of the pipe at different flow regimes: 1- $< J_l >$ =0.49 m/c; 2-2.01 m/c

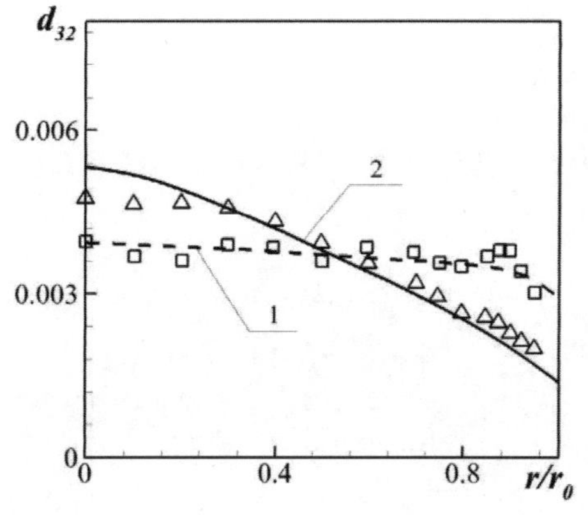

Figure 4. The change in the average diameter of bubbles d_{32} in the cross section of the pipe for different flow regimes: 1- $< J_l >$ =0.49 m/c; 2-2.01 m/c

4. Conlusions

The numerical model to describe the flow of gas bubbles in a turbulent flow is presented. Model satisfactorily describes the structure of the gas-liquid flow, and the local distribution of bubbles over the cross section in the tube. Presented the calculations of different modes of bubble flows show the complexity and mutual influence between the transfer of bubbles of variable sizes, the effects of coagulation, fragmentation and the effect of interfacial forces. Interfacial forces play a decisive role in the distribution of the main parameters (the concentration of bubbles, the speed of the carrier and dispersed phase, the specific interfacial surface area) in the cross section of the tube. Good agreement of simulation results with experimental data proves the correctness of the choice of these models.

References

[1] H A Jakobsen 2013 *Chemical reactor modelling. Multiphase reactive flows.* (Springer).
[2] R I Nigmatulin 1987 *Dynamics of multiphase flow.* (M.: Science).
[3] L E Sternin 1974 *Fundamentals gasdynamics of twophase flow in nozzles.* (M.: Machine enginerring).
[4] Serizawa A, Kataoka I, Michiyoshi I Turbulence structure of Air-Water bubbly flow. Parts I-III *J. Multiphase Flow* 1975 **2** 221-267.
[5] Hibiki T, Ishii T, Xiao Z Experimental and numerical study of downward bubbly flow in a pipe *Int. J. Heat Mass Transfer* 2006 **49** 3717-3727.

Journal of Physics: Conference Series **669** (2016) 012019 doi:10.1088/1742-6596/669/1/012019

Weak waves in multifractional liquids with bubbles

D A Gubaidullin[1,2], A A Nikiforov[1,2] and R N Gafiyatov[1,2]

[1] Institute of Mechanics and Engineering, Kazan Science Center, Russian Academy of Sciences, 2/31 Lobachevsky str., Kazan, 420111, Russia
[2] Kazan (Volga Region) Federal University, 18 Kremlyovskaya str., Kazan, 420008, Russia

E-mail: gafiyatov@mail.ru

Abstract. The propagation of weak waves in multifractional mixtures of liquid with vapor–gas and gas bubbles of different sizes and different compositions with phase transitions is studied. The dispersed phase consists of $N+M$ fractions having various gases in bubbles and different in the bubbles radii. Phase transitions accounted for N fractions. The total bubble volume concentration is small (less than 1%). The dispersion relation is derived and dispersion curves is built. Influence of the mass concentration is shown. It is shown that dispersion and dissipation of acoustic waves depends significantly on presence of different bubbles in fractions of the dispersed phase.

1. Introduction

A propagation of acoustic waves in liquid with gas bubbles is studied intensively both theoretically and experimentally. Nigmatulin [1], Nakoryakov et al. [2], Temkin [3] considered various basic aspects of acoustics of multiphase flows. Hao and Prosperetti [4] obtained the results of study of the vapor bubble dynamics in liquid. Commander and Prosperetti [5] presented a model of the propagation of plane waves with small amplitude in a mixture of liquid and of gas bubbles. This model is adequate for the frequencies far from resonance at the volume content of a disperse phase of 1-2%. Nigmatulin et al. [6] showed the need of allowing for the effect of liquid compressibility for problems of acoustics of bubbly liquids. Shagapov [7] considered a problem of propagation of small disturbances in liquids with polydisperse bubbles. Nigmatulin et al. [8] studied the small radial oscillations of vapor-gas bubbles in liquid under the influence of an acoustic field. They showed that capillary effects and phase transitions lead to the new resonant frequency of small bubbles. This frequency is different from the Minnaert frequency. Vakhitova and Shagapov [9] investigated the propagation of small disturbances in two-component two-phase bubbling media with a three-temperature model. They found that heat and mass transfer rather than an interphase friction mainly determine dispersion. Gubaidullin and Nikiforov [10] obtained a dispersion relation describing a propagation of harmonic spherical and cylindrical disturbances in monodisperse mixture of liquid with the vapor-gas bubbles. They showed strong dependence of an attenuation of pulse waves on a vapor concentration in bubbles. The propagation of acoustic waves in a two-fractional mixture of liquid with bubbles is investigated in [11-13]. The propagation of sound waves in two-fractional liquids with polydisperse bubbles is investigated in [14-17]. A mathematical model of the propagation of sound waves in the two-fractional mixture of liquid with polydisperse vapor-gas and gas bubbles with phase transitions is shown in [18].

The propagation of acoustic waves in multifractional mixtures of liquid with gas and vapor-gas bubbles of different sizes and different compositions is studied in [19].

In the present paper, a propagation of weak waves in multifractional liquids with bubbles is studied.

2. Dispersion relation

Linearized equations for one-dimensional disturbances in a multifractional mixture of liquid with bubbles are obtained from the general equations of motion for bubbly mixtures [1]. Solving this system this dispersion relation is obtained

$$\left(\frac{K_*}{\omega}\right)^2 = \frac{1}{C_f^2} + \frac{\rho_{10}}{p_0}\frac{\left(\sum_{i=1}^{M} H_{3i} + \sum_{j=1}^{N} H_{3j}\right)\left(\sum_{i=1}^{M} H_{1i} + \sum_{j=1}^{N} H_{1j}\right)}{\sum_{i=1}^{M}\left(\frac{m_i}{\tau_{T1i}} + H_{2i}\right) + \sum_{j=1}^{N}\left(\frac{m_j}{\tau_{T1j}} + H_{2j}\right) - i\omega} +$$

$$+\frac{\rho_{10}}{p_0}\sum_{i=1}^{M}\left(\frac{\alpha_{20i}}{N_{Ri}}\left(1-\frac{M_{4i}}{M_{3i}}\right)\right) + \frac{\rho_{10}}{p_0}\sum_{j=1}^{N}\left(\frac{\alpha_{20j}}{N_{Rj}}\left(1-\frac{M_{4j}}{M_{3j}}\right)\right) \quad (1)$$

Here C_f is the frozen velocity of sound ($C_f = C_1/\alpha_{10}$) and the following notations are also accepted:

$$H_{1j} = \frac{m_j}{\tau_{T1j}}\left(\frac{M_{1j}M'_{4j}}{M_{3j}} + M_{2j}\right),\ H_{1i} = \frac{m_i}{\tau_{T1i}}\left(\frac{M_{1i}M_{4i}}{M_{3i}} + M_{2i}\right),\ H_{2j} = \frac{m_j}{\tau_{T1j}}\frac{M_{1j}b_j}{M_{3j}},\ H_{2i} = \frac{m_i}{\tau_{T1i}}\frac{M_{1i}b_i}{M_{3i}},$$

$$H_{3j} = \frac{\alpha_{j0}}{N_{Rj}}\frac{b_j}{M_{3j}},\ H_{3i} = \frac{\alpha_{i0}}{N_{Ri}}\frac{b_i}{M_{3i}},\ M_{1j} = G_j - M_{2j} - \frac{L_{1j}N_{3j}}{L_{4j} - \delta N_{2j}},\ M_{2j} = \frac{N_{2j}L_{1j}}{N_{Rj}(L_{4j} - \delta N_{2j})},$$

$$M_{3j} = \frac{N_{3j}\delta - L_{2j}N_{3j}}{L_{4j} - \delta N_{2j}} + L_{3j} + M_{4j},\ M_{4j} = \frac{L_{4j} - L_{2j}N_{2j}}{N_{Rj}(L_{4j} - \delta N_{2j})},\ M_{1i} = -(N_{3i} + M_{2i}),\ M_{2i} = \frac{N_{2i}}{N_{Ri}},$$

$$M_{3i} = 1 + N_{3i}(1+b_i) + M_{4i},\ M_{4i} = \frac{1}{N_{Ri}}(1 + N_{2i}(1+b_i)),\ N_{1j} = \frac{i\omega\tau_{T1j}}{m_j} - 1,\ N_{1i} = 1,\ N_{2j} = i\omega\tau_{Tj} - 1,$$

$$N_{2i} = i\omega\tau_{Ti} - 1,\ N_{3j} = k_{2j}(c_j - R_j) - 1 + G_j,\ N_{3i} = k_{2i}(c_i - R_i) - 1,\ L_{1j} = E_j(1 - i\omega\tau_j),$$

$$L_{2j} = -\frac{l_0 k_{2j}}{(1-k_{Vj0})T_0} + \Delta R_j - L_{1j}(1+b_j),\ L_{3j} = 1 - G_j(1+b_j),\ L_{4j} = L_{1j} + \Delta R_j N_{2j},\ k_{2j} = \frac{i\omega\tau_{Tj}}{c_j},$$

$$k_{2i} = \frac{i\omega\tau_{Ti}}{c_i},\ b_j = \frac{c_1\tau_{Tj}}{c_j\tau_{T1j}},\ b_i = \frac{c_1\tau_{Ti}}{c_i\tau_{T1i}},\ N_{Rj} = \frac{-(i\omega)(a_{j0})^2 G_{Rj}\overset{\circ}{\rho}_{10}}{3(t_{Aj}G_{Rj}+1)p_0},\ N_{Ri} = \frac{-(i\omega)(a_{i0})^2 G_{Ri}\overset{\circ}{\rho}_{10}}{3(t_{Ai}G_{Ri}+1)p_0},$$

$$G_{Rj} = \frac{1}{t_{Rj}} - i\omega,\ G_{Ri} = \frac{1}{t_{Ri}} - i\omega,\ t_{Rj} = \frac{(a_{j0})^2}{4\nu_1},\ t_{Ri} = \frac{(a_{i0})^2}{4\nu_1},\ t_{Aj} = \frac{a_{j0}}{C_1(\alpha_{j0})^{1/3}},\ t_{Ai} = \frac{a_{i0}}{C_1(\alpha_{i0})^{1/3}},$$

$$m_{j0} = \frac{\overset{\circ}{\rho}_{j0}}{\overset{\circ}{\rho}_{10}},\ m_{i0} = \frac{\overset{\circ}{\rho}_{i0}}{\overset{\circ}{\rho}_{10}},\ m_j = \frac{\rho_{j0}}{\rho_{10}},\ m_i = \frac{\rho_{i0}}{\rho_{10}},\ i = \overline{1,M},\ j = \overline{1,N}.$$

Journal of Physics: Conference Series **669** (2016) 012019 doi:10.1088/1742-6596/669/1/012019

Dispersion relation (1) (i.e. the function of the complex wave number K_* on the frequency ω) determines a propagation of weak waves in multifractional mixtures of liquid with vapor-gas and gas bubbles (different initial radii, different initial volume contents and different thermal properties of fractions) with the interphase diffusion mass transfer.

3. Results

Figure 1 shows the phase velocity and attenuation coefficient in three-fraction mixture of water for the volume contents of fractions of $\alpha_{2N1} = \alpha_{2N2} = \alpha_{2M1} = 0.0033$, for the radius of vapor-air bubbles of $a_{0N1} = 1.5 \cdot 10^{-3}$ m, carbon dioxide with vapor bubbles – $a_{0N2} = 10^{-3}$ m, helium bubbles – $a_{0M1} = 0.5 \cdot 10^{-3}$ m, calculated by dispersion relation (1). In unperturbed state a pressure of a two-fractional mixture is $p_0 = 0.1$ MPa. Curves *1* are obtained for value of mass concentration of water vapor in bubbles is $k_{V0} = 0.1$, curves *2* – for value $k_{V0} = 0.5$, curves *3* – for value $k_{V0} = 0.9$.

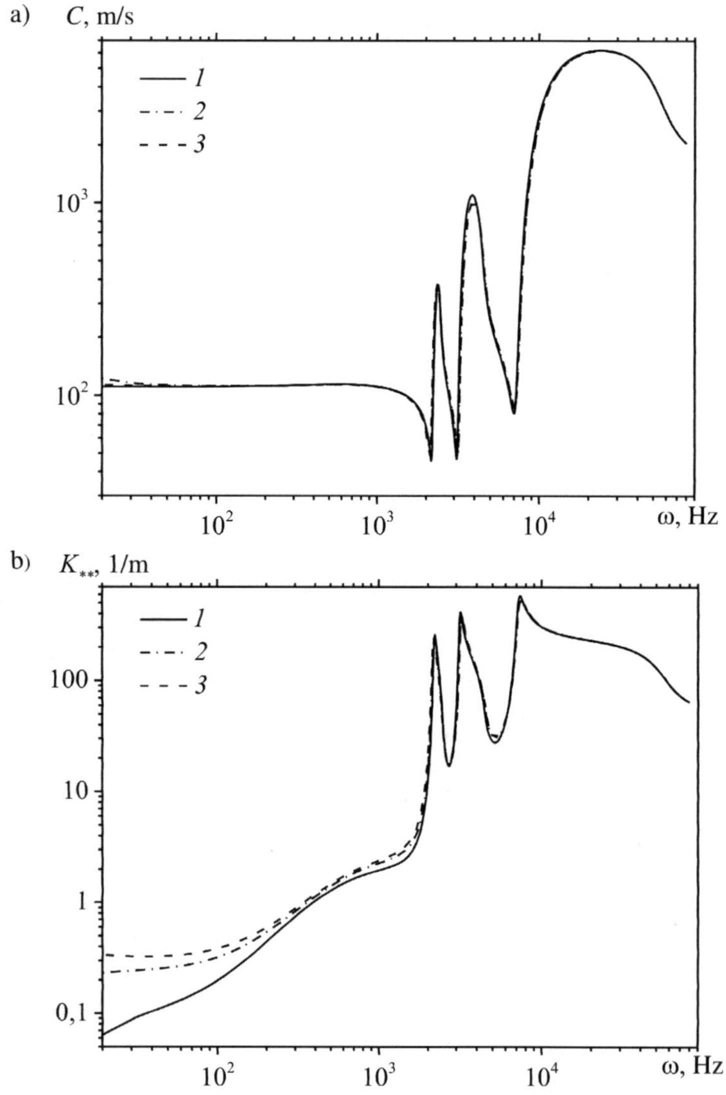

Figure 1. Frequency dependences of the phase velocity (a) and of the attenuation coefficient (b) for the three-fraction mixture of water for three different mass concentration: k_{V0}=0.1 (solid line), k_{V0}=0.5 (dash-dotted line), k_{V0}=0.9 (dashed line).

The difference of mass concentration of water vapor in bubbles almost has no effect on the phase velocity value (figure 1(a)). However, the presence of three different bubble sizes in the mixture is important significantly. This leads to the occurrence of the three minimum points of the phase velocity curve in the near of the resonance frequencies of the oscillations of bubbles and to a characteristic inflection of a phase velocity curve.

Figure 1(b) shows that increasing the mass concentration of vapor in bubbles leads to increasing the value of the attenuation coefficient in low-frequency region. For the same frequencies the difference is for four-five times. And also, three different bubble sizes leads to the occurrence of the three maximum values of the attenuation coefficient curves.

4. Conclusions

The propagation of weak waves in multifractional mixture of liquid with vapor-gas and gas bubbles taking into account the phase transitions is studied. In particular, it is shown that three fractions of bubbles with different radii cause the appearance of three local minimum values of the phase velocity and three local maxima in the dependence of the attenuation coefficient in the vicinity of the resonance frequencies of the natural vibrations of bubbles. It is shown that mass concentration of vapor in bubbles can significantly influence on dispersion curves.

Acknowledgments
This work was funded by the subsidy of the Russian Government to support the Program of Competitive Growth of Kazan Federal University among World's Leading Academic Centers.

References

[1] Nigmatulin R I 1991 *Dynamics of Multiphase Media* (New York: Hemisphere)
[2] Nakoryakov V E, Pokusaev B G and Shreiber I R 1993 *Wave Propagation in Gas-Liquid Media* (New York: CRC Press)
[3] Temkin S 2005 *Suspension Acoustics: An Introduction to the Physics of Suspensions* (New York: Cambridge University Press)
[4] Hao Y and Prosperetti A 1999 *Physics of Fluids* **11** 2008
[5] Commander K W and Prosperetti A 1989 *Journal of the Acoustical Society of America* **85** 732
[6] Nigmatulin R I, Shagapov V Sh and Vakhitova N K 1989 *USSR Academy of Sciences Doklady* **304** 1077
[7] Shagapov V Sh 1977 *Journal of Applied Mechanics and Technical Physics* **18** 77
[8] Nigmatulin R I, Khabeev N S and Nagiev F B 1981 *International Journal Heat and Mass Transfer* **24** 1033
[9] Vakhitova N K and Shagapov V Sh 1984 *Journal of Applied Mechanics and Technical Physics* **25** 682
[10] Gubaidullin D A and Nikiforov A A 2010 *High Temperature* **48** 170
[11] Gubaidullin D A, Nikifirov A A and Gafiyatov R N 2012 *High Temperature* **50** 250
[12] Gafiyatov R N, Gubaidullin D A and Nikifirov A A 2013 *Fluid Dynamics* **48** 366
[13] Gubaidullin D A and Gafiyatov R N 2014 *J. Phys.: Conf. Ser.* **567** 012020
[14] Nigmatulin R I, Gubaidullin D A and Fedorov Yu V 2012 *Doklady Physics* **57** 451
[15] Nigmatulin R I, Gubaidullin D A and Fedorov Yu V 2013 *Doklady Physics* **58** 261
[16] Gubaidullin D A and Fedorov Yu V 2013 *Journal of Applied Mathematics and Mechanics* **77** 532
[17] Gubaidullin D A, Gubaidullina D D and Fedorov Yu V 2014 *J. Phys.: Conf. Ser.* **567** 012011
[18] Gubaidullin D A and Fedorov Yu V 2015 *Fluid Dynamics* **50** 61
[19] Gubaidullin D A, Nikifirov A A and Gafiyatov R N 2015 *High Temperature* **53** 240

Journal of Physics: Conference Series **669** (2016) 012020 doi:10.1088/1742-6596/669/1/012020

The movement of a particle at nonlinear oscillations of high amplitude in the closed tube

D A Gubaidullin[1,2], L A Tkachenko[1,2] and R G Zaripov[1,2]

[1] Institute of Mechanics and Engineering, Kazan Science Center, Russian Academy of Sciences, 2/31 Lobachevsky str., Kazan 420111, Russia
[2] Kazan (Volga Region) Federal University, 18 Kremlyovskaya str., Kazan 420111, Russia

E-mail: luda_tkachenko@inbox.ru

Abstract. The movement of a flat particle at longitudinal oscillations of gas of high amplitude in the closed tube is experimentally investigated. It is shown that the oscillating particle on an axis moves from the closed end to the piston with increase in oscillation swing, and in near to a wall in the opposite direction. In the radial direction drift of the oscillating particle comes from an axis to a tube wall. Such behavior of a particle is caused by acoustic streaming of gas in a tube. It is revealed that the increase in frequency of excitement of gas or length of a tube lead to growth of oscillation swing of a particle and increase of its average speed.

1. Introduction

Research of wave processes in limited working environments represents one of actual problems. The main models of wave dynamics of multiphase environments and a number of results in this area are presented in [1]. Important applied value has studying of oscillations of environments with strongly nonlinear fronts of waves of pressure and their impact on disperse systems which examples are gas mixes with drops or solid particles. The processes occurring thus are of interest at an intensification of liquid spraying for the purpose of purification of flue gases on harmful productions, sedimentation of particles in various technical devices, an intensification of hashing and burning. Coagulation of drops and particles (drops of machine oil and a tobacco smoke [2], drops of oleic acid [3], particles of the smoke received from combustion of a stick of an incense [4] with diameters of 1-10 microns at nonlinear oscillations in the closed tube with the e first eigenfrequency is investigated. Results of experimental studies of the accelerated coagulation and sedimentation of drops of a aerosol di-ethyl-hexyl-sebacate (DEHS) with a diameter of drops of 0.863 microns in the closed and opened tubes near resonant frequency, twice smaller the first eigenfrequency (a subharmonic resonance) are considered in [5, 6], and the shock-free mode of oscillations of an aerosol at the first eigenfrequency is studied in [7-10]. The theoretical description of the accelerated coagulation and sedimentation of aerosols on walls of a tube presents considerable difficulties. Therefore when modeling such phenomena there is a problem of both theoretical, and experimental studying of the movement of separate particles at nonlinear oscillations of homogeneous gas. In [11] drift of the easy sphere from expanded polystyrene

[1] To whom any correspondence should be addressed.

with a diameter of 3.5 mm suspended in the axial direction on a thin wire in vertically located tube near the first eigenfrequency is analyzed. In work [12] the movement of a flat particle at nonlinear oscillations of gas in a tube and in an external field about the open end in the shock-free mode near the first eigenfrequency is in detail experimentally investigated.

The purpose of this work is research of drift of a flat particle in the closed tube at longitudinal oscillations of gas of high amplitude with small frequency excitation.

2. Experiment

Briefly we will describe setup and a technique of carrying out experiment. An experimental setup including a resonance tube, a system of excitation of oscillation, and a system for recording the process parameters. The tube was mounted horizontally. The tube consisted of identical sections with a length of 0.5 m and inside diameter $2R_0 = 0.048$ m. One section made of quartz glass was 1.2 m long. It is fixed to the closed end of the tube. The closed end of quartz section of a tube was accepted to the beginning of coordinates $x = 0$. The model of a particle represented a washer from expanded polystyrene with a diameter of 16.5 mm and 0.6 mm thick. The cylindrical polyethylene tubule was inserted into an opening of a washer. The full weight of a particle made 4.8 mg. Such particle was used for research of drift in the longitudinal direction along a tube axis. Other particle with a total weight of 2.5 mg had the following parameters: diameter is 8 mm, thickness is 0.4 mm. It was applied in experiments with the movement along a wall, at distance of 0.01 m from an internal surface, and radial the directions of a tube. Through a polyethylene tubule the scaffold with a diameter of 0.3 mm which stretched special arms in the longitudinal direction was passed. Existence of the tubule which is directing allowed avoiding undesirable cross oscillations of a particle in the course of its movement. For research of drift of a particle in the radial direction (with r = 0 on a tube axis) the special knot was made. It represented the cylinder 0.01 m high. Two medical needles 0.1 m long with distance between them 0.044 m were soldered to it. The cylinder was placed in a tube, adjoining to its wall. Through needles the scaffold with a diameter of 0.25 mm which passed through a polyethylene tubule of the tested particle was stretched with a tension. A standard compressor with the piston stroke $2l_0 = 0.086$ m and inside diameter of the cylinder $2R = 0.077$ m was used for the excitation of aerosol oscillation. The compressor cylinder was connected to the resonance tube via tapering adapter of height $h = 0.22$ m. For measurement of frequency of oscillations of the piston $\nu = \omega/2\pi$ where ω – cyclic frequency, the digital phototachometer AKTAKOM ATT-6000 was used. Frequency was measured with an accuracy of 0.05 Hz. Drift of a particle was fixed by a digital video camera "Panasonic" VSK 0631 (Japan) with a frequency of shooting of 25 frames per second. It was established on a support perpendicular to quartz section of a tube at distance of 0.5 m from it for increase in quality of the image and avoiding of oscillation when shooting and at turn. In case of drift of a particle in the radial direction shooting was conducted with distance of 0.05 m.

Measurements were taken in a tube by length L_0 = 2.7, 3.7, 4.7 m. In experiments with the movement of a particle along an axis it was established in section with x = 0.3 m. At research of dynamics of a particle near a wall she accommodated at x = 0.9 m. For measurements in the radial direction initial coordinates equaled: $x = 0.45$ m and $r = 4$ mm. Then the tube was densely closed by a cover. Tension on the electric motor via the autotransformer moved. It allowed to change smoothly turns of rotation of a rotor and, respectively, a frequency of oscillations of the piston. The video camera joined synchronously with the engine. Video filming fixed the movement of a particle at distance of 0.6 m along quartz section of a tube. In case of the movement of a particle in the radial direction video filming was conducted until a particle stop at some distance from a tube wall. Then the camera and the electric motor were cut out. Such experiment repeated several times for all values of frequencies and lengths of a tube. After each experiment the particle was returned in a starting position. For processing of results of video filming the video camera was connected to the computer. The received video images opened by means of Virtual Dub 1.5.3 where all video broke into shots, and time from the beginning of shooting corresponded to each shot. Transition from a shot to a shot allowed to record the provision of a particle in different time points.

3. Results

We will consider some experimental results received in the analysis of data of video filming. In Figure 1, *a* dependence of coordinate of a particle on time along a tube axis is presented. Points are represented by experimental data, the continuous line – approximation, the shaped line – average value of coordinate concerning which there are oscillations. Follows from Figure that the movement of a particle has oscillatory character. Thus time of the movement of a particle to the piston is more, than movement time by the closed end during one oscillation. It causes drift of a particle from the closed end of a tube towards the piston. In the experiments made for a case of a drift of a particle in a lengthwise direction near a wall back motion by the closed end of a tube is observed (Figure 1, *b*). And researches begin with the frequency of 8 Hz as with smaller frequencies driving of a particle is absent owing to a great influence of a sliding friction about a scaffold. Let's note similar nature of behavior of a particle for frequency period of the piston with a drift on a tube axis. Such behavior of a particle for all studied lengths of a tube and frequency excitation is caused by existence of an acoustic streaming [13, 14]. The toroidal-shaped vortex with the direction of flow of gas on an axis to the piston, and near a wall to the closed end is formed. The oscillation frequency of a particle equals to a frequency excitation of oscillations of gas. Similar dependences are observed at other oscillation frequencies of gas for all considered tube lengths.

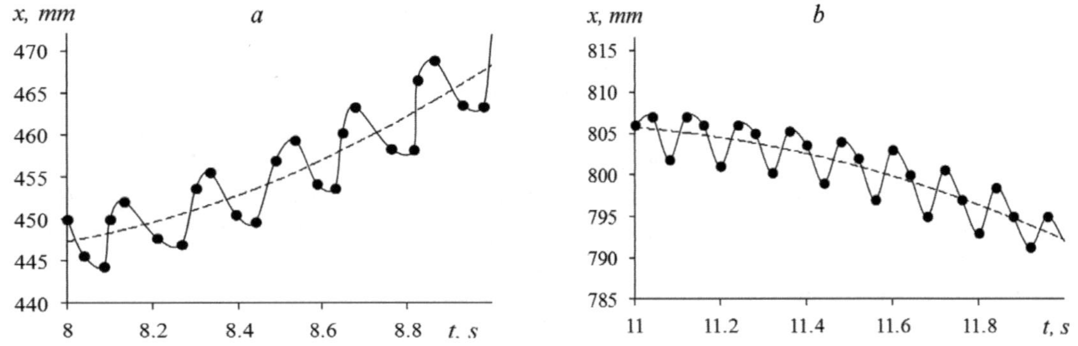

Figure 1. Dependence of coordinate of a particle along a tube length L_0 = 4.7 m from time: a) along an axis with a frequency of ν = 6 Hz, b) near a wall with a frequency of ν = 8 Hz. Points – the experimental dates, continuous and shaped lines – approximations.

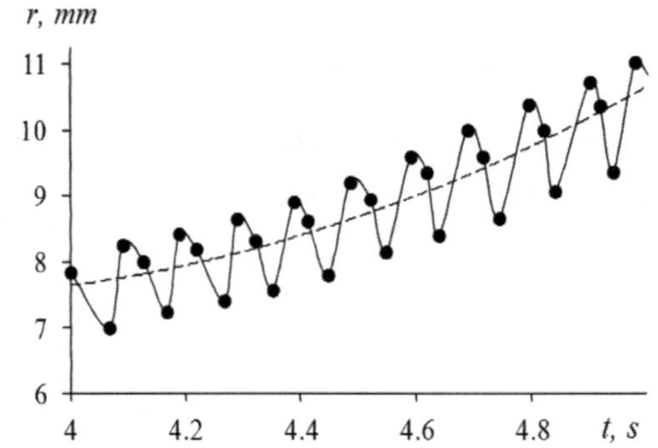

Figure 2. Dependence of coordinate of a particle in the radial direction from time for a tube length L_0 = 3.7 m with frequency ν = 10 Hz. Points – the experimental dates, continuous and shaped lines – approximations.

In Figure 2 dependence of coordinate of a particle is given in the radial direction from time. Thus the oscillating particle moves in the range from a point with coordinate $r = 4$ mm to some boundary point concerning which it oscillates with a constant amplitude without driving to a tube wall. In experiments this minimum interval equaled 7 mm, and maximal – 13 mm. The interval of driving of a particle from initial increased to a boundary point with growth of frequency.

From dependences of coordinate of a particle on time it is possible to obtain data on particle oscillation swing Δx (where $\Delta x = x_{max} - x_{min}$, x_{max} and x_{min} is the maximal and minimum values of coordinate for period oscillation of a particle concerning the centerline in Figure 1 and Figure 2). With driving of a particle along an axis from the closed end of a tube a range of its oscillations increases, reaching in the set mode of the value which is not exceeding the doubled piston oscillation amplitude. When driving a particle near a wall towards the closed end of a tube from which it is visible that the oscillation swing decreases. The increase in frequency accompanied with increase of intensity of oscillations of gas leads to height of a oscillation swing of a particle when driving in both directions. Thus the oscillation swing in near a wall is much less, than on a tube axis. It is caused by losses on a wall on viscosity, heat conductivity and change of conditions of a flow of a particle near an internal tube wall. Let's note that when driving a particle in the radial direction from an axis to a tube wall there is an increase in a oscillation swing from 0.01 mm to the maximal value in a boundary point which at $\nu = 10$ Hz equal 2 mm.

Also data on the average speed of a particle were obtained. For the quantitative analysis of average speed of the movement along an axis and near a wall data for the set distance (from coordinate from 0.75 m to 0.9 m) where the maximum a oscillation swing in both directions was observed got out. In Figure 3, a dependences of average speed of $V_a = 0.15/t$ (m/s) of a particle of an axis and near a wall from the frequency of oscillations of gas are presented. With increase of frequency excitation of gas the average speed of drift of a particle grows. It is visible that the specified dependence in case of the movement near a wall has nonlinear the character caused by losses on walls. Thus average speed near a tube wall, for example, at a frequency of $\nu = 10$ Hz is 3 times lower, than on an axis. It is possible to note that the increase in length of a tube with the set frequency leads to growth of average speed of a particle. In Figure 3, *b* dependence of average speed of a particle on frequency for a case of the movement of a particle would be given in the radial direction. It is visible that with increase in frequency of excitation of gas the average speed of the movement of a particle at distance of equal 7 mm increases, as well as at the movement of a particle in the longitudinal direction of a tube.

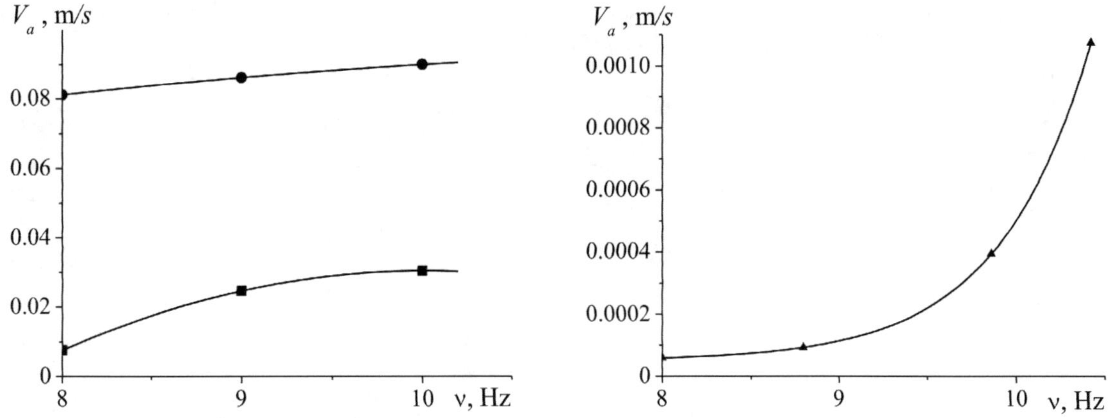

Figure 3. Dependence of average speed of a particle on the frequency of oscillations of gas for a tube length $L_0 = 3.7$m: a) • – along a tube axis, ■ – near a wall, continuous lines – polynomial approximations; b) in the radial direction, the continuous line – exponential approximation.

4. Conclusions

As a result of experimental studies of dynamics of a particle at longitudinal oscillations of gas in a tube it is revealed that the particle moves, making longitudinal oscillations. On an axis – from the closed end to the piston, and near a wall – in the opposite direction. This movement is caused by the acoustic streaming of gas in a tube which are followed by formation of a toroidal vortex. In the radial direction the oscillating particle moves from an axis to a tube wall to a boundary point. It is revealed that with the fixed frequency with the movement of a particle on an axis from the closed end of a tube its oscillations swing increases, reaching the value which isn't exceeding the doubled amplitude of oscillations of the piston. In case of drift of a particle near a wall towards the closed end of a tube the specified swing on the contrary decreases. At the movement of a particle in the radial direction of a tube there is an increase in oscillations swing to the maximum value in a boundary point. It is established that the increase in frequency of excitement of gas leads to growth of oscillations swing of a particle for all directions and to growth of its average speed of the movement.

Acknowledgments
This work was funded by the subsidy of the Russian Government to support the Program of Competitive Growth of Kazan Federal University among World's Leading Academic Centers.

References

[1] Nigmatulin R I 1990 *Dynamics of Multiphase Media* (New York: Hemisphere)
[2] Gulyaev A I and Kuznetsov V M 1963 *Sov. Phys. Acoust.* **8** 236
[3] Temkin S 1970 *Phys. Fluids* **13** 1639
[4] Shuster K and Fichman M, Goldshtein A and Gutfinger C 2002 *Phys. Fluids* **14** 1802
[5] Gubaidullin D A, Zaripov R G, Galiullin R G, Galiullina E R and Tkachenko L A 2004 *High Temperature* **42** 794
[6] Gubaidullin D A, Zaripov R G, Tkachenko L A 2012 *Journal of Engineering Physics and Thermophysics* **85** 259
[7] Gubaidullin D A, Zaripov R G and Tkachenko L A 2012 *High Temperature* **50** 564
[8] Gubaidullin D A, Zaripov R G and Tkachenko L A 2013 *High Temperature* **51** 955
[9] Gubaidullin D A, Zaripov R G and Tkachenko L A 2013 *Doklady Physics* **58** 392
[10] Gubaidullin D A, Zaripov R G and Tkachenko L A 2014 *High Temperature* **52** 895
[11] Goldshtein A, Shuster K, Vainshtein P, Fichman M and Gutfinger C 1998 *J. Fluid Mech.* **360** 1
[12] Gubaidullin D A, Zaripov R G, Tkachenko L A and Shaidullin L R 2015 *Journal of Engineering Physics and Thermophysics* **88** 871
[13] Ilgamov M A, Zaripov R G, Galiullin R G, Repin V B 1996 *Appl. Mechanics Reviews* **49** 137
[14] Galiullin R G, Timokhina L A, Filipov S E 2001 *Acoustical Physics* **47** 529

Journal of Physics: Conference Series **669** (2016) 012021 doi:10.1088/1742-6596/669/1/012021

Computation of shock waves in media with an interphase boundary by the CIP-CUP method on an adaptive grid

T S Guseva

Institute of Mechanics and Engineering, Kazan Science Center, Russian Academy of Sciences, 420111 Kazan, Russian Federation

E-mail: ts.guseva@mail.ru

Abstract. A numerical technique of computing shock waves in compressible media with movable deforming interphase boundaries including those of the gas-liquid type has been realized. The approach without explicit separation of the interphase boundary is applied. The CIP-CUP method is used for integrating the equations of gas dynamics. An adaptive grid of special kind (the soroban-grid) is utilized. Some results of testing the technique using one- and two-dimensional problems are given. Results of computation of impact of a jet on a thin liquid layer on a wall are presented.

1. Introduction

Shock processes in media with deforming interphase boundaries arise, in particular, at high-speed impacts of jets and drops on the surfaces of bodies. In the case the interphase boundary is deformed quickly and strongly, the intensity of the shock waves is moderate the CIP-CUP method (Constrained Interpolation Profile - Combined Unified Procedure) is quite effective [1]. One feature of this method is an opportunity of calculating the dynamics of the media being in contact without explicit separation of the contact boundaries (including those of the gas-liquid type) [2]. A numerical technique based on the CIP-CUP method with the staggered Eulerian grids was realized and tested in [3]. In problems with shock waves and fast changing contact boundaries the adaptive grids are more preferable, and are sometimes necessary. However the CIP-CUP method has a number of features reducing its efficiency or complicating its application on the structured adaptive (non-uniform) grids. In [4], a new way of adaptive discretization of a computational domain using the soroban-grid ("soroban" in Japanese means an abacus), and also a corresponding modification of the CIP-CUP method were proposed. In the present paper, some results of computing shock waves in the liquid and gas media being in contact are given, which show possibility of effective application of this modification to calculating high-speed jet impact on the body surface including the case with a liquid layer covering it.

2. Mathematical model and calculation procedure

An example of two-dimensional soroban-grid is given in figure 1. The grid points are located on parallel straight lines and are not connected among themselves by any elements like cells. The use of staggered grids is complicated, but there are no restrictions which could be caused by deformations of the grid. On each time step a new grid not connected with the old one is constructed. The distance between the lines, their number, and also the distance between the grid points on each line and their number can change and be different for different lines.

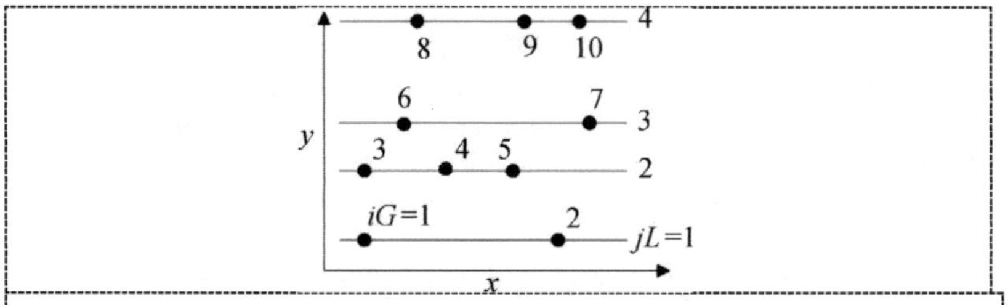

Figure 1. View of a two-dimensional soroban-grid.

Adaptation of the grid is determined by a monitor function which, in the one-dimensional case, can be written as $M(x,t)=\left(1+\alpha f_x^{\,2}\right)^{1/2}+\beta\left|f_{xx}\right|$ where f is a parameter of the solution, for example, the density. In the two-dimensional case the monitor function is then written as $M(x,y,t)=\left(1+\alpha\left(f_x^{\,2}+f_y^{\,2}\right)\right)^{1/2}+\beta\left(\left|f_{xx}\right|+\left|f_{yy}\right|\right)$.

Dynamics of the liquid and gas media being in contact without allowing for the effects of viscosity, heat conductivity and phase changes is described by a set of equations of the form

$$f_t+\mathbf{u}\cdot\nabla f=G \tag{1}$$

where $f\equiv\rho$, $\mathbf{u}$, p, φ, accordingly, $G\equiv-\rho\nabla\cdot\mathbf{u}$, $-\rho^{-1}\nabla p$, $-\rho C_S^{\,2}\nabla\cdot\mathbf{u}$, 0. Here ρ is the density, $\mathbf{u}$ the velocity, p the pressure, $C_S=\varphi C_{S1}+(1-\varphi)C_{S2}$ the sound speed, $C_{Si}=\left(\Gamma_i(p+B_i)/\rho\right)^{1/2}$, Γ_1, B_1 are the constants of the Tait equation of state for the liquid and $\Gamma_2=\gamma$ is the isentropic exponent, $B_2=0$ for the gas. The identifier $\varphi=1$ in the liquid area and $\varphi=0$ in the gas area.

In the CIP-CUP method equations (1) are split into the advection and non-advection parts. The CIP method belonging to semi-Lagrangian ones is applied to calculating each of the advection equations making up the convective part. On the new step $n+1$ the solution of the advection equation at the grid point of the new grid with coordinates $\mathbf{x}^{n+1}$ can be approximately presented as $f^*=f^n\left(\mathbf{x}^{n+1}-\mathbf{u}(\mathbf{x}^{n+1})\Delta t^n\right)$ where the argument is a departure point, i.e. the location of the "Lagrangian particle" which during Δt^n will arrive at the grid point $\mathbf{x}^{n+1}$. If the location of the departure point does not coincide with the grid point $\mathbf{x}^n$ of the old grid at which f^n is already known, interpolation is applied. Thus, the displacement of the grid points and recalculation of the values from the old grid to the new one are combined in one procedure. Monotonicity, small numerical diffusion and dispersion of the CIP method are determined by effective interpolation [2].

The non-advection part of system (1) is reduced to the equation

$$\frac{p^{**}-p^*}{\rho^* C_S^{*2}\Delta t^{n2}}=\nabla\cdot\left(\frac{\nabla p^{**}}{\rho^*}\right)-\frac{\nabla\cdot\mathbf{u}^*}{\Delta t^n},$$

which is solved by the SOR method in this case. The CIP interpolation is applied to determining the values of parameters at the nodes of the stencil of the finite difference scheme if they do not coincide with the grid points.

In solving problems with shock waves the CIP-CUP method is used in combination with the artificial viscosity [4]

$$\mathbf{u}^{n+1}=\mathbf{u}^{**}-\Delta t^n\,\nabla q_v/\rho^*,\quad p^{n+1}=p^{**}-\Delta t^n(\kappa-1)q_v\nabla\cdot\mathbf{u}^*,$$

$$q_v=\rho^*\left(-c_{v1}C_S^*\Delta\mathbf{U}+c_{v2}0.5(\kappa+1)\Delta\mathbf{U}^2\right),\quad \Delta\mathbf{U}=\min\left(0,\lambda\nabla\cdot\mathbf{u}^*\right),$$

c_{v1}, c_{v2} are the coefficients of the linear and quadratic viscosity of the media, $\kappa=\Gamma_1$ in the liquid and $\kappa=\Gamma_2$ in the gas, $\lambda=(\Delta x\,\Delta y)^{1/2}$, Δx, Δy are the local spatial grid steps.

3. Computational results

3.1. Riemann problem with discontinuity on the gas-liquid boundary

At the initial moment the gas (γ=1.4) and the liquid (Γ=7.15, B=3072 bar) are at rest, while the pressure is discontinuous on the contact boundary at x_0=360 m. In the liquid area ($x > x_0$): φ=1, p=14088 bar, ρ=1.216ρ_0 (ρ_0=10^3 kg/m^3), in the gas area ($x < x_0$): φ=0, p=p_0=1 bar, ρ=$10^{-3}\rho_0$. Figure 2 presents comparison of the results of calculation on the grid with Δx_{min}=Δx/2 and Δx_{max}=2Δx (Δx =5.7 m, α=3, β=0.5) with the exact solution. The calculations were carried out with c_{v1}=c_{v2}=0 in the liquid and with c_{v1}=c_{v2}=0.7 in the gas.

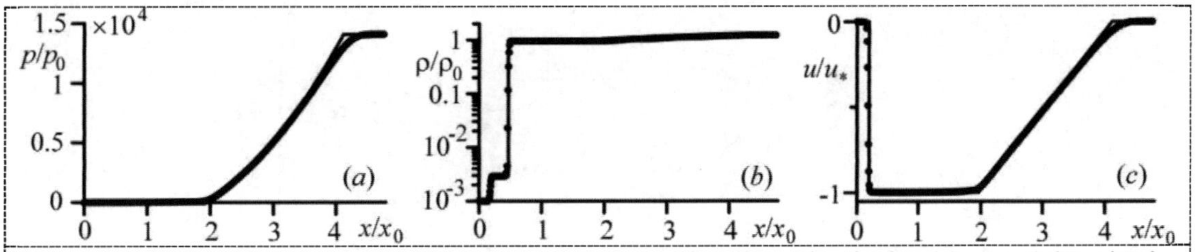

Figure 2. Spatial distributions of the pressure (*a*), density (*b*) and velocity (*c*) at t=0.35 s in the Riemann problem with discontinuity on the gas-liquid boundary, u_* = 540 m/s. The exact solution is given by solid lines, the numerical solution is shown by symbols.

3.2. Cylindrical explosion in a liquid

At the initial moment inside of the cylindrical domain $(x-x_0)^2+(y-y_0)^2<R_c^2$ (x_0=y_0=1500 m, R_c=297 m) there is gas: φ=0, γ=1.4, p=10^4 bar, ρ=720 kg/m^3, and outside it there is liquid: φ=1, Γ=7.15, B=3072 bar, p=p_0=1 bar, ρ_0=10^3 kg/m^3, the both media are at rest.

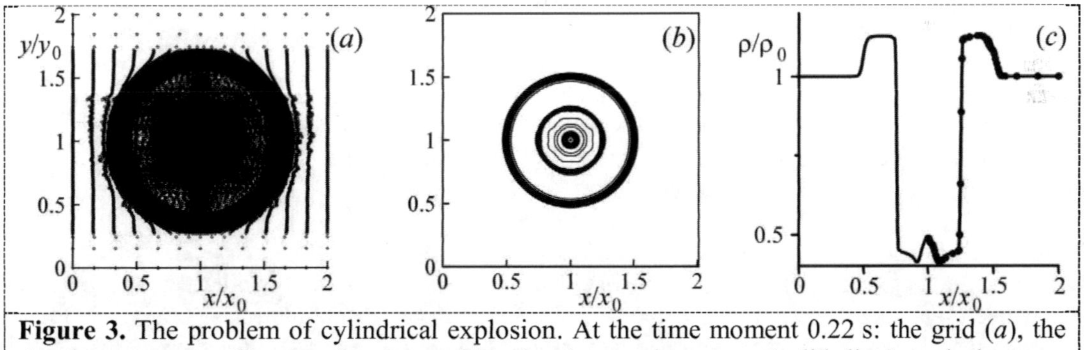

Figure 3. The problem of cylindrical explosion. At the time moment 0.22 s: the grid (*a*), the density contours (*b*), the density-profiles in the section y=y_0 (solid line) and the one-dimensional axisymmetric solution (symbols) (*c*).

The problem was solved in the two-dimensional planar statement, and also in the one-dimensional statement with axial symmetry. The initial distribution of the parameters in the both cases was set on the uniform grid with the spacing Δ_0=7.5 m. The coefficients of artificial viscosity in the liquid are c_{v1}=1.2, c_{v2}=0.4. Figure 3 shows some results of computation on the soroban-grid with Δ_{min}=Δ/4, Δ_{max}=8Δ, Δ=30 m (α=1.1, β=0.73).

3.3. Impact of a jet on a liquid layer on a wall

Impact of an axisymmetric liquid jet on a rigid wall covered with a thin layer of the same liquid is considered (figure 4*a*). In the liquid: φ=1, Γ=7.15, B=3072 bar, in the gas: φ=0, γ=1.33. The end of the jet is hemispherical, the velosity of the jet is V=250 m/s, its radius is R=12.7 μm, the liquid and gas pressure is p=1.25 bar, the liquid density is $\rho=10^3$ kg/m^3, the gas density is 0.34 kg/m^3, the thickness of the liquid layer on the wall is $6.25\cdot10^{-3}R$.

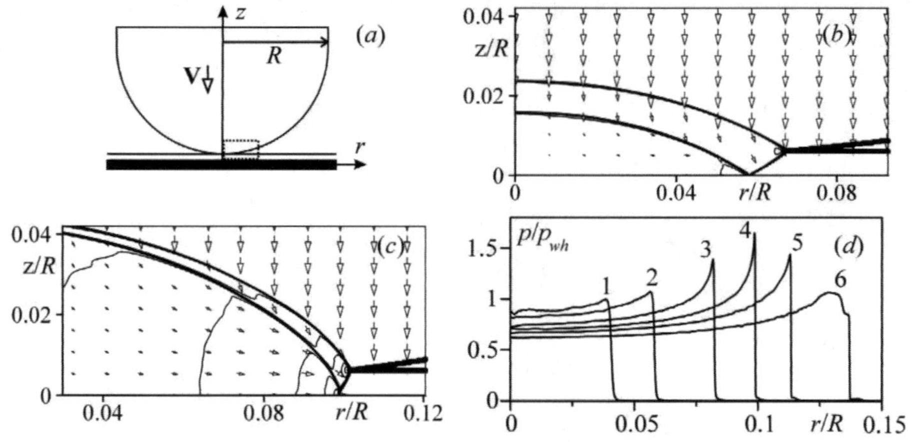

Figure 4. Impact of a jet on a liquid layer on a wall: the scheme (*a*), the interphase surface (bold lines), the pressure contours (thin lines), the velocity vectors at moments 2 and 4 in the half of the axial section in the vicinity of the end of the jet (*b*, *c*); the pressure profiles on the wall (*d*), p_{wh}=5 kbar is the waterhammer pressure under the given conditions. Moments 1–6 correspond to t/τ = 0.01, 0.017, 0.029, 0.039, 0.049, 0.065, $\tau = R / C_{S1}$ = 8.6 ns.

Figure 4 presents the results of computation on the grid with the minimum step $\Delta_{min}=4.4\cdot10^{-5}R$ and the maximum step $\Delta_{max}=6.3\cdot10^{-3}R$. In the liquid c_{v1}=0.6, c_{v2}=0.3, in the gas $c_{v1}=c_{v2}$=0.9. During impact of the jet on the layer two shock waves arise: one moves up the jet, another propagates in the layer to the wall and reflects from it (figure 4*b*). At the moment the second wave reaches the wall the pressure in the center of the area of its reflection (r=0) is approximately equal to the waterhammer pressure p_{wh} and then it decreases. On the boundary of the reflection area the pressure profiles have a growing maximum due to the growth in the pressure on the edge of contact between the jet and the layer (figure 4*d*). Interaction of the shock wave with the wall is separated into the stages of regular (figure 4*b*) and irregular (figure 4*c*) reflection.

4. Conclusion

A numerical technique of computing shock waves in media with moving boundaries of the gas-liquid type has been realized. The approach without explicit separation of the interphase boundary, the CIP-CUP method and the adaptive soroban-grid were applied. The one- and two-dimensional tests confirm the reliability of the created technique. Some results of computation of impact of a jet on a liquid layer on a wall are given.

Acknowledgments

The author would like to thank Prof. Alexander A Aganin for useful discussions. This work was supported by the Russian Foundation for Basic Research (project No 14-01-97004 r povolzhje a).

References

[1] Yabe T and Wang PY 1991 *J. Phys. Soc. Japan* **60** 2105–08
[2] Yabe T, Xiao F and Utsumi T 2001 *J. Comput. Phys.* **169** 556–93
[3] Aganin A A and Guseva T S 2012 *Uchenye Zapiski Kazanskogo Universiteta. Seriya Fiziko-Matematicheskie Nauki* **154** 74–99 (in Russian)
[4] Takizawa K, Yabe T, Tsugawa Y, Tezduyar T E and Mizoe H 2007 *Comput. Mech.* **40** 167–83

Low temperature plasma RF capacitive discharge in helium at atmospheric pressure

A Hakki, I Fayrushin, N Kashapov
Kazan Federal University
420008, KremlyovskayaSt, bdg.18, Kazan, Russia

E-mail: hakki.adnan@gmail.com

Abstract. The paper describes Low temperature plasma RF capacitive discharge in helium at atmospheric pressure. The circuit has been done, to obtain output currentabout 90mA,and the maximum power was 100W, The frequency of the discharging was f = 40MHz.Twolamps (ГУ-50) were used in power supply. Helium consumption was about 1.5l/m.

Key words. Low temperature plasma, RF capacitive discharge, atmospheric pressure

1.Introduction

Plasma power less than100Whas many applications such as in medical, industrial, searching etc[1-5]. This kind of plasma has a very simple design, very large operating range of1MHz to100MHz, very simple cooling system, not expensive final design.

2. Experimental Setup

Fig.1 demonstrates the installation setup of RF discharge. This installation contains a RF power supply 40MHz, matching circuit, upper electrode, lower electrode, glass cover with inlet gas, specimen [6].

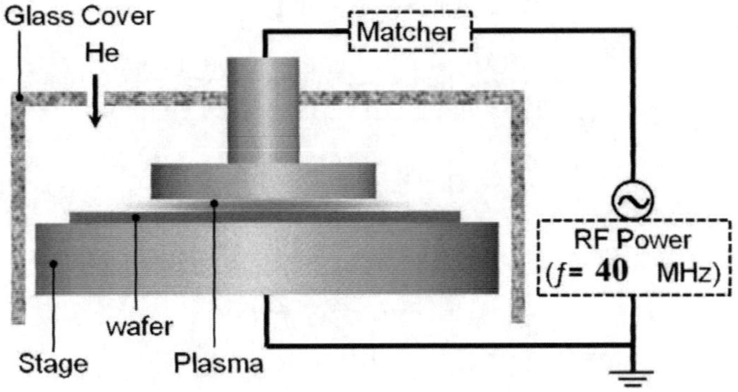

Figure 1. Installation of RF discharge

3. Results and discussion

Fig.2 demonstrates the RF power supply 40MHzThe output frequency is given by [7, 8]:

$$f = \frac{1}{4l\sqrt{L_1 C_1}}$$

Where $l = 0.5m$length of cable, L_1, C_1 its inductance and capacitance,

$$L_1 = 4\mu H \quad C_1 = 40pF$$

Then

$$f = 40MHz$$

Type of lamps VL1.1, VL1.2 is (ГУ-50). The maximum power of this power supply is 100 W.

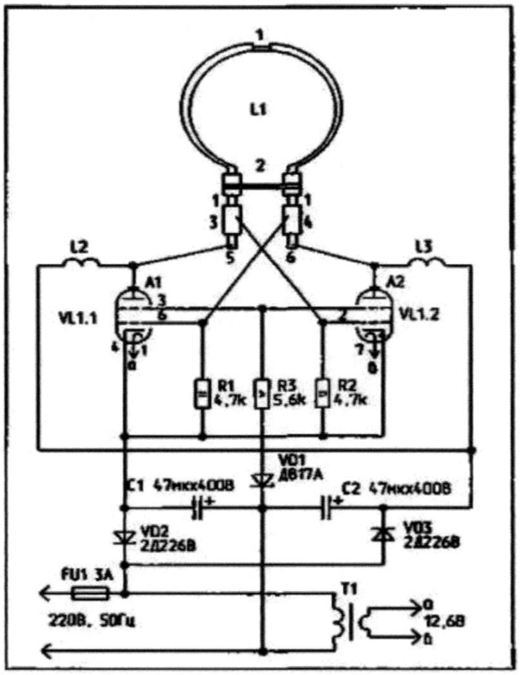

Figure 2.RF power supply 40MHz

Fig.3 demonstrates the electric schematic of the installation. We have added a layer of mica as a barrier, L1, L Sare coaxial cables(75Ω), C and L are variable elements for matching, where:

$$L = 2\mu H \quad C = 8pF$$

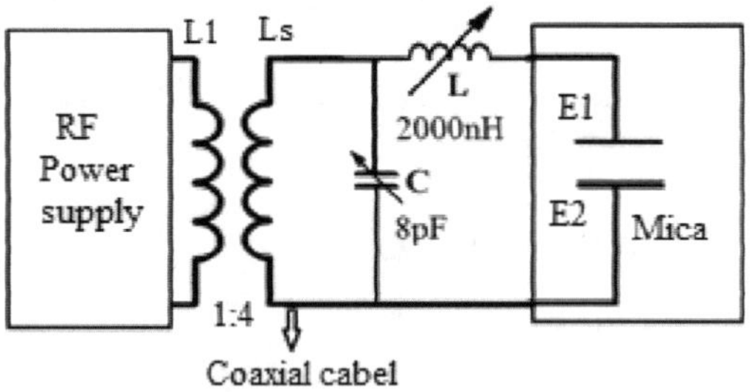

Figure 3. Electric schematic

Fig.4 demonstrates the output current measurement circuit using Rogowski coil [9, 10].

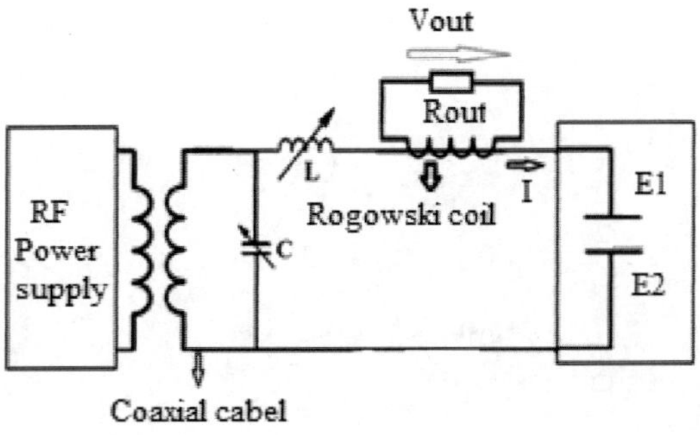

Figure 4. Output current measurement circuit

The output currentis given by:

$$V_{out} = K * i$$

Where K isthe sensitivity of the Rogowski coil, and it is defined as:

$$K = M * \frac{R_{out}}{L_c},$$

Where$M$is the mutual inductance of the Rogowski coil, and it is defined as:

$$M = \mu_0 A n,$$

Then we can write the sensitivity oft he Rogowski coil as [11-13]:

$$K = \mu_0 A n * \frac{R_{out}}{L_c},$$

Where μ_0 is the magnetic constant, A area of one loop, n number of turns, R_{out} the output resistance. Thus, the load current is determined by the formula:

$$i = \frac{1}{k} * V_{out}$$

Finally the output current is given by:

$$i_{out} = \frac{1}{k} * V_{out} = \frac{1}{0.5} * V_{out} = 2 * V_{out}$$

Fig.5 shows the Rogowski coil.

Journal of Physics: Conference Series **669** (2016) 012022 doi:10.1088/1742-6596/669/1/012022

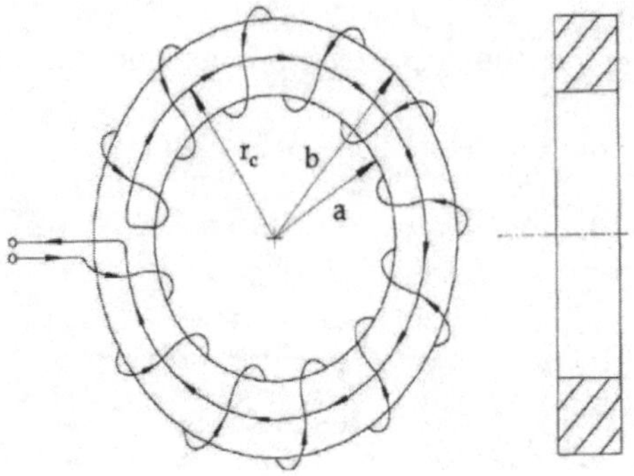

Figure 5. Rogowski coil

Fig.6 demonstrates the output current and voltage

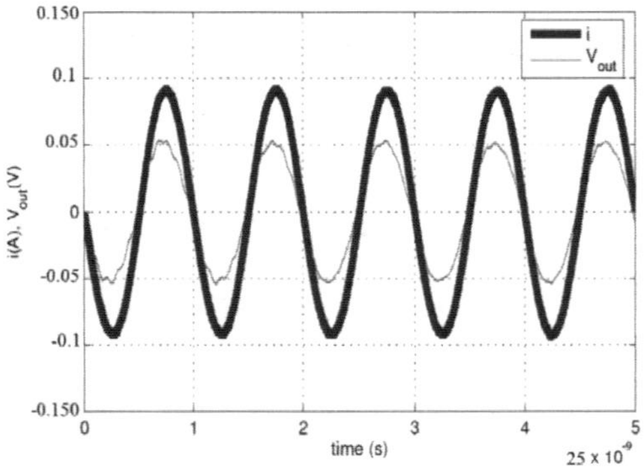

Figure 6. Outputcurrent and voltage

Dimensions of the electrodes

The table below demonstrates the dimensions of the lower electrode and the upper electrode.
The lower electrode:

Diameter (mm)	Thickness (mm)
100	8

The upper electrode:

Diameter (mm)	Thickness (mm)
90	8

Notice that the both electrodes made of aluminum, and they are cooling by air only.

Parameters of the Rogowski coil

The table below demonstrates the selected Parameters of the Rogowski coil.

Parameters	The value
a	13 mm
B	50 mm
A	$1850mm^2$
Lc3,083 μH	
Rc0,274**Ω**	
Rout	10 **Ω**
N	12 turns
K	0.5 V/A

The experimental parameters

The table below demonstrates all the experimental parameters and their conditions.

Conditions Parameters	
ReactivegasHe	
Gas consumption	0.5 –1.5l/min
Power Supply Maximum Power Frequency	RF 100W 40MHz
Maximum load current	90 mA
The gapbetween electrode3 –10mm	
Chamber pressureAtmospheric	

3. Conclusion

Low temperature plasma RF capacitive discharge in helium at atmospheric pressure was constructed and calibrated . The maximum output current was about 90mA at frequency 40MHz.

The current was measured using Rogowski coil method. With this exceptionally valuable and low priced method fast signals can be measured and recorded in an oscilloscope.

References

[1] Yamamura K, Takiguchi T, Ueda M, Deng H, Hattori AN, Zettsu N (2011) Plasma assisted Polishing of Single Crystal SiC for obtaining Atomically Flat Strain-free Surface. Annals of CIRP 60:571–574.
[2] Dautov G, Dautov I, Fayrushin I and Kashapov N 2013 *J. Phys.: Conf. Series* **479** 012001
[3] Kashapov L N, Kashapov N F and Kashapov R N 2013 *J. Phys.: Conf. Series* **479** 012003
[4] Saifutdinov A I, Fadeev S A, Saifutdinova A A and Kashapov N F 2015 *JETP Lett.* **102** 637–642
[5] Israfilov Z K and Kashapov N F 1991 *Journal of Engineering Physics* **60** 364–368
[6] M. A. Lieberman and A. J. Lichtenberg, Principles of Plasma Discharges and Materials Processing ~Wiley, New York, 1995!, p. 175, Eq. 6.6.7.
[7] E. Kawamura, M.A. Lieberman, and A.J. Lichtenberg, "Capacitive Discharges Driven by Combined DC/RF Sources", J. Vac. Sci. Technol. A, 25, 1456–74, Sept. 2007

[8] Raizer YP, Shneider MN, Yatsenko NA. Radio-frequencycapacitive discharges. Boca Raton, FL: CRC Press;1995.
[9] W. F. Ray and C. R. Hewson, \High performance Rogowski current transducers", IEEE - IAS Conf. Proc., Sept. 2000.
[10] W. Rogowski and W. Steinhaus, \Die Messungder magnetische Spannung", Arch Electrotech, vol. 1, pp. 141{150, 1912.
[11] W. F. Ray, \Wide bandwidth Rogowski current transducers. part i: The rogowski coil", EPE Journal, vol. 3, no. 1, pp. 51{59, March 1993.
[12] J. D. Ramboz, \Machinable Rogowski coil. Design and calibration", IEEE Transactions on Instru- mentation and Measurement, vol. 45, no. 2, pp.511{515, April 1996.
[13] D. A. Ward and J. La T. Exon, \Using Rogowski coils for transient current measurements", Engineering Science and Education Journal, pp. 105{113, June 1993.

Frequencies of gas oscillations in a pipe with a concentrated heat source

O V Iovleva, V M Larionov, E V Semenova

Kazan Federal University, 18 Kremlyovskaya str., Kazan 420008, Russian Federation

E-mail: techphyskfu@kpfu.ru

Abstract. It is known that the location of the heat source significantly affects the frequency of acoustic oscillations in the channels. The case of a step change in the sound speed is investigated. In this article, linear distribution of sound speed in hot gas is considered. The well-known equations are used to calculate frequencies of the gas oscillations. The analysis shows that the movement of the flame from the down up in an open tube causes a nonmonotonic change in the resonant frequency. The calculation results are in good agreement with the experimental data.

1. Introduction

Acoustic gas oscillations were investigated in several works: in channels [1–3] and in combustion chambers [4–6]. One such laboratory installations is a vertical pipe with open ends. In some section is located concentrated heat source, such as a flame (Figure 1).

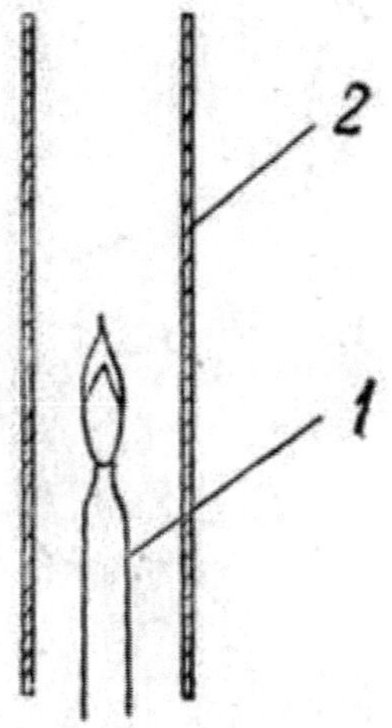

Figure 1. A diagram of the setup. 1 – burner 2 – pipe.

The air enters the pipe, having a constant temperature. After the heat source the mixture of air and burnt gases at a high temperature. Then, heat transfer to the pipe walls causes a gradual decrease of the gas temperature and therefore the sound speed. There was obtained a generalized equation which allows one to calculate the frequency of the gas oscillations in a pipe with any boundary conditions [7,8]. The effect of location of heat source on the resonant frequencies of an open pipe is investigated in this paper.

2. Resonance gas oscillations in a pipe with flame

The equation to calculate the frequency of longitudinal gas oscillations in a tube with any boundary conditions, is as follows:

$$-\frac{b}{2\omega}+\beta tg\left[\frac{\omega\beta}{b}ln\left(1-\frac{bx^*}{a}\right)-\varphi_2\right]+\frac{c_2^*}{c_1}\tan\left(\frac{\omega x^*}{c_1}+\varphi_1\right)=0. \quad (1)$$

where a, b - the coefficients of the linear distribution of sound, ω - cyclic frequency, $\beta=\left[1-(b/(2\omega))^2\right]^{1/2}$, $c_2(x)=a-bx$, $c_2^*=c_2(x^*)$, x^* - the distance from the entrance pipe to the rear of the source.

For pipe that is open at the ends, replacing the actual length of pipe effective $l^*=l+0.61R$, R - the radius of the pipe, believing $Y_0=0=Y_l$, get

$$\varphi_1=0,$$

$$\varphi_2=\frac{\omega\beta}{b}ln\left(1-\frac{bl^*}{a}\right)-arctg\left(\frac{b}{2\omega\beta}\right).$$

We measured the resonant frequencies of acoustic oscillations of gas in a pipe, open at the ends, with a length of 0.91 m with an internal diameter of 0.036 m. A mixture feed pipe 3 m long with an internal diameter of 0.005 m and propane–air flames were located on the pipe axis. Figure 2 shows experimental and theoretical data for the first two resonant frequencies under the condition $c^*=497\ m/s, b=165\ s^{-1}$.

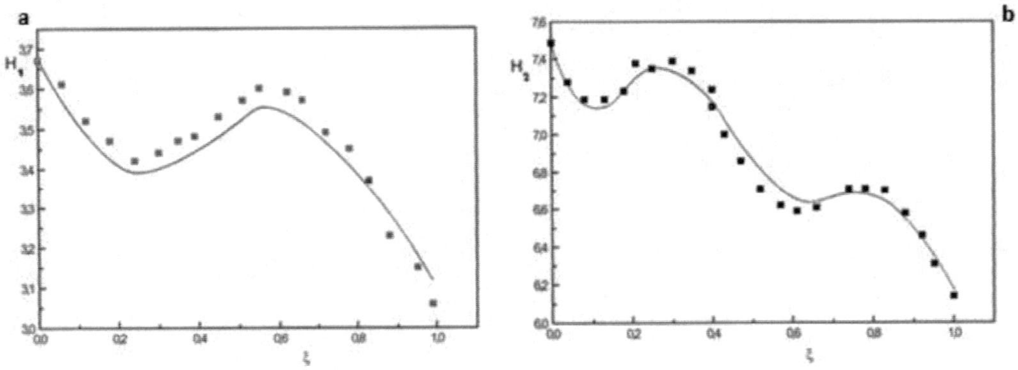

Figure 2. The dependence of the resonant frequencies from the flame location in the pipe: a - the first frequency; b - the second frequency; $H=\omega l/c_1$; $\xi=x^*/l$; line – theory; points – experiment.

These values have been calculated. The measured distribution of the mean gas temperature along the pipe was used. Experimental data and calculation results are consistent. The movement of the flame from the inlet to the outlet pipe causes the non-monotonic change in frequency. Curves have minimums and maximums. The dependence of the first frequency has one minimum and one maximum. The dependence of the second frequency has two minimums and two maximums.

3. Conclusion

Linear analysis of the equations of resonant frequencies in the pipe that is open at the ends with a concentrated heat source was carried out. In the case of a linear decrease of the sound speed in the hot part of the flow resonance frequency depend on the location of the heat source in the pipe. Curves have minimums and maximums. For comparison, if there is no sound speed gradient, the heat source moving from input to output causes the a continuous decrease of the resonant frequency.

The work is performed according to the Russian Government Program of Competitive Growth of Kazan Federal University.

References

[1] Ermakov R A, Galiullin R G, Larionov V M, Nikolaev A N 2008 *Chemical and Petroleum Engineering* **44** 387-392
[2] Ermakov R A, Sadykov A F, Galiullin R G, Larionov V M 2011 *Chemical and Petroleum Engineering* **46** 749-751
[3] Fadeev S A, Kashapov N F, Larionov V M *2014 IOP Conf. Ser.: Materials Science and Engineering* **69** 012006
[4] Akhmadullin A N, Ahmethanov E N, Iovleva O V, Mitrofanov G A 2013 *J. Phys.: Conf. Ser.* **479** 012004
[5] Bykovets A P, Larionov V M, Marchukov E Yu 1992 *Izvestiya Vysshikh Uchebnykh Zavedenij. Aviatsionnaya Tekhnika* (3) 71-74
[6] Larionov V M, Iovleva O V 2003 *Izvestiya Vysshikh Uchebnykh Zavedenij. Aviatsionnaya Tekhnika* (4) 48-51
[7] Iovleva O V, Larionov V M 2007 *Russian Aeronautics* **50** 303-308
[8] Semenova E V, Larionov V M, Kazakova E I 2014 *J. Phys.: Conf. Ser.* **567** 012033

Journal of Physics: Conference Series **669** (2016) 012024 doi:10.1088/1742-6596/669/1/012024

Study of processes of steels surfaces modification with highly concentrated energy flows

R M Khisamutdinov [1], V V Zvezdin[1], I H Israfilov[1], Ruz R Saubanov [1], A A Spirin[2], R R Rakhimov[1]

[1]Naberezhnye Chelny institute (branch) of KFU, Naberezhnye Chelny, Russia
[2] Ford Sollers holding LLC, Naberezhnye Chelny, Russia

E-mail:irmaris@yandex.ru, rafisih88@mail.ru

Abstract. The peculiarities of steels surface layer modification under the influence of highly concentrated energies which increase the parts and tools quality have been studied . Here are the results of microstructure studies of tool steels and steel 45 samples for the analysis of strengthened steel and surface layer formation conditions. The conditions for the required quality of modified surfaces achievement using nitriding, hardening and cladding.

1. Introduction

The modern machine engineering is characterized by the complex operating conditions for machines related to vibrations, high level of effective voltage, broad temperature range, aggressive environments, etc. Specific requirements for metals surface layer must be complied with. According to the statistics, the majority of machining parts and tools (85-90%) fail due to wearing of separate parts surfaces [1].

Among the various means of metal surface layer quality improvement, the main ones are cementation, nitride cementation (carbonizing), carbonitriding, nitriding, etc.

Such courses as laser, plasma, etc, surface modification are expanding at the moment. Surface strengthening of operating areas of tools and parts is of pretentious nature both in energy parameters accuracy of the concentrated energy source, and in time-space condition of the zone where it influences the surface [2, 3].

Currently, there are numerous various methods of surface modification [7]. Their use requires adjustment of output parameters of technological complexes and time-space features so that the required surface layer quality parameters are stabilized. In the laboratory the processes of ion-plasma nitriding of parts surface, its laser and plasma hardening were studied.

This work contains the results of metallographic studies of surfaces after plasma nitriding, laser and plasma cladding and hardening. This relates to the most perspective and modern methods of hardening technologies with an opportunity to control and stabilize the tools and parts surfaces.

2. Ion-plasma nitriding of tool steels

Creation of surface layer with higher hardness, wear resistance, improved fatigue resistance and corrosion resistance is a perspective and rapidly developing course. The finished products which have already passed the mechanical and final heat treatment are subjected to nitriding.

The main purpose of this study is to control and manage the technological process parameters. The microstructure of the tool steel 4Х5МФС after impulse plasma flow treatment in the nitrogen environment is shown on the Figure 1.

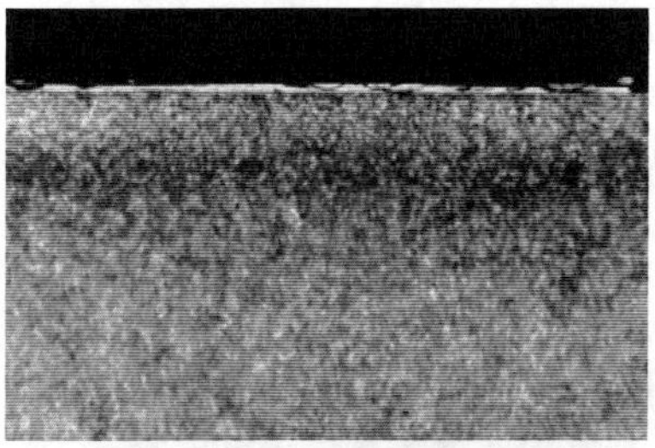

Figure 1. Microstructure of nitrided layer of the tool steel 4Х5МФС.
The cores of the sample – tempered martensite, carbides, from the sample surface – nonerodible nitride zone ~5,4 micron thick; further – transition diffusion layer with nitride phase flecks

After the nitriding process not only the hardness and wear resistance, but also corrosion resistance of metals increases which is especially important during extrusion dies usage [4]. This is why the purpose of the studies is to define the technological parameters influence on the microstructure values variability and metals microstructure change in the nitride atmosphere under the influence of impulse plasma flow and continuous laser irradiation.

For experimental studies an impulse plasma technological complex (IPTC) has been developed. The basis for it is a developed plasmatron with an alternating current with a power consumption up to 35 kW [5]. During this experiment, the plasma flow with a diameter of 5 to 10 mm is directed to the sample positioned in the operating booth in nitrogen gas environment. The treatment (processing) is being done almost without surface melting decreasing the heating load at the die.

A nitride layer is observed on the sample surface, this layer consists of nitride zone in the shape of a white nonerodible stripe, inner nitriding zone which consists of nitrogenous solid solution with micro-hardness starting from 1238 $HV_{0,3}$ (0,03mm from the surface) (Fig.1).

3. Laser and plasma hardening and cladding of steel 45

Experimental studies have been conducted in respect to metals surface hardening in the nitrogen environment under flux and with alloyed powders of higher hardness. These include powder ПГ – СР4 – ОМ with micro-hardness of 55 HRC and flux АН-43 providing cladded metal acquisition with an optimal manganese, silicone and other alloy elements as well as with an optimal sulphur and phosphorus content. The Fig. 2 and 3 show steel 45 surface microstructure with a powder ПГ – СР4 – ОМ cladding through plasma flow and laser irradiation.

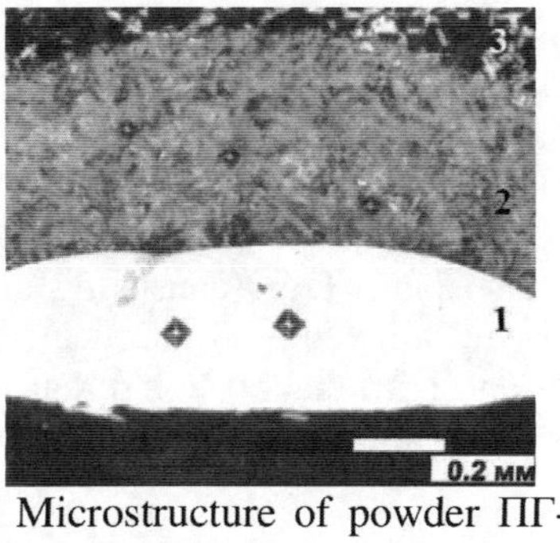

Figure 2. Microstructure of powder ПГ-СР4-ОМ layer cladded on steel 45 in nitrogen environment with a plasma flow

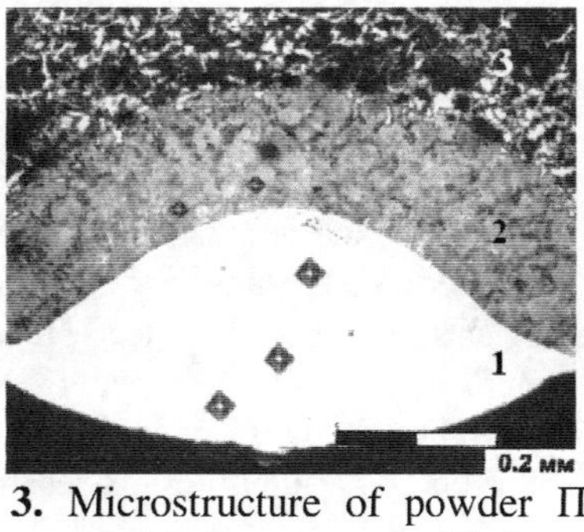

Figure 3. Microstructure of powder ПГ-СР4-ОМ layer cladded on steel 45 in nitrogen environment with a laser irradiation

Energy concentration is about 10^5 W/cm^2, thermal impact zone in the metal acquires a segment form. Laser irradiation 0,9 mm in diameter is directed to the sample as well, transition speed is 20 mm/sec.

On Figures 2, 3 (light area) a microstructure of powder ПГ – СР4 – ОМ is observed, in the upper area — unchanged structure of steel 45. A steel 45 with powder ПГ – CP4 – OM interaction zone can be seen between them which is characterized by diffusion processes taking place in the transition zone. Micro-hardness of the powder layer (light area) cladded with plasma flow in the nitrogen environment is 514 $HV_{0,1}$, and middle area after heat treatment acquired structure of a needle martensite with a micro-hardness of 1003 $HV_{0,1}$. During the laser treatment is 920 $HV_{0,1}$ and 1003 $HV_{0,1}$ respectively. Micro-hardness was measured according to the GOST 9450-76.

Figures 4, 5 show microstructures of steel 45 samples with hardened plasma zones and laser irradiation under flux AH - 43.

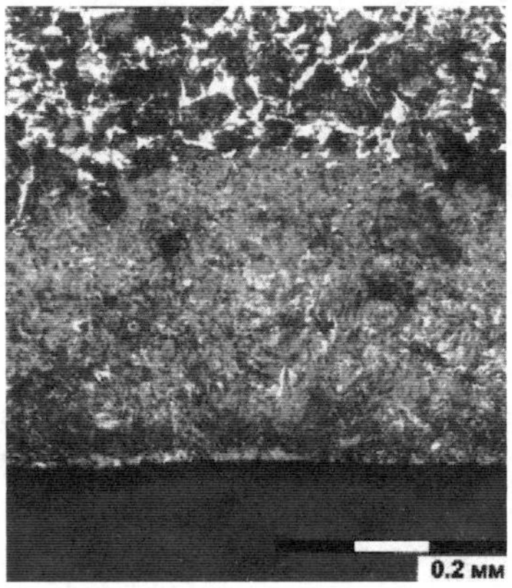

Figure 4. Microstructure of hardened under flux (AH – 43) zone of steel 45 with plasma flow in nitrogen environment

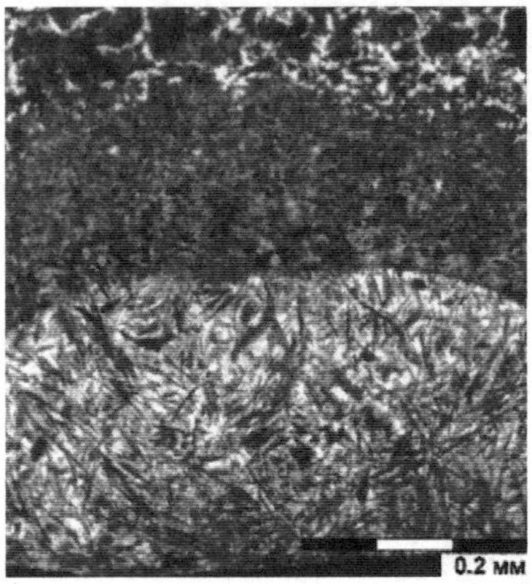

Figure 5. Microstructure of hardened under flux (AH – 43) zone of steel 45 with laser irradiation in nitrogen environment

For the comparative analysis, microstructures and micro-hardnesses of steel 45 processed with laser irradiation in nitrogen environment under flux AH - 43 were studied in the same modes (Fig. 5). Subsurface layers micro-hardness value of martensite layer of steel 45, processed under flux is 1003 $HV_{0,1}$, and with powder ПГ–CP4–OM layer cladding these values equal 1097 $HV_{0,1}$.

Heat treatment of steel 45 with plasma flow in nitrogen environment under flux (AH – 43) shows ground needle microstructure of martensite layer (Fig. 3) in comparison with the microstructure of steel 45 – ПГ – CP4 – OM samples which characterizes micro-hardness reduction up to 824 $HV_{0,1}$. Furthermore, a nitride layer has been identified from the sample surface, this layer consists of nitride zone in the form of a white nonerodible stripe 0,007mm thick under which a diffusion zone is located. The previously conducted analysis results for these modes show that a layer consisting of e-phase (Fe_2N) forms on the steel 45 surface. The diffusion zone consists of alloyed with nitrogen matrix and incorporated nitrides of alloy elements. During hardening with laser irradiation, micro-hardness was 1003 $HV_{0,1}$. Higher hardness is explained by high cooling speed of the thermal impact zone due to high density of energy and higher processing speed.

4. Laser hardening of cutter teeth made of steel P18K5F2

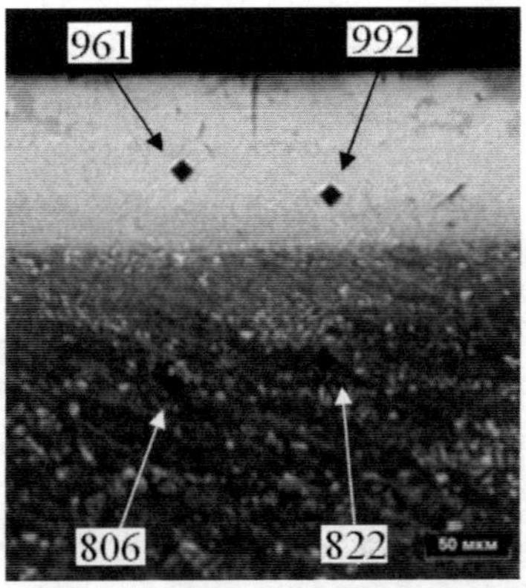

Figure 6. Microstructure of laser hardened layer on the tool teeth tops with micro-hardness measurements results in $HV_{0,1}$, (x 500)
Chemical composition of worm cutter metal complies with the steel P18K5F2 GOST 19265-73.
The part microstructure is a latent needle martensite of grade 1 GOST 8233-56, carbides.
Carbide heterogeneity of the part structure corresponds to the Grade 1A of the scale 1 which complies with the requirements of the GOST 19265-73.

At the operating surface of teeth edges a surface layer hardened with a laser irradiation with nonerodible structure and carbides is observed.

Part micro-hardness is: main cutter metal - 806 – 852 $HV_{0,1}$; hardened with laser irradiation la - 961 – 992 $HV_{0,1}$.

5. Conclusion

The study findings show that the technological process of the surface layer modification is an effective method of steel wear resistance increase for extrusion dies and micro-hardness of the tool experiencing various wear types. Based on the analysis of results, main parameters of plasma flow and laser irradiation can be derived which effectively influence the technological process quality. This allows for conducting precision processing for parts and tools. This process requires use of high-accuracy lase irradiation and plasma flow positioning system [6].

References

[1] Yunusov F S, Khisamutdinov R M Shape-forming tools producing surfaces accuracy increase (monograph) // KNRTU-KAI. - 2008. – p 212.

[2] Grigoriants A G, Shiganov I N, Misyurov A I Technological processes of laser treatment: Textbook for higher education institutions/ Edited by A. Grigoriants. — M.: Publishing house of Bauman MSTU, 2006. — p 664: ill.

[3] Saubanov Ruz R, Rakhimov R R, Zvezdin V V, Israfilov I Kh Method of nanostructured modification of parts surface layer with concentrated energy flows / Materials of the 13th international science-practice conference, section 3- Saint-Petersburg; Publishing house of Polytechnic university, 2011. - pp 389-394.

[4] Arzamasov B N, Bratukhin A G, Eliseev Y S, Panayoti T A Ion chemical thermal processing of alloys in gas environments. - M: Publishing house of Bauman MSTU, 1999. – p 400.

[5] Saubanov Ruz R, Zvezdin V V, Israfilov I Kh, Rakhimov R R, Saubanov Rusl R Invention patent №2558713, Russia, Россия, IASC H05H1/24; Application of 11.03.2014; Published: 10.08.2015; Priority 11.03.2014. Bul. №22. Impulse alternating current plasma generator device.

[6] Zvezdin V V, Khamadeev A V, Zagirov R G, Shangaraev I R Laser irradiation positioning in relation to the welding joint as a technological process quality indicator / KNRTU-KAI Bulletin, 2008. — № 3. — pp 17-21.

[7] Portnov S M, Kisaev R A, Kuznetsov I N, Israfilov I Kh, Zvezdin V V, Nugumanova A I System of parts surface heat treatment process with concentrated energy flows / «Global science potential» Saint-Petersburg: - №8. - 2011.- pp 95-100.

Investigation of crater form and dimensions evolution during laser treating of materials

A Yu Ivanov[1], A V Kapytski[2], N V Zharkiy[1] and N A Yerema[1]

[1] Grodno State University, 22 Ozheshko Street, Grodno 230023 Belarus
[2] Grodno State Medical University, 78 Gorky Street, Grodno 230023 Belarus

e-mail: ion_ne@mail.ru

Abstract. Acoustic emission of a zone of the destruction formed during influence of pulse laser radiation on a surface of metal is considered. Dependence of the time form of acoustic fluctuations on parameters of an irradiated material and the law of increase in depth of a crater was estimated. It is revealed, that at action on a surface of the copper sample of a laser impulse duration ~ 20 μs time of growth of a zone of destruction makes approximately 40 μs. It is shown also that having detected a diffraction picture appearing on incidence of SHF-radiation on the surface of solid and having achieved during the experiment the best coincidence of the experimental and calculated using the computer model data one can define the character of the zone of destruction on the surface of structure being under treatment relief.

1. Introduction

Laser treating of solids is accompanied by several phenomena. One of these phenomena is an acoustic emission of irradiated matter destructed zone. The experiments and theoretical investigations show the existence of simple connection between the parameters of emitted acoustic waves and the characteristics of destructed zone [1]. Thus estimating the properties of the acoustic wave one can estimate the parameters of destructed zone [2]. The authors of mentioned works have obtained their results using the model of a loaded zone radiating waves into elastic medium. Calculated with the help of this model temporal dependences of pressure of an elastic wave correspond satisfactory to the experimental data only beginning from the second quarter of the pressure of elastic wave oscillations because of the acoustic wave generation during the growth of the zone of non-elastic deformation.

Our aim was to investigate emission of the acoustic waves generated on the laser treating of solids during the laser radiation action on the metal surface and estimation of the dependence of temporal parameters of the pressure of generated elastic wave and the temporal dependences of the geometrical properties of irreversibly deformed zone on the surface of the irradiated sample. The second purpose of research was to develop the new method of determination of geometrical parameters of destructed zone in real time scale, based on the features of SHF-radiation diffraction on the different objects with complicated form are perspective.

2. Acoustic emission of laser destruction zone

Investigating the acoustic emission let us use the model of a loaded zone radiating waves into elastic medium. Corresponding to this model let us consider the destructed zone as a spherical segment with the curvature radius R, depth d and diameter $2r_1$. z-ax of the coordinate system is directed along the laser beam. It is important that the parameters of the irreversibly deformed zone are changing on the time: $R=R(t)$, $d=d(t)$, $r_1=r_1(t)$; here t is the time.

The displacement vector in an elastic zone consists of a longitudinal and a transversal components, $\vec{A}=\vec{A}_l+\vec{A}_t$, and each of them can be described by corresponding wave equation. Because of the presence of the media board direct in the elastic wave generating zone the solution of the wave equations system we shall search as a sum of the volume and the surface components

$$\vec{A}_l=\vec{A}_{lO}+\vec{A}_{l\Pi}=\nabla\psi_O+\nabla\psi_\Pi\ ,$$
$$\vec{A}_t=\vec{A}_{tO}+A_{t\Pi}=\vec{A}_{t0}+rot(\vec{B})\ .$$

With regard to the symmetry of our problem, $\vec{A}_{t0}=0$, scalar potential $\psi_0(\omega)=-\tilde{A}(\omega)\cdot\dfrac{\exp(-ik_l r)}{r}$, $\psi_\Pi(\omega)=-Z_0(k_R\rho)\cdot(\tilde{B}(\omega)\cdot\exp(-\chi_l z)+\tilde{Q}(\omega)\cdot\exp(\chi_l z))$, $B_\varphi(\omega)=Z_1(k_R\rho)\cdot(\tilde{D}(\omega)\cdot\exp(-\chi_t z)+\tilde{S}(\omega)\cdot\exp(\chi_t z))$, $B_\rho=B_z=0$.

Here ω is a frequency, $k_\ell=\omega/c_\ell$, c_ℓ and c_t – longitudinal and transversal sound velocities,

$\tilde{A}(\omega)$ – wave amplitude, $k_R=\omega/c_R$, c_R – surface wave velocity, $\chi_i=(k_R^2-k_i^2)^{1/2}$, $\tilde{B}(\omega)$, $\tilde{D}(\omega)$, $\tilde{Q}(\omega)$, $\tilde{S}(\omega)$, – oscillation amplitudes, $Z_i(x)$ – spherical function. When $\rho\to 0$ and

$z\to\infty$ the result must remain finite, therefore $Z_i(x)=J_i(x)$ (Bessel function), $\tilde{Q}(\omega)=\tilde{S}(\omega)=0$, and

$$\vec{A}(\omega)=A(\omega)\frac{\vec{r}}{r^3}(1+ik_l r)\exp(-ik_l(r-R))+B(\omega)(\vec{\rho}_0 k_R J_1(k_R\rho)+\vec{z}_0\chi_l J_0(k_R\rho))\times$$
$$\times\exp(-\chi_l(z-h))+D(\omega)(\vec{\rho}_0\chi_t J_1(k_R\rho)+\vec{z}_0 k_R J_0(k_R\rho))\exp(-\chi_t(z-h)) \qquad (1),$$

where $A(\omega)=\tilde{A}(\omega)\cdot\exp(-ik_l R)$, $B(\omega)=\tilde{B}(\omega)\cdot\exp(-\chi_l h)$, $D(\omega)=\tilde{D}(\omega)\cdot\exp(-\chi_t h)$.

Let us consider that on the surfaces $r=R$ and $z=h$ temporal dependence of pressure in the plasma cloud is

$$P/_{r=R}=p(t).$$

On the surface of spherical segment $R=R(t)$ $\sigma_{rr}=-p(t)$, $\sigma_{r\theta}=\sigma_{r\varphi}=0$; on the surface $z=h$ $\sigma_{zz}=p(t)$, $\sigma_{\rho z}=0$, $\sigma_{z\varphi}=0$. Here σ_{ij} are the components of the stress tensor, r, θ, φ are the coordinates of spherical system.

Out of the spherical segment $R=R(t)$ medium is elastic, and so [3]

$$\left[\lambda\left(\frac{\partial A_r}{\partial r}+2\frac{A_r}{r}+\frac{1}{r}\frac{\partial A_\theta}{\partial\theta}+\frac{A_\theta}{r}ctg\,\theta\right)+2\mu\frac{\partial A_r}{\partial r}\right]_{r=R(t)}=-p(t),$$
$$\left(\frac{\partial A_\rho}{\partial z}+\frac{\partial A_z}{\partial\rho}\right)_{z=h(t),\rho=\rho_1(t)}=0, \qquad (2)$$

$$\left[\lambda\left(\frac{\partial A_\rho}{\partial\rho}+\frac{A_\rho}{\rho}+\frac{\partial A_z}{\partial z}\right)+2\mu\frac{\partial A_z}{\partial z}\right]_{z=h(t),\,\rho=\rho_1(t)}=-p(t).$$

Here A_i are the components of the displacement vector $\vec{A}$, λ, μ are the Lame coefficients.

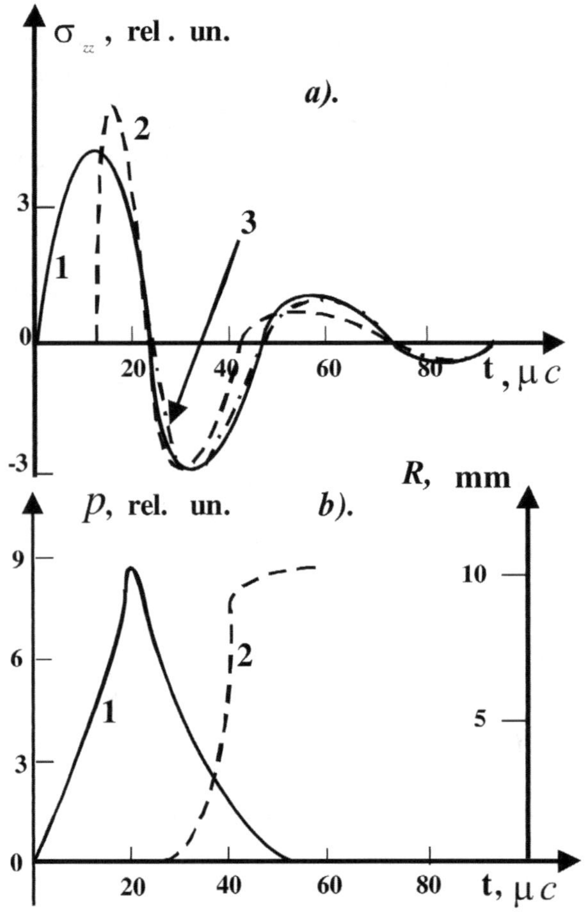

Figure 1. *a).* Temporal dependences of pressure of acoustic wave on the action of laser pulse with duration of 20 μs on cupper sample: 1 – experimental dependence [2]; 2 – dependence calculated without account of crater growth [4]; 3 – dependence calculated with account of crater growth.
b). 1 – temporal dependences of pressure of plasma cloud on the border of irreversibly deformed zone; 2 – temporal dependences of curvature radius.

Substituting A_i from the equation (1) to the system (2) we can calculate for each temporal moment $A(\omega,R,d,\rho_1)$, $B(\omega,R,d,\rho_1)$, $C(\omega,R,d,\rho_1)$, $\sigma_{zz}(\omega,R,d)$ and

$$\sigma_{zz}(t)=\int_{-\infty}^{+\infty}\sigma_{zz}(\omega,R,d)Exp[i\omega t]d\omega=\int_{-\infty}^{+\infty}\sigma_{zz}(\omega,R(t),d(t))Exp[i\omega t]d\omega=\int_{-\infty}^{+\infty}\sigma_{zz}(\omega,t)Exp[i\omega t]d\omega.$$

The results of the calculations with $R(t) = R_{max} \cdot \exp\left(\frac{t^2}{\tau_0^2}\right)$ for $t<0$ and $R(t) = R_{max}$ for $t>0$ are presented on the figure 1; $\tau_0 = 40\ \mu s$.

Evidently that at action on a surface of the copper sample of a laser pulse duration ~ 20 μs time of growth of a zone of destruction makes approximately 40 μs, that will well be coordinated with time of existence of plasma formation at a surface of the target exposed to laser-plasma processing (~ 50 μs). Use of model of the loaded area with the moving borders radiating acoustic waves in the elastic medium allows to solve the important practical problem – the definition of a law of time growth of a zone of irreversible deformations on a surface of the sample exposed to pulse laser-plasma processing.

3. Use of SHF-diapason electromagnetic waves diffraction for the crater evolution diagnostics

The second purpose of research was to develop the new method of determination of geometrical parameters of destructed zone in real time scale. In this research the modeling and calculations were realized with the help of the program set CST Microwave Studio – a system of SHF-devices modeling based on the method of approximation and the method of definite integrals in temporal area [5].

The pictures of diffraction of SHF-radiation on the models of the structures with different dielectric penetrability periodicity including periodical structures (figure 2). Parameters of structure represented at fig. 2, are following. 9x9 right-angled insulators with dimensions 1x1x9 мм; dielectric penetrabilities ε_1=100, ε_2=200, ε_3=300. The diffracted radiation direction diagrams (figure 3) was obtained in SHF-diapason on the frequency f = 10.0*10^9 s^{-1}.

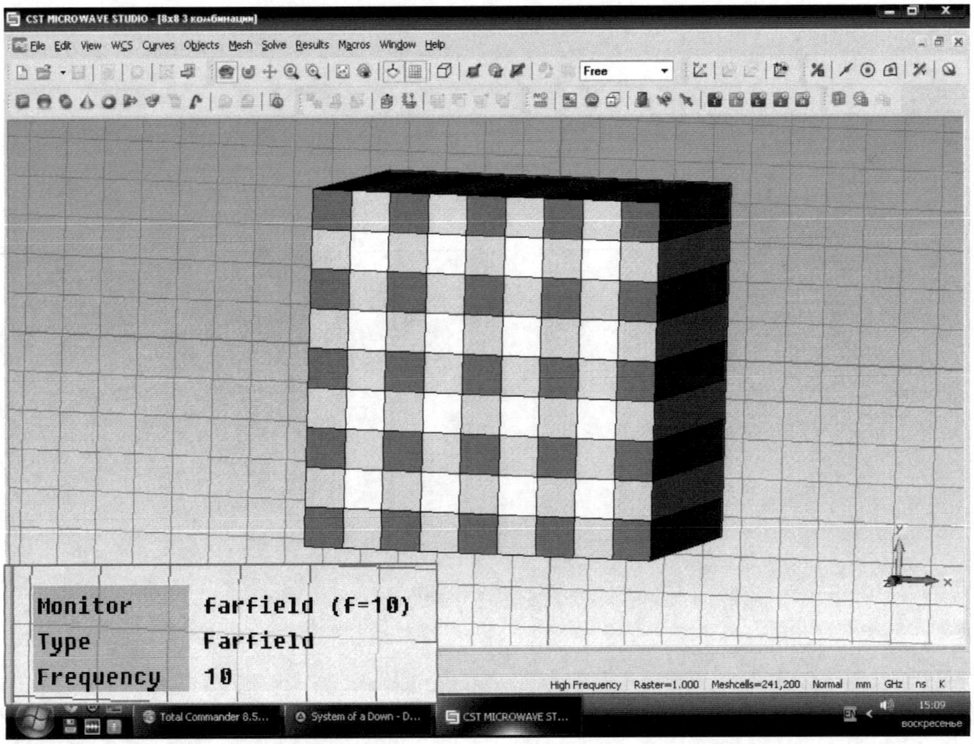

Figure 2. Computer model of periodical structure in CST Microwave Studio product.

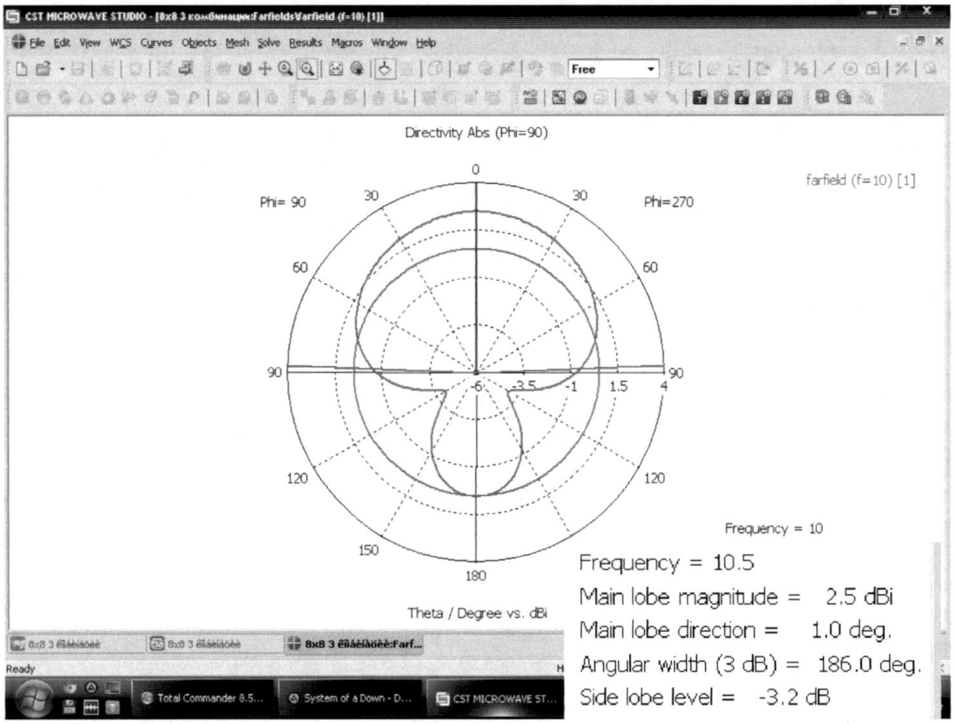

Figure 3. Direction diagram of described structure obtained on the frequency f = $10.5*10^9$ s^{-1}.

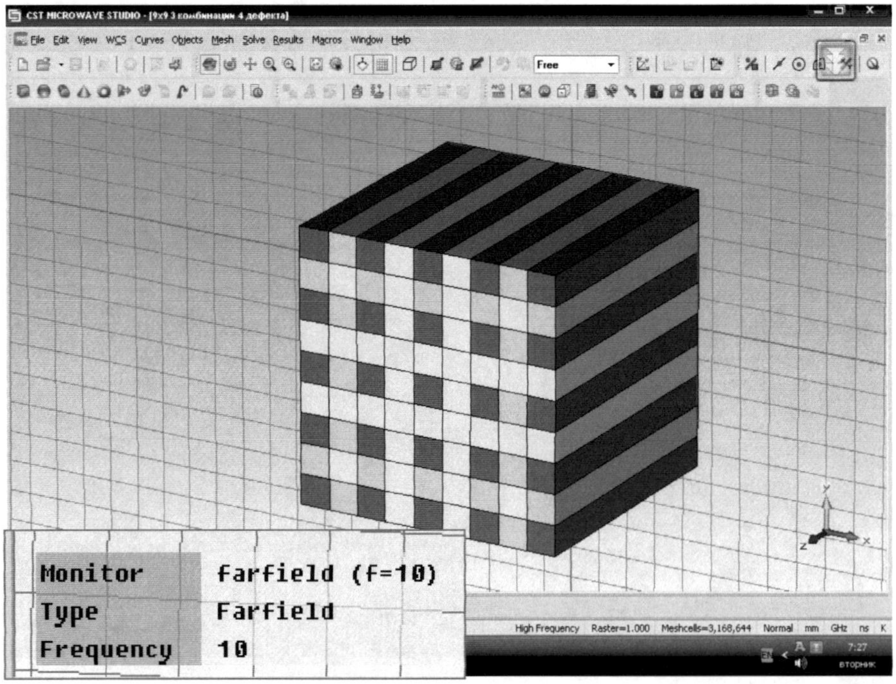

Figure 4. Computer model of periodical structure with some types of unperiodicity and defects in CST Microwave Studio product.

Figure 4 represents a of periodical structure with some types of unperiodicity and defects.

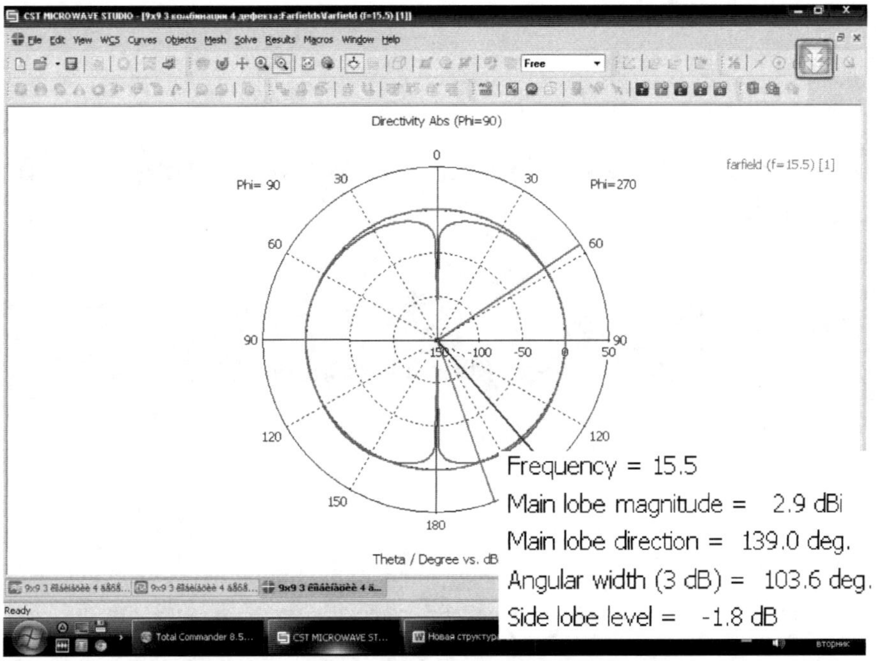

Figure 5. Direction diagram of described periodical structure with some types of unperiodicity and defects obtained on the frequency f = $15.5*10^9$ s^{-1}.

Figure 5 represents the diffracted radiation direction diagram of periodical structure with some types of unperiodicity and defects obtained in SHF-diapason on the frequency f = $10.0*10^9$ s^{-1}.

The results of modeling showed a considerable difference of the diffracted radiation direction diagrams obtained using some different objects. Using the distinctive parameters of diffractive reflection curve one can solve with the definite precision the reverse problem of character of defects of the surface of structure being under treating determination. The SHF-radiation admission spectrums for the described structures are also obtained. They are also different.

4. Conclusion

So having detected a diffraction picture appearing on incidence of SHF-radiation on the surface of solid and having achieved during the experiment the best coincidence of the experimental and calculated using the computer model data one can define the character of the zone of destruction on the surface of structure being under treatment relief.

References

[1] Ashmarin I.I., Bykovskii Yu A, Zysin Ya Yu, et al. KvantovayaElektron. (Moscow) 15 393 (1988) [Sov. J. Quantum Electron. 18 250(1988)].
[2] A.Yu. Ivanov. Acoustical diagnostics of process of laser treating of materials / – Grodno: GrGU, 2007. – 280 p.
[3] L.D.Landau, E.M.Lifshits. Theory of elasticity / – M: Nauka, 1987. – 246 p.
[4] A.Yu. Ivanov, V.I.Nedolugov. KvantovayaElektron. (Moscow) 16, 801 (1989) [Sov. J. Quantum Electron. 19 668 (1989)].
[5] CST Studio Suite 2009. Workflow&Solver Overview. [Electronic resource]. – 2013. – The access regime: http://www.cst.com. – The access date: 07.09.2014.

Influence of electric field on crater formation during laser treating of metals

A Yu Ivanov[1], **S V Vasil'ev**[1]

[1] Grodno State University, 22 Ozheshko Street, Grodno 230023 Belarus

e-mail: ion_ne@mail.ru

Abstract. It is shown that by varying the external electric field with different polarity from 0 to 10^6 V/m^{-1} in the course of laser processing with the mean radiation flux density ~10^6 W/cm^{-2} the change in the evolution features of the plasma torch at the surface of some metals (Cu, Al, Sn, Pb) at early stages is quantitative rather than qualitative. At the same time the characteristic size of the target material droplets, carried out from the irradiated zone, becomes essentially (by several times) smaller as the amplitude of the external electric field strength grows, independently of its polarity.

1. Introduction

In connection with permanent expansion of the field of technological applications of lasers new problems arise, the solution of which requires investigation of processes taking place under the action of light fluxes on the surface of solids. Such studies are of particular importance for choosing the most efficient regimes of laser processing of materials, including laser pattern cutting, perforation, welding of materials, modifying their surface properties, and laser film deposition, as well as for developing new methods for control of laser technological processes. In addition, such studies are of separate scientific interest. It is urgent to study the mechanisms of changing the surface relief of a solid in the course of exposing it to pulsed laser radiation under different external conditions, particularly, in the presence of external electric fields. The study of processes that occur in a steam-plasma cloud arising near the surface of the irradiated sample are also of great interest since they affect most significantly the material processing.

The aim of the present paper is to study the influence of electric fields of different strength (from 0 to 10^6 V/m^{-1}) on the spatial and temporal evolution of the laser plasma arising under the action of millisecond laser pulses at the surface of metals (copper, aluminum, tin, lead) and on the mechanisms of formation of the surface relief of the irradiated samples.

2. Experimental equipment

The radiation of the GOR-100M ruby laser (λ = 0.694 mm) operating in the free oscillation regime (pulse duration ~1.2 mc, figure 1) passed through the focusing system and was directed through the hole in the first electrode onto the sample that served as the second electrode and was mounted in air at a pressure of 10^5 Pa. The radiation spot diameter on the sample with sharp edges was varied in the course of the experiments from 1 to 2 mm. From the front face of the glass wedge a part (4 %) of laser radiation was directed into the IMO-2N energy meter, whose entrance window was located in the focal plane of the lens. The energy of the laser pulses varied from 5 to 60 J. The FEK-14 coaxial photodetector, the signal from which was coupled to the S8-13 oscilloscope, was used to record the temporal shape of the laser pulse. The voltage was applied to the electrodes from the source, built on the basis of the UN 9/27-13 voltage multiplier of the TVS-110 unit. The source allowed the voltage variation within 25 kV and its stabilization in the course of the experiment.

To study the spatial and temporal evolution of the laser plasma torch in the course of laser radiation action on the sample, we used the method of high-speed holographic motion-picture recording [1]. The sample was placed in one of the arms of a Mach – Zehnder interferometer, which was illuminated with the radiation of the ruby laser (l = 0.694 μm) operating in the free oscillation regime. The pulse duration of the radiation amounted to 400 mc. The transverse mode selection in the probing laser was accomplished using the aperture, placed in the cavity, and the longitudinal mode selection was provided by the Fabry – Perot cavity standard used as the output mirror. The probing radiation after the collimator was a parallel light beam with the diameter up to 3 cm, which allowed observation of the steam-plasma cloud development. The interferometer was attached to the SFR-1M high-speed recording camera, in which the plane of the film was conjugate with the meridian section of the laser beam, acting on the sample, by means of the objective.

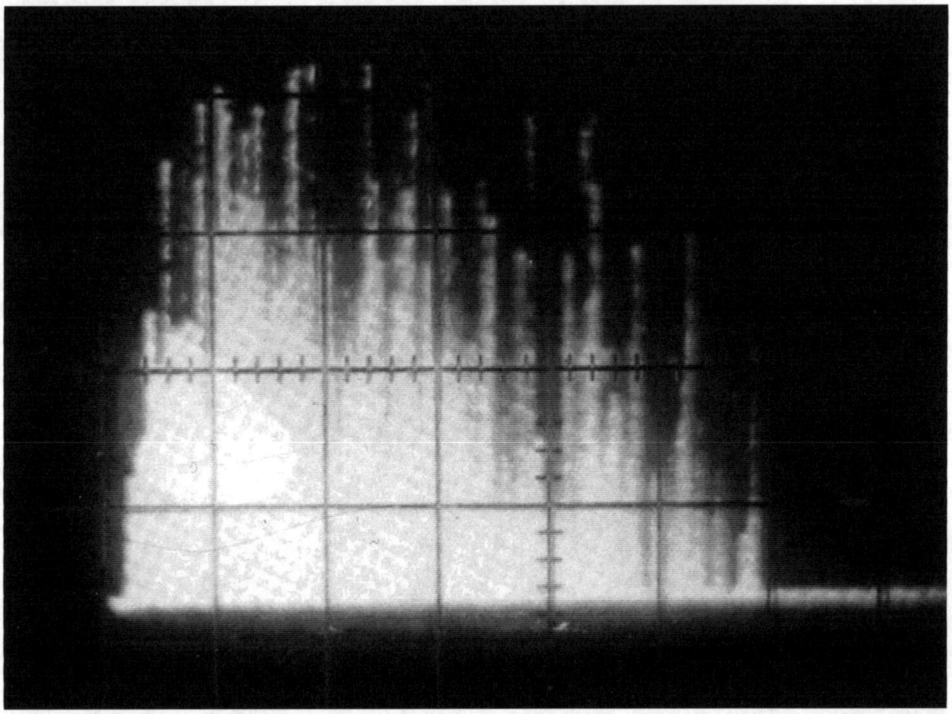

Figure 1. – Oscillogram of the radiation pulse from the GOR-100M laser. The scanning rate is 200 m/s div^{-1}.

The high-speed camera operated in the time magnifier regime. The described setup allowed recording of time-resolved holograms of the focused image of the laser plasma torch. Separate holographic frames provided temporal resolution no worse than 0.8 mc (the single frame exposure time) and the spatial resolution in the object field ~ 50 mm. The diffraction efficiency of the holograms allowed one to reconstruct and record interference and shadow pictures of the studied process under the stationary conditions.

3. Experimental results and discussion

The interferograms reconstructed from the holograms recorded at different instants in the course of high-speed holographic motion-picture shooting clearly illustrates both the initial stage of the laser plume development and the plasma flow around the electrode at different directions of the external electric field strength vector. The data on the distribution of concentration of free electrons in the plasma of an evaporated metal at different instants, obtained by processing the interferograms [2] show that although the energy distribution over the laser radiation focusing spot is not uniform, the lines of equal concentration are practically smooth, which is an evidence of relatively uniform ionization of the eroded substance steams. It is essential that, despite a substantial increase in the plasma formation over time, the mean electron concentration in the plume remains practically unchanged and even slightly grows, which may be associated both with a constant increase in the mass of emitted substance and with secondary ionization of plasma by the laser radiation. Note, that the presence of an external electric field weakly affects the concentration of electrons in the laser plasma plume.

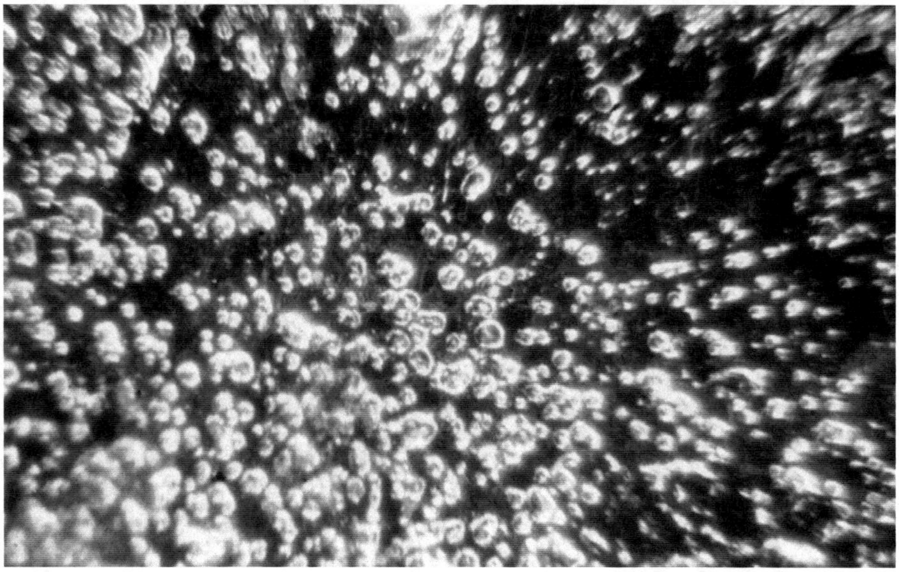

Figure 2. – Photographs of the microscopic surface relief of the crater outer zone.

When either positive or negative potential is applied to the sample, many small droplets appear on its surface after the laser action (figure 2). In particular, at the laser pulse energy 20 J, the diameter of the focusing spot 2 mm, and the electric field strength 10^6 V/cm we observed ejection of droplets having the mean characteristic size less than 0.1 mm to the distance up to 2 cm from the crater centrum. The maximal characteristic size of drops was ~ 0.4 mm. In the absence of the external electric field the mean size of the droplets was ~ 0.4 mm. The droplets were seen at the distance up to ~ 1 cm from the crater centrum.

In accordance with the results presented above, the dynamics of the processes on the surface of a sample, placed in an external electric field with the strength from 0 to 10^6 V/m and subject to the action of the pulsed laser radiation with the parameters mentioned above, is thought to be the following. The primary plasma formation and the initial stage of the laser plume development, in principle, do not differ from those observed in the absence of the external electric field. The metal is melted and evaporated. As a result of local formation of steam and plasma, the erosion plume begins to form with the fine-dispersed liquid-drop phase. Note, that the bulk evaporation is promoted by the gases, diluted in the metal, and by the spatiotemporal non-uniformity of the laser radiation. At a radiation flux density 10^6 – 10^7 W/cm^2 the bulk evaporation is typical of all metals used in the experiments [3]. Obviously, the presence of the external electric field affects (increases or decreases depending on the direction of the field strength vector) the velocity of motion of the plasma front and causes some distortion of the plasma cloud shape. It is essential that the mentioned differences (at the considered parameters of laser radiation) are observed only at the initial stage of the laser plume development, because after the steam-plasma cloud reaches the electrode an electric breakdown (short-circuit) occurs, and the external field in the interelectrode gap disappears.

Consider now the motion of the molten metal droplets in the steam-plasma cloud. In our opinion, the significant difference in the characteristic size of droplets, observed on the surface of the irradiated sample in the presence of the external electric field (independent of the direction of the field strength vector) and in the absence of the field, is a manifestation of the following mechanism of droplet formation. It is known that at the surface of a liquid (including a liquid metal) the formation of gravity-capillary waves [4] is possible under the action of various perturbations. Undoubtedly, the examples of such perturbations are the spatially non-uniform evaporation of the target material due to non-uniform heating caused by non-uniform energy distribution over the focusing spot, the non-uniform primary plasma formation [5, 6] caused by roughness of the irradiated sample surface, and, in the first place, the slop of the molten metal initiated by each spike of laser radiation, acting on the exposed sample [7].

Using the method presented in [4], one can show that at insignificant thickness of the molten metal layer (confirmed by the view of the 'outer' (directed) zone of the crater, particularly, the absence of fillets of significant height at the crater edge) the dispersion equation for the gravity-capillary waves takes the form

$$\omega^2 = \frac{\alpha k^3}{\rho} + g k - \frac{k E_0 E'}{4\pi \rho \xi'}\Big|_{z=0} \qquad (1),$$

where α is the surface tension coefficient of the molten metal; ρ is the metal density; g is the free fall acceleration; k is the magnitude of the wave vector of the gravity-capillary wave; E_0 is the electric field strength at the surface $z=0$ of the molten metal (the z axis is perpendicular to the surface of the irradiated sample, directed towards the laser radiation source and parallel to the vector $\vec{E}_0$); $E' = -\partial\varphi'/\partial z$ is the perturbation of the electric field in the space surrounding the molten metal; ξ' is the small displacement of the surface of the liquid in the z axis direction in the gravity-capillary wave.

Because for the uniform field E_0 the potential is $\varphi = -E_0 \cdot z$ (the potential at the metal surface is considered to be zero), the displacement of the mentioned surface by the small quantity ξ' leads to a small distortion of the potential:

$$\varphi'\big|_{z=0} = E_0 \cdot \xi' \qquad (2).$$

In our case the maximal concentration of electrons in the plasma formation does not exceed ~10^{18} cm^{-3}, which corresponds to the change in the dielectric constant of the medium ε by approximately 10^{-5}. Therefore, near the metal surface $\varepsilon \cong 1$ and with the boundary condition (2) taken into account

$$\varphi' = E_0 \xi' e^{-kz}.$$

In this case, the dispersion equation for gravity-capillary waves takes the form

$$\omega^2 = \frac{\alpha k^3}{\rho} + g k - \frac{k^2 E_0^2}{4\pi\rho}.$$

Because the frequency of the gravity-capillary waves ω is determined by the temporal characteristics of the abovementioned perturbations and, therefore, does not depend on the strength of the electric field E_0, the growth of the magnitude E_0 (independent of the direction of the vector $\vec{E}_0$) should cause the increase in the magnitude of the wave vector $k = \frac{2\pi}{\Lambda}$ and the decrease in the wavelength Λ of the gravity-capillary wave. If we assume that the droplets are 'torn away' by the plasma flow from the 'tops' of the gravity-capillary wave and, therefore, their characteristic size is proportional to Λ, then it becomes clear why in the presence of the external electric field (of any direction) the observed mean size of the droplets becomes essentially reduced.

The escaped droplets possess the charge of the same sign as the sample. That is why the droplets begin to move with acceleration towards the second electrode. However, since the maximal initial velocity of the outgoing droplets under the analogous conditions [8] is ~ 45 m/c, i.e., an order of magnitude smaller than the velocity of steam-plasma cloud spreading, the droplets do not reach the electrode before the moment of the breakdown in the interelectrode gap. In what follows (in the absence of the external electric field) the droplets move under the action of the same forces as in [8] and, therefore, in the way, described in [8]. In this case, having acquired at the stage of accelerated motion in the electric field the velocity, exceeding the initial one, the droplets may fly to a greater distance along the surface of the irradiated sample than in the absence of the electric field, which is observed in the experiment. Moreover, having moved to a greater distance from the sample surface and, therefore, being affected by the plasma for longer time before returning to the surface, the droplets may be split into finer parts than in the absence of the external field.

4. Conclusion

The studies performed have shown that under the action of laser radiation with the mean radiation flux density ~ 10^6 W/cm^{-2} at the surface of some metals (Cu, Al, Sn, Pb) in the external electric field with different polarity and the strength up to 10^6 V/m the characteristic size of the target substance droplets, carried out of the irradiated zone, decreases by several times with increasing external electric field strength. Probably, this is due to a change in the wavelength of the gravity-capillary wave, excited on the molten metal surface. The observed effect offers the possibility to control the size of the metallic droplets in the course of laser deposition of thin films.

References

[1] Barikhin B.A., Ivanov A.Yu., Nedolugov V.I. Kvantovaya Elektron., 17, 1477 (1990) [Sov. J. Quantum Electron., 20, 1386 (1990)].

[2] Ostrovskii Yu.I., Butusov M.M., Ostrovskaya G.V. Golograficheskaya interferometriya (Holographic Interferometry) (Moscow: Nauka, 1977).

[3]. Goncharov V.K., Kontsevoy V.L., Puzyrev M.V. Kvantovaya Elektron ., 22 , 249 (1995) [Quantum Elelctron., 25, 232 (1995)].

[4] Rabinovich M.I., Trubetskov D.I. Vvedenie v teoriyu kolebanii i voln (Introduction tothe theory of oscillations and waves) (Moscow: Nauka, 1984).

[5] Dorofeev I.A., Libenson M.N. Opt. Spektrosk., **76**, 73 (1994) [Opt. Spectrosc., **76**, 66 (1994)].

[6] Chumakov A.N. et al. *Kvantovaya Elektron.*, **21**, 773 (1994) [*Quantum Electron.*, **24**, 718 (1994)].

[7] Bosak N.A., Vasil'ev S.V., Ivanov A.Yu., Min'ko L.Ya., NedolugovV.I., Chumakov A.N Kvantovaya Elektron ., 27, 69(1999) [Quantum Elelctron., 29, 69(1999)].

[8] Vasil'ev S.V., Ivanov A.Yu., Lyalikov A.M. Kvantovaya Elektron., 22, 830(1995) [Quantum Elelctron., 25, 799(1995)].

Journal of Physics: Conference Series **669** (2016) 012027 doi:10.1088/1742-6596/669/1/012027

Hydrogen transport through stainless steel under plasma irradiation

A A Airapetov, L B Begrambekov, A S Kaplevsky and Ya A Sadovskiy
National Research Nuclear University MEPhI, Moscow, 115409 Kashirskoe shosse 31, Russia Federation
E-mail: alexk1989@mail.ru

Abstract. The paper presents the results of investigation of gas exchange through stainless steel surface of the plasma chamber under irradiation with hydrogen atoms in oxygen atmosphere or oxygen contaminated hydrogen plasma. Dependence of this process on various irradiation parameters, such as the metal temperature, energy of irradiating ions, gas composition of plasma are studied. It is shown, that desorption from stainless steel is activated with the increase of the plasma chamber walls temperature and energy of irradiating ions. Hydrogen release occurs also under irradiation of the walls by helium and argon plasmas added with oxygen, however the amount of released hydrogen is several times lower than in the case of irradiation with oxygen contaminated deuterium plasma.

1. Introduction.

Stainless steel is the main structural material of plasma and fusion devices. Hydrogen trapping in the plasma facing components of the plasma and fusion devices as well as it unexpected release are serious problem for their sustainable operation. For the reason of safe operation the retention of radioactive tritium in structural elements of International Thermonuclear Experimental Reactor (ITER) is limited by 700 grams. Therefore, investigation of methods of outgassing of ITER vacuum chamber at the time between main discharges of ITER is actual task. It was shown in our previous work [1], that hydrogen desorption and deuterium trapping occurs during irradiation of the plasma chamber wall with deuterium atoms in the oxygen contaminated atmosphere and with deuterium plasma with oxygen addition. The present work is devoted to study of peculiarities of hydrogen transport through the stainless steel walls surface under irradiation in different experimental conditions. In particular, the paper presents the dependences of the features of the plasma chamber walls outgassing on irradiating ion energy, oxygen concentration in irradiating species, temperature of the walls and working gas composition.

2. Experimental.

The Multipurpose Investigation Complex for Mass Analysis (MICMA) [1] is used in the experiments. This complex is an automated device intended for experiments concerning the interaction of hydrogen atoms, ion beams, and gas discharge plasma with solids and for analyses of the processes of trapping and retention of gases in metals via thermal desorption spectrometry (TDS). The experiments are conducted in the MICMA plasma chamber made from stainless steel of the type 12Cr18Ni10Ti (0.12% C, 18% Cr, 10% Ni, < 1%Ti). The plasma chamber volume is ~2×10^{-3} m^3. The surface area of the plasma contacting elements is ~0.1 m^2. The residual gas composition is $\geq$ 95%H_2O and $\leq$ 5%H_2. Its pressure is $\leq 1\times10^{-4}$ Pa. The working gas pressure is 6.6×10^{-1} Pa. The heated tungsten cathode and the anode allows ignition of a gas discharge in the plasma chamber. The plasma density is $\approx$ (2-3) $\cdot 10^{10}$ cm^{-3} at all stages of experiments, and the plasma electron temperature is $\approx$ 7 eV. The mean energy of the plasma ions irradiating chamber walls in these experiments is about 10 eV. The mass-spectrometer connected to the plasma chamber is used for analysis of the gas composition in the chamber during the experiments.

The experimental procedure is similar to that used in the paper [1]. At the first stage, the plasma chamber is filled by deuterium with an addition of oxygen (deuterium/oxygen mixture). The tungsten cathode is heated, deuterium atoms appeared in the gas due to dissociation of deuterium molecules on the tungsten surface. The plasma chamber walls are irradiated with deuterium atoms and oxygen. Then the plasma is ignited in the same gas mixture, and ions and atoms of oxygen added plasma irradiate the chamber walls. On the second stage, deuterium is replaced by hydrogen in the working gas, and hydrogen atoms in the hydrogen/oxygen mixture irradiate the walls.

In the first series of experiments, the plasma chamber walls are heated up to 200 °C and irradiated in accordance with above described scenario. The experiments are made with the oxygen concentration in the working gas from 0.5 to 30%. In the second series, the surface of the plasma chamber walls is irradiated with 50 eV/at and 100 eV/at ions of deuterium plasma with 10% oxygen addition. The plasma chamber walls are cooled and their temperature do not exceed 80°C during the experiment. The third experimental series includes irradiation of the plasma chamber walls by ions of helium and argon plasma with 10% oxygen addition. The walls temperature do not exceed 40°C. Finally (fourth series), the plasma chamber walls are irradiated with ions of glow discharge switched in deuterium with oxygen addition at the pressure 52 Pa.

3. Results and discussion.

Irradiation of plasma chamber heated up to 200 °C by deuterium atoms in the deuterium/oxygen mixture and then by the ions of the plasma ignited in the same gas mixture leads to the similar changes in the working gas composition as in the similar experiments made in the paper [1] with the chamber walls at 40 °C. Namely, the oxygen concentration drops down sharply, and hydrogen contained molecules HD, HDO, H_2O, D_2O H_2 HD including hydrogen released from the chamber wall appear in the working gas. Their concentrations, i.e. hydrogen desorption from the walls increase with increasing oxygen content in the working gas. When deuterium in the working gas is replaced by hydrogen, and hydrogen atoms in hydrogen/oxygen mixture irradiate the chamber walls, gas modification in the plasma chamber point out on the release of deuterium trapped in the walls in the preceded irradiation conditions.

The plots shown in figures 1a and 1b illustrate, as an example, the changes of the compositions of the working gas during all the steps of the experiment performed in the D_2+2 % O_2 mixture.

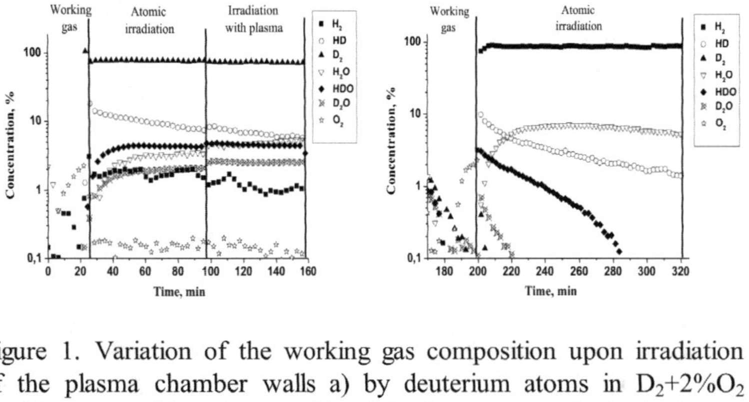

Figure 1. Variation of the working gas composition upon irradiation of the plasma chamber walls a) by deuterium atoms in D_2+2%O_2 working gas and by plasma initiated in the same gas; b) by hydrogen atoms in D_2 +2% O_2 working gas.

The modifications of gas composition in the plasma chamber during all the stages of experiments are registered by mass-spectrometer. The results of both these measurements and of the rates of pumping of different gases by the MICMA vacuum system allow calculation of the amount of hydrogen released from- and deuterium trapped in the plasma chamber walls during its irradiation in D/O mixture. The table 1 compares the hydrogen release and deuterium trapping by the chamber walls at 40 and 200 °C. It is seen, that the increase of the walls temperature up to 200°C leads to on average a 20% increase of the hydrogen release from the walls. At the same time, trapping of deuterium is not practically changed. The amount of released hydrogen is two or more times bigger than the amount of trapped deuterium for all oxygen concentrations and temperatures of the walls. This means that under irradiation with deuterium atoms in oxygen atmosphere or oxygen contaminated hydrogen plasma the amount of hydrogen isotopes in the walls decreases, i.e. outgassing of the walls occurs. One can mention that higher both oxygen concentration in the working gas and wall temperature, are better conditions for both plasma chamber walls outgasing (the (N_H–N_D) value is higher) and detritiation of the walls of thermal nuclear devices (the N_H is higher).

Table 1. Parameters of gas exchange processes through the surface of the plasma chamber walls irradiated during 40 minutes with deuterium atoms in the working gases of different compositions.

Oxygen concentration (at. %) in the D/O mixture	Hydrogen (N_H) release from the chamber walls, (10^{21} at./m^2)		Deuterium (N_D) trapping in the chamber walls, (10^{21} at./m^2)		The net outgassing (N_H-N_D) T= 40°C, (10^{21} at./m^2)	The net outgassing (N_H-N_D) T= 200°C, (10^{21} at./m^2)
	40°C	200°C	40°C	200°C		
0.5	0.6	0.8	0.3	0.4	0.3	0.4
2	1.0	1.9	0.5	0.5	0.5	1.4
10	1.7	2.4	0.9	0.7	0.8	1.7
20	2.0	2.4	1.1	0.9	0.9	1.5
30	2.9	3.4	1.5	1.4	1.4	2.0

The table 2 lists the parameters of hydrogen transport through the plasma chamber walls during irradiation with ions of different energies. The data show that irradiation with 50 eV/at deuterium ions provides the best regime for degassing and for detritiation of the plasma chamber walls.

Table 2. Parameters of gas exchange processes at the surface of the plasma chamber walls irradiated during 40 minutes in working gas D_2 + 10 at. % O_2 by deuterium ions with 10, 50 and 100 eV/at energies.

Ion energy, eV/at	Hydrogen (N_H) release from the chamber wall, (10^{21} at./m^2)	Deuterium (N_D) trapping in the chamber wall, (10^{21} at./m^2)	The net outgassing (N_H-N_D), (10^{21} at./m^2)
10	1.7	0.9	0.8
50	1.8	0.4	1.4
100	1.3	0.3	1.0

Heating of the cathode and ignition of discharge in the mixtures Ar +10 % O_2 and He +10 % O_2 lead to increase of concentration in the plasma chamber of H_2 and H_2O containing hydrogen released from the plasma chamber walls. As an example Figure 2 shows gas composition changes after heating of the cathode and ignition of discharge in He+10% O_2.

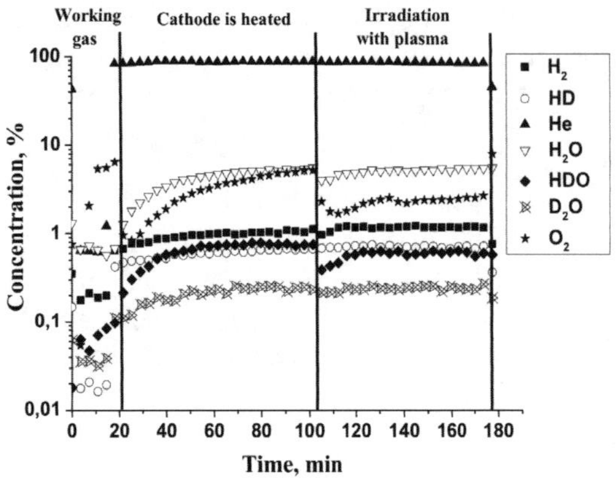

Figure 2. Variation of the working gas composition after heating of the cathode and ignition of the discharge in He +10 % O_2.

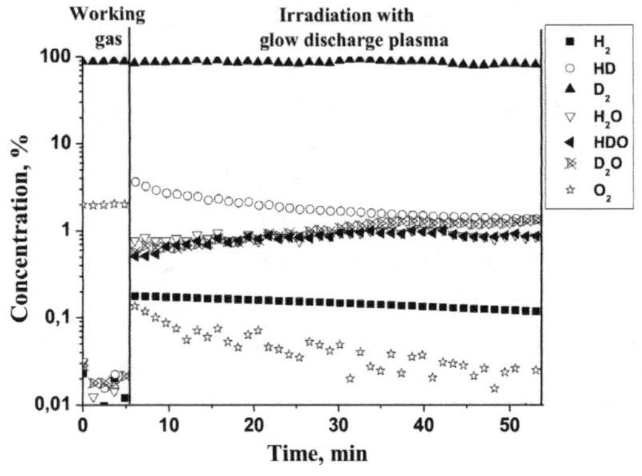

Figure 3. Variation of the working gas composition after ignition of the glow discharge in D_2 + 2 % O_2.

The Table 3 allows comparison of hydrogen release from the chamber walls under irradiation by deuterium, argon and helium plasmas added by oxygen. It is seen that irradiation by deuterium discharge with oxygen addition leads to the highest hydrogen release ($1.7 \cdot 10^{21}$ at./m^2). However, hydrogen release in this experiment is accompanied with deuterium trapping. That is why, the net chamber walls outgasing appears to be equal to that of the experiment with helium plasma. Conclusion could be made that for outgassing both irradiation regimes (deuterium and helium plasmas with oxygen addition) can be used with similar efficiency.

Table 3. The amount of released hydrogen from the plasma chamber walls after irradiation during 40 minutes with plasma of different composition.

Working gas composition	Hydrogen (N_H) release from the chamber walls, (10^{21} at./m^2)	Deuterium (N_D) trapping in the chamber walls, (10^{21} at./m^2)	The net outgassing (N_H - N_D), (10^{21} at./m^2)
D_2 + 10 ат. % O_2	1.7	0.9	0.8
He + 10 ат. % O_2	0.8	-	0.8
Ar + 10 ат. % O_2	0.2	-	0.2

The table 4 comprises the hydrogen release and deuterium trapping during the chamber walls irradiation by plasma with heated cathode and by glow discharge plasma. One can believe that in both conditions, the similar processes take place (see Figure 3), but in the glow discharge plasma they are less intensive.

Table 4. Parameters of gas exchange processes through the surface of the chamber walls during 40 minute irradiation by plasma of activated by heated cathode (HCP) and glow discharge plasma (GDP).

Plasma type	Working gas	Hydrogen (N_H) release from the chamber wall, (10^{21} at./m^2)	Deuterium (N_D) trapping in the chamber wall, (10^{21} at./m^2)	The net outgassing (N_H - N_D), (10^{21} at./m^2)
HCP	D_2+2 ат.% O_2	1.0	0.5	0.5
GDP	D_2+2 ат.% O_2	0.7	0.4	0.3

4. Summary

The paper presents the peculiarities of hydrogen transport through stainless steel 12Cr18Ni10Ti under irradiation by deuterium atoms in the deuterium/oxygen mixture and by ions and atoms of plasmas ignited in different gases (D_2, H_2, He, Ar) with oxygen addition. The features of the plasma chamber walls outgassing in dependences on irradiating ion energy, temperature of the walls and working gas composition are measured and analyzed. It is shown that the increase of the plasma chamber walls temperature from 40°C to 200°C leads to 20% growth of the hydrogen release, at the same time deuterium retention does not change. Ions and atoms of glow discharge in deuterium initiate the same processes on stainless steel surface like in the case of discharge with heated cathode, however in the glow discharge plasma they are more intensive. Plasma ignited in the mixture He/O_2 seem can be used successfully for the stainless steel outgaqssing. Conclusion is made that in our experimental conditions the irradiation of the chamber walls at 200°C with atoms in D_2+30 at.% O_2 gas mixture and with plasma ignited in the same working gas is the best regime for both plasma chamber walls outgasing and detritiation of the walls.

Acknowledgments

This work was supported by Competitiveness Growth Program of the Federal Autonomous Educational Institution of Higher Professional Education National Research Nuclear University MEPhI and by the Ministry of Education and Science of the Russia Federation (agreement contract № 14.575.21.0049 of 27.06.14).

References

[1] L. Begrambekov, A. Grunin, A. Kaplevsky, et al. *J. of Surf. Inv. Xray, Sunch. and Neut. Tech.*, **9**, 1 (2015)190-195.

Journal of Physics: Conference Series **669** (2016) 012028 doi:10.1088/1742-6596/669/1/012028

Changes of thermal balance in a plasma-electrolyte system according to the shape of the applied voltage

L N Kashapov[1,2], N F Kashapov[1] and R N Kashapov[2,3,4], V D Semushin[2]
[1] Institute of Physics, Kazan Federal University, Kazan, 420008, Russia
[2] Engineering Institute, Kazan Federal University, Kazan, 420008, Russia
[3] Laboratory of Radiation Physics, Kazan Physical-Technical Institute of the Kazan Scientific Center of the Russian Academy of Sciences, Kazan, 420029, Russia
[4] Kazan State Medical University, Kazan, 420012, Russia

E-mail: kashramil.88@mail.ru

Abstract. The work is aimed at studying the changes of the heat balance in the system of the electrode-plasma-gas / vapor-electrolyte in various forms of applied voltage. It was established that in contrast to the voltage, which is obtained after full-wave rectification, film boiling occurs by using a smoothing capacitive filter. This is because in the pulsing mode voltage value has zero value at the beginning and end of the pulse, also at that time takes place cooling of the electrode at the contact with the electrolyte. In this paper describes a heat exchange in two different modes of plasma-electrolytic treatment.

Introduction

Modern materials science has two main lines of development - is changing volume characteristics of the materials or modification of the surface layers. Surface modification is more attractive and necessary to the economic position [1]. One of actively developing methods for surface treatment is a plasma-electrolytic treatment. Interest in this method is caused by wide possibilities of this method (cleaning, hardening, diffusion saturation, formation of microrelief coating, synthesis of nanoparticles etc.) and its environmental performance [2, 3, 4]. The basis of plasma electrolytic treatment is a discharge burning between liquid and solid electrodes. During the processing, occurs the electrochemical reactions, heat impact, the formation of streams of electrons, ions, photons, and their effect on the surface. As known one of the factors, strongly influence on the properties of the materials, is thermal influence [5]. This was the necessity of carrying out works aimed at the research of the heat balance of the system. It should be noted that the heat balance of the system during cleaning, hardening, saturation, and formation of the diffusion surface microrelief would vary. Data processing methods can be divided into two groups, which are classified by the form of the applied voltage. Therefore, the aim of this work was to study the influence of the shape of the applied voltage to the thermal balance of the system electrode-plasma-gas / vapor-electrolyte.

Experimental

The experimental setup is designed for the study of plasma-electrolyte discharge in a range of parameters voltage $U = 0 \div 250$ B, the current $I = 0 \div 50$ A, pH = 1-12 and T = 20-110 ^{0}C. Functional schemes of setup is shown in Figure 1. Installation is intended for research of electrical and power characteristics of the discharge, including the research of the potential distribution in the electrode gap in the treatment process.

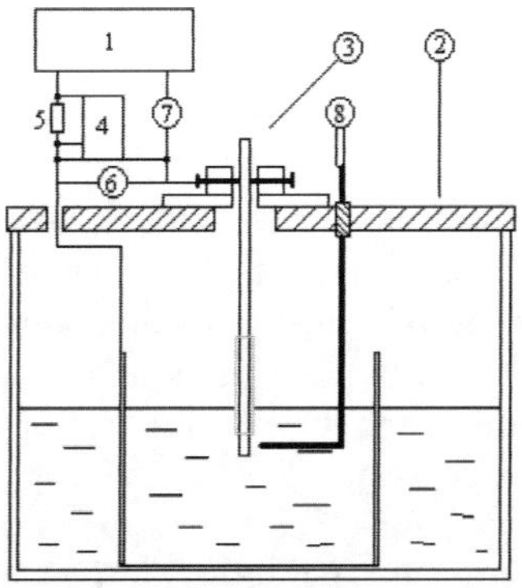

Fig. 1. Functional scheme of the experimental setup plasma-electrolytic treatment

Consider the installation of which is presented in Figure 1. It consists of a current source 1, the electrolytic bath 2, the electrode system 3, the oscilloscope 4, the additional resistance 5, the voltmeter 6, the ammeter 7, thermocouple 8. Power supply 1 supplies a regulated DC voltage to the electrode system 3. The depth of immersion of the anode in an electrolyte solution controlled by using the electrode system. With an oscilloscope 4 was controlled the shape of the applied voltage and current with a voltmeter and an ammeter measures the voltage and current of the discharge.

The magnitude of the contact area of the anode with the electrolyte controlled by applying to the electrode surface of the dielectric protective coating, leaving open on strictly defined part. During the research anode area was S_A = 0.1 - 0.65 cm^2, cathode area was S_K = 10 - 34cm^2. A plasma-electrolytic treatment process takes place, subject to inequality of surface areas of electrodes $S_K > 5\ S_A$. In previous studies it was established that at the same areas of the electrodes discharge burns on the cathode. The gradual reduction of the area of the anode relative to the cathode initially leads to the discharge quenching at the cathode, and then to the appearance of micro-discharges at the anode. These data, about the condition of the generation of electrical discharges at the electrodes immersed in the electrolyte, in agreement with the data of [2].

To determine the amount of heat that goes into heating the electrolytic cathode used laboratory mercury thermometer, measuring lab cylinder of 0.25 liters, laboratory flask 1 liter and stopwatch.

Measurements were taken during the discharge by the following procedure:

1) After the ignition of the discharge, the electrolyte circulating through the discharge zone, started to heat up. After 30 seconds, the temperature stops changing and sets the thermal equilibrium.

2) After 40 seconds after ignition of the discharge, began measuring. The drain hose is directed into a laboratory flask (1 liter) to collect the water. Flowing water washes thermometer showing the temperature of the water in real time. The start and the end of water intake recorded stopwatch. The exact amount of water collected in the flask was determined dimensional laboratory cylinder capacity of 0.25 liter.

3) In the workbook recorded values of voltage and current of discharge and obtained at those discharge parameters the thermometer readings, the stopwatch readings and the volume of collected water. Calorimetric measurements for each value of the discharge power were repeated ten times to average the obtained values.

The energy of an electric current, which is inserted into the gas discharge, calculated by the formula:

$$Q_э = U \cdot I \cdot t \text{ [J]},$$

Where (U) - voltage across the electrodes, (I) - discharge current, (t) - time of current flow.

The amount of heat that spent on the heating of the electrolyte, calculated by the formula:

$$Q_B = c_B \cdot m_B \cdot \Delta T \text{ [J]},$$

Where c_B = 4190 (J / kg ° C) - specific heat of water, m_B(kg) - the mass of water heated in a minute, ΔT (° C) - the heating temperature of the electrolyte.

The proportion of discharge energy consumed for the heating of the electrolyte is calculated as:

$$\eta=(Q_B/Q_э)\ 100\%$$

Experiments have shown, film boiling is observed only when using capacitive smoothing filter. Nature of the electrolyte has a strong influence. For solutions of alkalis film boiling is observed at lower voltages than aqueous solutions of acids and salts. This is due to the higher viscosity of alkali. Consider the release of heat, in this case, there is a four-phase system, which contains a steam-gas shell, in which the discharge is burning and which is separating the anode from the electrolyte. In this shell releases heat under the influence of the electric current. We single out three heat flux: heat flux from the shell in an electrolyte solution, the heat flow which is spent on evaporation of of the electrolyte, the heat flow directed from the anode shell. The volumetric power of the heat source in a conductive shell, we write down with the influence of the space charge on flow of current through the shell in strong electric fields. The current density of anions is equal $j = K\ \rho\ E^{1/2}$. Experiments have shown that the gradient of the temperature distribution of cylindrical samples is 5 degrees per 1 mm. Measurement error of 15 degrees. It gives the right to assume that the samples up to 10 mm is homogeneously heated. We neglect the loss of heat by evaporation of the liquid and write heat balance: $jU=Q_A+Q_B$, Q_A and Q_B - the heat flux density at the anode and an electrolyte.

When using film boiling ripple voltage is not observed. This is because in the pulsing mode voltage value at the beginning and the end of the pulse has zero value; at that time there is a cooling of the electrode in contact with the electrolyte. Another distinctive feature is the active dissolution of anode in terms of its passivation of friable film, which constantly forms at all current densities. The appearance of the passivation layer is likely occurs because of interaction between the metal an electrode with oxygen which allocated on the electrode. As a result, take place screening and passivation of the surface of the electrode. In addition, the discharge can be carried out in some of vials. These bubbles are plasma zone with a low electrical resistance. During the discharge volume of the heated bubble due to viscous resistance of the liquid are not rapidly increase and this is leading to a rapid rise of pressure in the vial and its rebound from the electrode surface. Then discharge burns on another favorable location.

Conclusions

It was established that in contrast to the voltage, which is obtained after full-wave rectification, film boiling occurs by using a smoothing capacitive filter. This is because in the pulsing mode voltage value has zero value at the beginning and end of the pulse, also at that time takes place cooling of the electrode at the contact with the electrolyte. In this paper describes a heat exchange in two different modes of plasma-electrolytic treatment.

This work was funded by the subsidy allocated to Kazan Federal University for the state assignment in the sphere of scientific activities.

References

[1] Rybalko A, Sahin O. Mesyats A. Some features of micro-arc oxidation at high current densities. Metal processing, number 2 (56), 2010, pp 30-38.

[2] L. Kashapov, N. Kashapov and R. Kashapov, Journal of Physics: Conference Series, Volume 479, Issue 1, article id. 012011 (2013).

[3] L. Kashapov, N. Kashapov and R. Kashapov, Journal of Physics: Conference Series, Volume 479, Issue 1, article id. 012003 (2013)

[4] L. Kashapov, N. Kashapov and R. Kashapov, Journal of Physics: Conference Series 2014, 567, 012025

[5] D. Denisov, N. Kashapov and R. Kashapov, IOP Conference Series: Materials Science and Engineering 2015, 86, 012005

Journal of Physics: Conference Series **669** (2016) 012029 doi:10.1088/1742-6596/669/1/012029

Plasma electrolytic treatment of products after selective laser melting

L N Kashapov[1,2], N F Kashapov[1] and R N Kashapov[2,3,4], D G Denisov[2]
[1] Institute of Physics, Kazan Federal University, Kazan, 420008, Russia
[2] Engineering Institute, Kazan Federal University, Kazan, 420008, Russia
[3] Laboratory of Radiation Physics, Kazan Physical-Technical Institute of the Kazan Scientific Center of the Russian Academy of Sciences, Kazan, 420029, Russia
[4] Kazan State Medical University, Kazan,420012,Russia

E-mail: kashramil.88@mail.ru

Abstract. . The aim of the work was to study the possibilities of plasma electrolytic treatment for cleaning surfaces of metal products obtained by the SLM-technology. We found that the most effective cleaning from the large alloy particles occurs in the "hydrodynamic" mode, when the occurrence of hydrodynamic pulses observed. Further smoothing of irregularities eliminated by a stable burning of discharge in vapor shell. Analysis the morphology of the surface of difficult specialized products, such as crown conical gears, after plasma hydrodynamic treatment showed efficiency and advantages in comparison to conventional methods of final cleaning such as shot blasting.

Introduction

Additive manufacturing technologies are increasingly used in industry and move from the area of prototyping in small-scale production of functional objects. This is especially true for selective laser melting (SLM) and selective laser sintering (SLS). SLS technology was created and developed by Carl Deckard and Joe Beamanat the University of Texas at Austin in the mid 80s of the twentieth century for the manufacture of plastic products. SLM-technology is the next step of SLS, as it allows obtaining metal products by melting metal powders. It was on the developed in 1995 in Fraunhofer Institute for Laser Technology ILT in Aachen, Germany [1]. The melting of the metal powder is achieved by using a more powerful laser compared to SLS-technology. In these capacities observed intensive heating of the product, which may lead to deformation. To prevent overheating phenomena was used the system of substrate which support hanging elements of products and conduct heat to the platform of SLM-installation. Now more and more efforts is aimed at the development of new compositions of alloys of powders and study of the effect of particle size on the properties of the resulting products. Powders of instrumental and stainless steels most commonly used in SLM-production. After receiving of fabricated metal products need to remove mechanically the substrate and to clean the surface from the sintered powder particles and drops of molten metal. In this way, finishing mechanical machining is necessary step in obtaining high quality products with a desired surface roughness. One of these widespread methods is sandblasting. This method has several drawbacks, namely the creation of a large amount of dust, which must be removed with special suction devices, the gradual destruction of the nozzle, the flow of sand, the presence of micro- and nanoparticles of aluminum oxide on the surface of the metal, which deteriorate the adhesion of the coating. This causes the necessity of searching a new method of removing the excess of fused metal particles and polishing the surface without changing the basic geometry of the product. In recent years the increasing distribution began receiving plasma electrolytic method of treatment [2,3]. One of its features is the ability to generate hydraulic impacts in the liquid [4, 5]. The aim of this work was to research of process of plasma electrolytic treatment of steel products, which are obtained by selective laser melting, and determine the possibility of replacing the sand blasting.

Experimental

In experiments on selective laser melting was used the installation Realizer SLM 50. This machine has a table top mounting and is very well suited for research purposes, as it has a small working area 70 mm in diameter and 40 mm in height. Working chamber of the machine filled with high-purity argon environment. To start the machine must be loaded three-dimensional model of the product in the form of stl-file. Using the software are generated the automatic supportы, manual control is also possible. SLA-unit is equipped with Nd: YAG fiber laser, with maximum power 100 W and 1.064 μm wavelength. It is possible to adjust the laser power to 100 W and a scan speed of 200mm/s. Laser spot 0,2 - 0,4 μm. The thickness of the layer of construction was 30 μm. The energy density varied from 1.5 to 2,5 J/mm^2. In the experiments the powder steel 316 L (X2CrNiMo17-12-3) was used, particle diameter of 20 - 40 μm.

Research the dependence of the relative density of a product from the applied energy density of laser spot showed that when the energy density of 1.5 J/mm^2 relative density was ~ 93% and increases up to the maximum 99% at a power of 2,5 J/mm^2. Metallographic research has shown that the relative density of 93% - 98% have a pore size of 50 - 250 μm. With the increase in the relative density of defects decreases the average size of 5 μm and is at 99%. The surface morphology of the obtained samples were investigated by scanning electronic microscope EVO 50 CarlZeissAG. Figure 1 shows images of the surface of products produced by SLM.

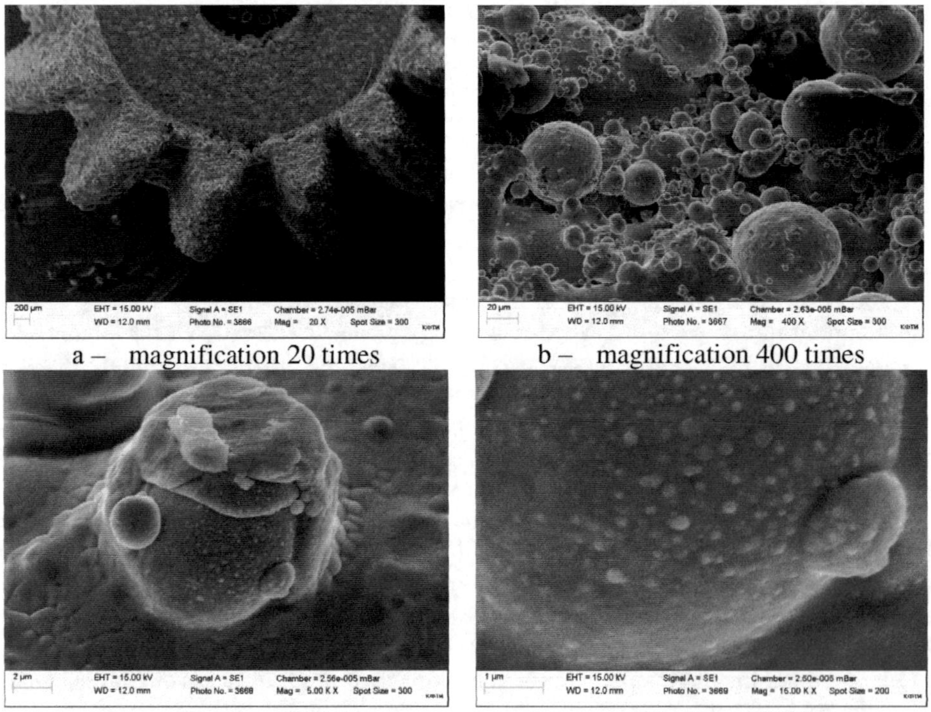

a – magnification 20 times

b – magnification 400 times

c – magnification 5000 times

d – magnification 15000 times

Fig. 1. SEM images of the surface of products produced by SLM.

To initiate plasma electrolyte discharge on the surface of the workpiece used an experimental setup, which described in [4]. In this paper investigated the influence of ripple voltage. In the process of initiating, the discharge can occur intense spattering of the electrode, and processing area will be different at different times. Therefore, the main focus was on the initiation of the discharge on prongs of gears which is 10 mm lower than the surface of the electrolyte. The area of the workpiece was 20 times less than the area of the electrode. An electrolytic bath filled with different electrolytes - aqueous solution of Na2CO3 or aqueous NaCl solution concentration of 5% by weight. The main operating parameters of the electrode system was the voltage on the electrodes, the magnitude of the discharge current and the current density on the electrodes. Changing the shape of the current and voltage at the time of discharge ignition was determined by an oscilloscope FLUKE 105 SCOPEMETERSERIESII, timebase varied from 5 ns to 60 s. The measurement of voltage at the anode was carried out with a digital multimeter device APPA 105, relative measurement error is 0.1%. Using a digital multimeter APPA 305 temperature measurement of the electrolytic cathode was carried out using chromel-copel thermocouple. The temperature of the electrolyte before the experiment and then measured by a mercury thermometer.

For comparison, treatment was conducted by the standard method with using a stream of particles of aluminum oxide 99.5% Al2O3 and SiO2, less than 0.06% grit 110 - 150 microns (Cobra, Renfert) on the installation APO-5U (Aveyron, Ekaterinburg). Figure 2 shows the image of the surface of the obtained samples.

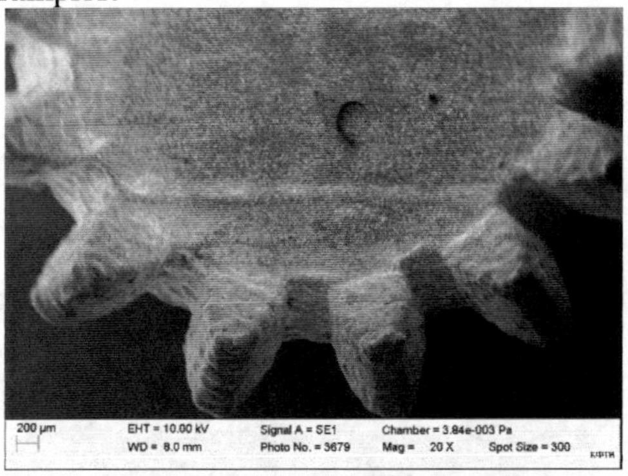

a – magnification 20 times

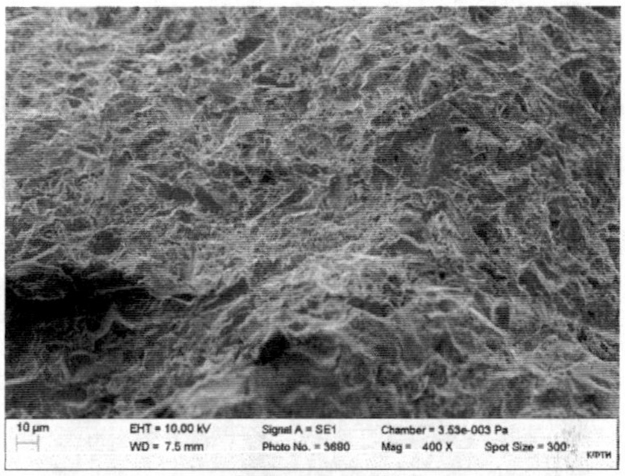

b – magnification 400 times

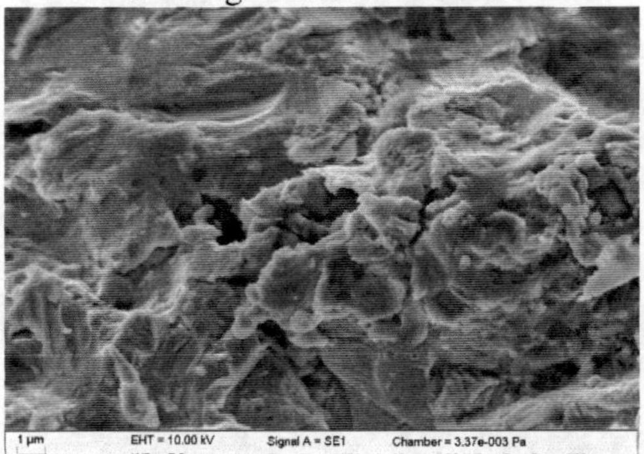

c – magnification 5000 times

d– magnification 15000 times

Fig. 2. SEM images of the surface of the SLM-products after blast cleaning.

Before the experiment of the plasma-electrolytic treatment and cleaning, the current-voltage characteristic of discharge burning on workpiece in an aqueous solution of Na2CO3 was obtained, presented in Figure 3. Watching the glow of discharge occurs at a voltage of 130 V. Before that, at 30 V are recorded the occurrence of noise that increases with increasing voltage and goes to the appearance of the hydraulic pulses at a voltage of 60V. The highest intensity of hydraulic impulses observed in the voltage 120 - 140 V. Further, with increasing the voltage reduction in current observed, but the glow of the discharge increases. At the same time stable combustion takes place in the of the discharge voltages above 190 V. Plasma electrolytic treatment and cleaning were carried out in the hydrodynamic regime, when the discharge was ignited on the product-anode and observed the occurrence of water hammer. When initiating of the discharge on the product-cathode there is a sharp rise in temperature, which is difficult to control and in its turn may lead to melting of the gear prongs.

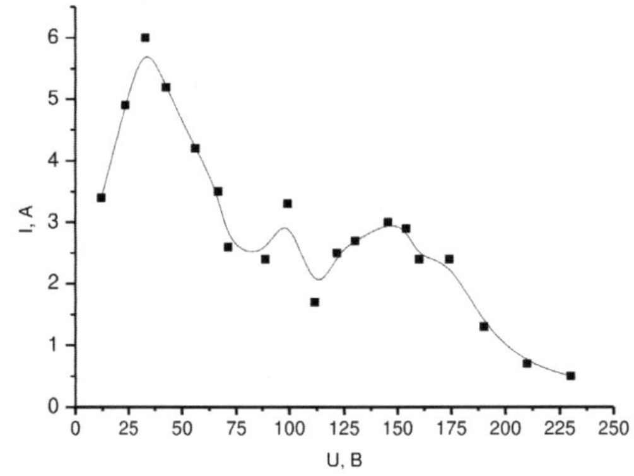

Fig. 3. The current-voltage characteristics of plasma electrolytic process of steel 316L.

After treatment were obtained two kinds of samples, the first processed in Na2CO3, represented in Figures 4a (magnification 20 times), 4b (magnification 400 times) and 4c (magnification 5000 times), the second processed in NaCl - 4d, 4e and 4f. The treatment in both instances was conducted at a voltage of 130 V, in the case of Na2CO3 solution under the action of the hydrodynamic impulses are removed sintered powder particles is about 75% of the irregularities, but the fused powder particles remained (Figure 4b).

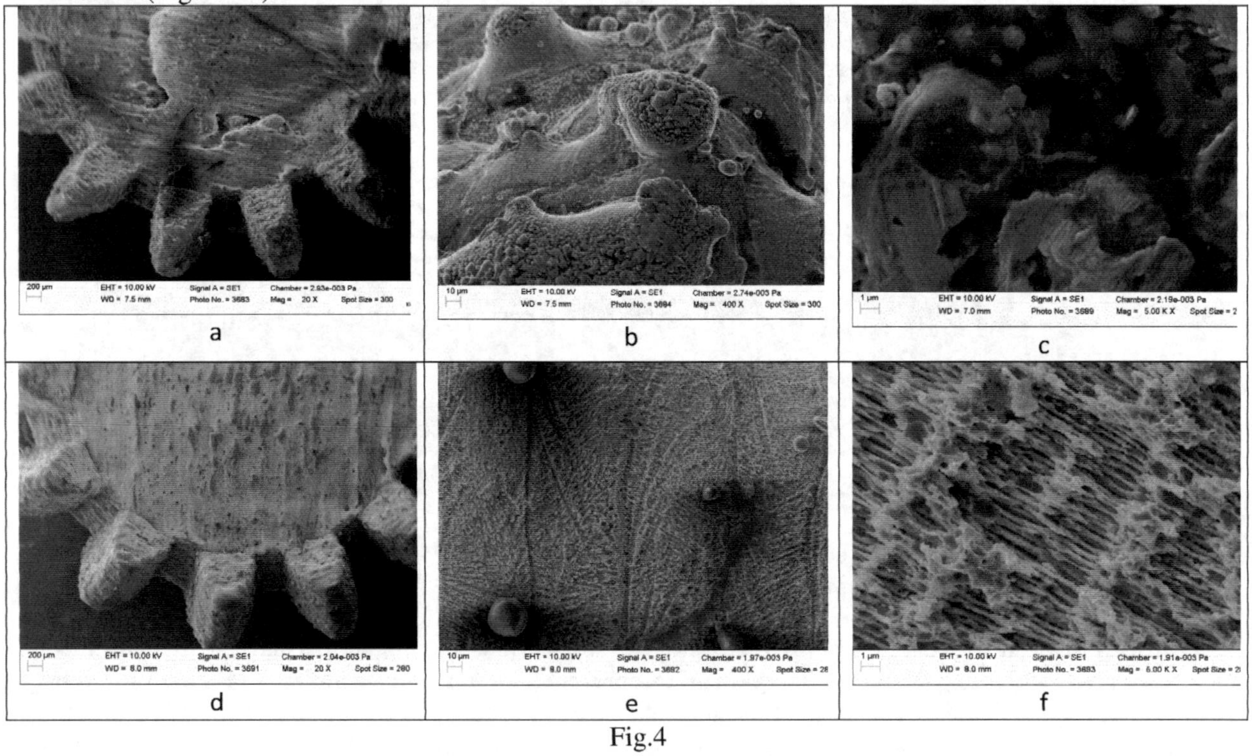

Fig.4

In the case of treatment in a solution of sodium chloride, in addition to the impact of the hydrodynamic impulses, also anodic dissolution of surface occurs, which leads to a more efficient cleaning process. Namely, the separation of larger particles under the influence of hydrodynamic impacts and smoothing fused powder particles. This is clearly seen when compared to sand blasting treatment (Figures 2a and 4d are similar).

Conclusions

We found that the most effective cleaning from the large alloy particles occurs in the "hydrodynamic" mode, when the occurrence of hydrodynamic pulses observed. Further smoothing of irregularities eliminated by a stable burning of discharge in vapor shell. Analysis the morphology of the surface of difficult specialized products, such as crown conical gears, after plasma hydrodynamic treatment showed efficiency and advantages in comparison to conventional methods of final cleaning such as shot blasting.

The work is performed according to the Russian Government Program of Competitive Growth of Kazan Federal University.

References

[1] L Lu, JYH Fuh, YS Wong. Laser-Induced materials and processes for rapid prototyping. Boston: Kluwer Academic publishers; 2001.
[2] L. Kashapov, N. Kashapov and R. Kashapov, 2013 *J. Phys.: Conf. Series* **479** 012011
[3] Kashapov L N, Kashapov N F and Kashapov R N 2013 *J. Phys.: Conf. Series* **479** 012003
[4] Kashapov L, Kashapov N and Kashapov R 2014 *J. Phys.: Conf. Series* **567** 012025
[5] Denisov D, Kashapov N and Kashapov R 2015 *IOP Conference Series: Materials Science and Engineering* **86** 012005

Study of thermal and electrical parameters of workpieces during spray coating by electrolytic plasma jet

A A Khafizov[1], Yu I Shakirov, R A Valiev, R I Valiev and G M Khafizova

Kazan Federal University Naberezhnye Chelny Institute, 68/19 Mira Street, Naberezhnye Chelny, 423810, Russian Federation

E-mail: Almazok75@yandex.ru

Abstract. In this paper the results are presented of thermal and electrical parameters of products in the system bottom layer – intermediate layer when applying protective coatings of ferromagnetic powder by plasma spray produced in an electric discharge with a liquid cathode, on steel samples. Temperature distribution and gradients in coating and intermediate coating were examined. Detailed descriptions of spray coating with ferromagnetic powder by plasma jet obtained in electrical discharge with liquid cathode and the apparatus for obtaining thereof is provided. Problem has been solved by using of Fourier analysis. Initial data for calculations is provided. Results of numerical analysis are provided as temporal functions of temperature in contiguity between coating and intermediate coating as well as temporal function of the value $Q=q-\varphi$; where q is density of heat current directed to the free surface of intermediate coating, φ is density of heat current in contiguity between coating and intermediate coating. The analysis of data given shows that in the systems of contact heat exchange bottom layer-intermediate layer with close values of the thermophysical characteristics of constituting materials is observed a slow increase of the temperature of the contact as a function of time .

1. Introduction

Durability and surface hardening of protective coating applied using plasma spray produced in the electric discharge containing a liquid cathode [1], and its adhesion to the bottom layer is largely dependent on the presence of residual stresses in the coating – bottom layer system. Residual tension in welded joints is extensively studied worldwide [2]. Residual tension has considerable influence on operational properties of machine parts subjected to cyclic mechanical and thermal action.

Nowadays spray coating is widely used in manufacturing of those machine parts [3]. Advantages of this process include quality and precision of resulting surface but this process also creates stretching residual tension. To reduce the level of residual stresses one may by applying an additional intermediate layer to the bottom layer [4]. Since the bottom layer and the intermediate layer have different thermal properties, then with the thermal effect on the system bottom layer – intermediate layer the thermal stresses arise in coating [5].

Therefore, in the development of the coating technology (bottom layer + intermediate layer) the information becomes important about the temperature distribution and temperature gradients in the bottom layer and the intermediate layer, and in contact area.

[1] To whom any correspondence should be addressed.

Journal of Physics: Conference Series **669** (2016) 012030 doi:10.1088/1742-6596/669/1/012030

2. Statement of the problem

In this paper the results are presented of thermal and electrical parameters of products in the system bottom layer – intermediate layer when applying protective coatings of ferromagnetic powder by plasma spray (Figure 1) produced in an electric discharge with a liquid cathode, on steel samples [6 – 8]. Ferromagnetic powder is obtained in the high-voltage electric discharge with voltage U = 1000 – 1300 V, current I = 0.6 –10 A, between the anode of steel and liquid cathode at an interelectrode gap S = 5 – 10 mm, at atmospheric pressure.

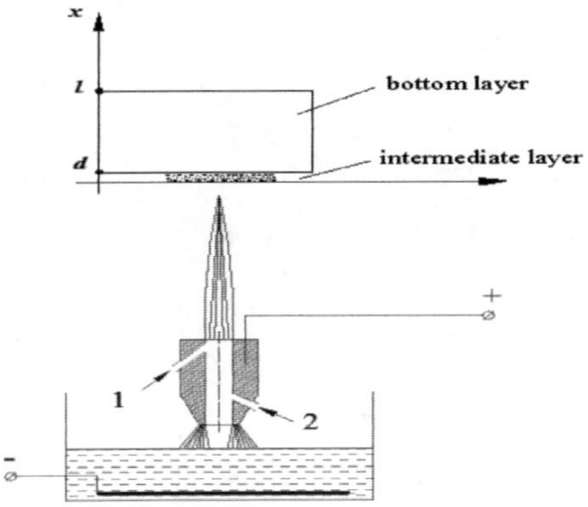

Figure 1. Diagram of the deposition process of the powder by the plasma spray: 1 – feed pipe of sprayed powder; 2 – orifice gas delivery port.

Thermal effect of grains of the coating on the free surface of the intermediate layer is accepted equivalent to heating of the surface by heat flow of given power q = const. The temperature t_0 of the bottom layer and of the intermediate layer at the initial moment are the same, and the thickness of the bottom layer (l – d) >> d – the thickness of the intermediate layer.

We can assume that the distribution of temperature fields in this system is carried out by heat conduction as follows:

$$\frac{\partial t_1(x,\tau)}{\partial \tau} = a_1 \cdot \frac{\partial^2 t_1(x,\tau)}{\partial x^2} \quad (0 < x < d, \quad \tau > 0), \tag{1}$$

$$\frac{\partial t_2(x,\tau)}{\partial \tau} = a_2 \cdot \frac{\partial^2 t_2(x,\tau)}{\partial x^2} \quad (d < x < l, \quad \tau > 0), \tag{2}$$

$$t_1(x,0) = t_2(x,0) = t_0 \tag{3}$$

$$t_1(d,\tau) = t_2(d,\tau) \tag{4}$$

$$-\lambda_1 \cdot \frac{\partial t_1(d,\tau)}{\partial x} = -\lambda_2 \cdot \frac{\partial t_2(d,\tau)}{\partial x} = \varphi(\tau) \tag{5}$$

$$-\lambda_1 \cdot \frac{\partial t_1(0,\tau)}{\partial x} = q \tag{6}$$

$$\frac{\partial t_2(l,\tau)}{\partial x} = 0 \tag{7}$$

where $t_1(x,\tau)$ $t_2(x,\tau)$ are the temperatures of respectively intermediate layer and bottom layer at a point x at a moment τ ; d – the thickness of the intermediate layer, *l-d* – the thickness of the bottom layer, a_j and $\lambda_j (j = 1,2)$ – are respectively the temperature conductivity and heat conductivity coefficients; $q = const$ – heat flow of given power; $\varphi(\tau)$ – the density of heat flow in contact of bottom layer with intermediate layer; t_0 – initial temperature.

The challenge is to find the temperatures $t_j(x,\tau), j = 1,2$, and the density $\varphi(\tau)$ from (1) – (7).

3. Research method

In solving the problem (1) – (7) we use the Fourier method [9]. First, we find the solution of equation (1) satisfying the conditions of (3), (5), (6), temporally considering $\varphi(\tau)$ as known. This solution we obtain in the form

$$t_1(x,\tau) = \sum_{k=0}^{+\infty} T_k(\tau) X_k(x) + t_1^*(x,\tau), \quad 0 \le x \le d, \tau > 0, \tag{8}$$

where the functions $t_1^*(x,\tau)$, $X_k(x)$, $T_k(\tau)$ are defined by following formulas:

$$t_1^*(x,\tau) = \frac{q - \varphi(\tau)}{2d\lambda_1} \cdot x^2 - \frac{q}{\lambda_1} \cdot x;$$

$$X_k(x) = \cos(\mu_k x), \mu_k = \left(\frac{\pi \cdot k}{d}\right)^2, k = 0,1...,$$

$$T_k(\tau) = \begin{cases} e^{-a_1 \mu_k \tau} \left[\int_0^\tau b_k(t) e^{a_1 \mu_k t} dt + a_k \right], k = 1,2..., \\ \int_0^\tau b_0(t) dt + a_0, \quad k = 0; \end{cases}$$

$$b_k(\tau) = \begin{cases} \dfrac{2(-1)^k}{\lambda_1 \cdot \mu_k \cdot d} \cdot \varphi'(\tau), \quad k = 1,2..., \\ \dfrac{a_1(q - \varphi(\tau))}{d \cdot \lambda_1} + \dfrac{d}{6\lambda_1} \cdot \varphi'(\tau), \quad k = 0; \end{cases} \qquad a_k = \begin{cases} -\dfrac{2q}{\lambda_1 \cdot d \cdot \mu_k}, \quad k = 1,2..., \\ t_0 + \dfrac{q \cdot d}{3\lambda_1}, \quad k = 0; \end{cases}$$

In the derivation of the decision (8) we have taken into account that $\varphi(0) = 0$.

Similarly, we find the solution of equation (2) satisfying the conditions of (3), (5), (7), which has the form:

$$t_2(x,\tau) = \sum_{k=0}^{+\infty} \tilde{T}_k(\tau) \cdot \tilde{X}_k(x) + t_2^*(x,\tau), \tag{9}$$

where

$$t_2^*(x,\tau) = \frac{\varphi(\tau)}{2(l-d)\lambda_2}(x^2 - 2lx),$$

$$\tilde{X}_k(x) = \cos[\tilde{\mu}_k(x - d)], \quad \tilde{\mu}_k = \left(\frac{\pi \cdot k}{l - d}\right)^2, \quad k = 0, 1, 2, ...;$$

$$\tilde{T}_k(\tau) = \begin{cases} e^{-a_2\tilde{\mu}_k\tau}\left[\int\limits_0^\tau \tilde{b}_k(t)e^{a_2\tilde{\mu}_k t}dt\right], & k = 1,\ 2, \ldots \\ \int\limits_0^\tau \tilde{b}_0(t)dt + t_0, & k = 0; \end{cases}$$

$$\tilde{b}_k(\tau) = \begin{cases} \dfrac{(-2)\varphi'(\tau)}{\lambda_2(l-d)\tilde{\mu}_k}, & k = 1,\ 2,\ \ldots, \\ \dfrac{a_2}{\lambda_2(l-d)}\varphi(\tau) + \dfrac{2l^3 - 3ld^2 + d^3}{6\lambda_2(l-d)^2}\cdot\varphi'(\tau), & k = 0; \end{cases}$$

Now the solutions (8), (9) we substitute in the condition (4). As a result, to determine the density $\varphi(\tau)$ we obtain the Volterra integral equation of the first kind [10] of the form

$$\int\limits_0^\tau \left\{\theta\left[\frac{\pi a_1}{d^2}(\tau - t)\right] + \frac{\beta_0}{l-d}\theta\left[\frac{\pi a_2}{(l-d)^2}(\tau - t)\right]\right\}\varphi(t)dt = \psi(\tau), \tag{10}$$

where $\theta(x)$ - theta function, defined by the formula

$$\theta(x) = 1 + 2\sum_{k=1}^{+\infty} e^{-\pi k^2 x},$$

$$\psi(\tau) = q(\tau) - \frac{d^2 q}{6\lambda_1} - \frac{2d^2 q}{a_1\pi^2}\sum_{k=1}^{+\infty}\frac{(-1)^k \cdot e^{-a_1\mu_k\tau}}{k^2},$$

$$\beta_0 = \frac{d\lambda_1 a_2}{a_1\lambda_2}$$

The integral equation (10) we solve computationally [11]. Assume $\tau \in [0,T]$, $\tau_j = j\cdot h$, $j = \overline{1,N}$, $h = T/N$. To find the values of $\varphi(\tau_j)$ in junction points τ_j, $j = \overline{1,N}$, we obtain the recurrent equations (11)

$$\varphi(\tau_{j+1}) = \frac{\left\{\psi_m(\tau_{j+1}) - h\left[\varphi(\tau_1)\cdot k_m(\tau_{j+1};\frac{\tau_1}{2}) + \varphi(\tau_2)\cdot k_m(\tau_{j+1};\frac{\tau_1+\tau_2}{2}) + \ldots + \varphi(\tau_j)\cdot k_m(\tau_{j+1};\frac{\tau_{j-1}+\tau_j}{2})\right]\right\}}{\left[h\cdot k_m(\tau_{j+1};\frac{\tau_j+\tau_{j+1}}{2})\right]}$$

$$j = 0,1,\ldots,N-1,$$

where

$$k_m(\tau;t) = \theta_m\left[\frac{\pi a_1}{d^2}\cdot(\tau - t)\right] + \frac{\beta_0}{l-d}\theta_m\left[\frac{\pi a_2}{(l-d)^2}\cdot(\tau - t)\right],$$

$$\theta_m(x) = 1 + 2\sum_{k=1}^{m} e^{-\pi k^2 x},$$

$$\psi_m(\tau) = q\tau - \frac{d^2 q}{6\lambda_1} - \frac{2d^2 q}{a_1 \pi^2} \sum_{k=1}^{m} \frac{(-1)^k e^{-a_1 \mu_k \tau}}{k^2},$$

m is reasonably large natural number.

4. Results and discussion

Computational scheme (11) is implemented with the following input data:

$$a_1 = 12{,}5 \cdot 10^{-6} m^2/s, \; \lambda_1 = 20 W/m \cdot K$$
$$a_2 = 12{,}8 \cdot 10^{-6} m^2/s, \; \lambda_2 = 46 W/m \cdot K$$
$$m = 100, \; d = 10^{-3} m, \; q = 4 \cdot 10^7 W/m^2, \; t_0 = 20^\circ C.$$

Results of numerical computations are shown in Figures 2, 3.

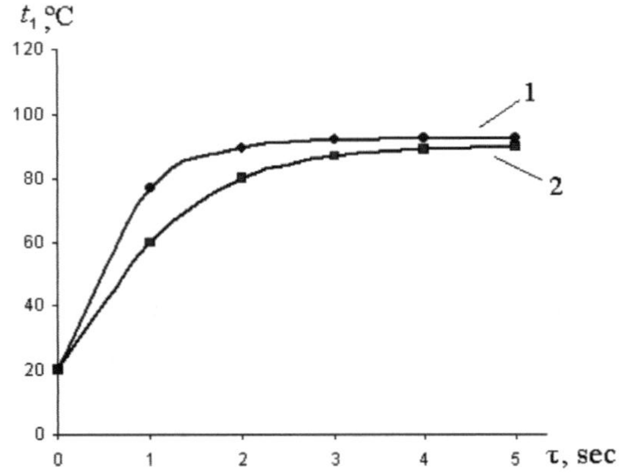

Figure 2. Variation with time τ of temperature t_1 in the contact area of intermediate layer and bottom layer: 1 – when l = 1 cm; 2 – when l = 2 cm.

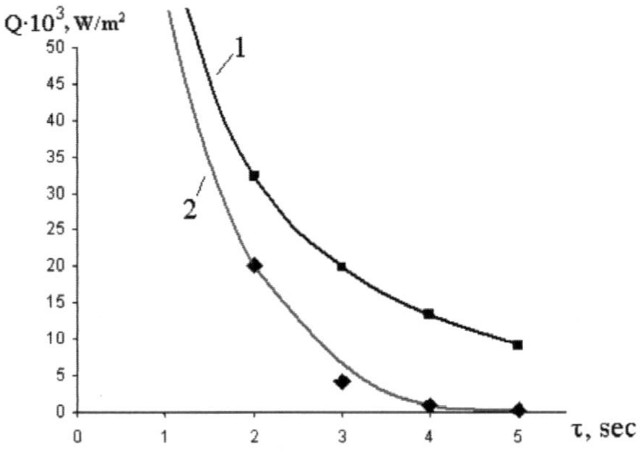

Figure 3. Time law of variable Q=q-φ (q is the density of heat flow falling on free surface of the intermediate layer; φ is the density of heat flow in the contact area of intermediate layer and bottom layer): 1 –when l = 1 cm; 2 – when l = 2 cm.

The analysis of data given shows that in the systems of contact heat exchange bottom layer-intermediate layer with close values of the thermophysical characteristics of constituting materials is observed a slow increase of the temperature of the contact as a function of time t_1 . In such systems, heat flow density in the region of contact of the intermediate layer with the bottom layer (φ) differs slightly from the density of the flow (q) causing heating of the free surface of the intermediate layer, and with time φ and q tend to equalization (corresponding to falling curves Q=q-φ in Figure 3). The specified time laws $t_1(\tau)$ and $Q(\tau)$ have caused by the condition that, in this case, the thickness of the bottom layer is much greater than the thickness of the intermediate layer.

5. Conclusion

Thus, in the process of applying protective coatings on the two-component facilities the solution of the problem of the regulated control of thermal parameters is directly dependent on the appropriate selection of components of a system in terms of optimizing the ratio between their sizes and thermal properties.

References

[1] Shakirov Yu I, Valiev R I, Khafizov AA and Shakirova G Yu 2011 Multichannel plasma installation with electrolytic cathode. *Scientific and technical Journal Auto Industry* **2** 36–38

[2] Belobragin Yu A, Guchek N E, Lebedev A M, Filin V I and Shcerbina V I 1987 Issledovanie vlijanija rezhimov jelektrojerrozionnohimicheskoj obrabotki na raspredelenie ostatochnyh naprjazhenij v poverhnostnom sloe *Jelektronnaja obrabotka materialov*. **1** 13–15

[3] Dautov G Y, Nail K F, Viktor L, Renat G Z, Raphael G T, Fayrushin I and Kashapov R N 2013 Low Temperature Plasma in the Processes of Functional Coating Preparation *J. of Phys.: Conf. Ser.* **479** 79.

[4] Matveev N V, Krasnov A N and Miloserdov I V 1985 Residual stresses in the coating of titanium nitride precipitated in vacuo *Problems of strength* **5** 90–93.

[5] Mrochek Zh A, Antonenko A B and Vershina A K 1987 The distribution of temperature and heat flow at the contact of the bottom layer with the intermediate layer during the electric arc application of protective coatings *Electronic processing materials* **1** 23–26

[6] Khafizov A A Valiev R I Shakirov Yu I and Valiev R A 2014 Steel surface modification with plasma spraying electrothermal installation using a liquid electrode. *J. of Phys.: Conf. ser.* **567** 012026

[7] Khafizov A A, Shakirov B Yu and Shakirov Yu I 2014 Surface hardening of components by automatic plasma electrothermal installation with molten cathode *IOP Conf. Ser.: Materials Science and Engineering* **69** 012019

[8] Valiev R A, Shakirov Yu I and Iliukhin A N 2014 Phase and structural conditions of low-temperature plasma interaction products with steel. *J. of Phys.: Conf. ser.* **567** 012040

[9] Sobolev S L 1966 *Equations of Mathematical Physics* (Moscow: Nauka) p 444

[10] Volterra V 1082 *The theory of functionals, integral and integro-differential equations* (Moscow: Nauka) p 304

[11] Bakhvalov N S, Zhidkov N P and Kobel'kov G M 1987 *The numerical techniques* (Moscow: Nauka) p 600

Features of deformation of metal body surfaces under impact of a water jet

A A Aganin and N A Khismatullina

Institute of Mechanics and Engineering, Kazan Science Center, Russian Academy of Sciences, 2/31 Lobachevsky str., Kazan 420111, Russia

E-mail: nailya_hism@mail.ru

Abstract. The paper presents a mathematical model and computational results on dynamics of a perfect elastic-plastic body under the load arising during impact of a high-velocity liquid jet with the hemispherical end. The body is simulated by the isotropic linearly-elastic semi-space, its plastic state is described by the von Mises condition. The dependence of features of the body surface deformation on the body material is studied. The problem is considered in the axisymmetric statement. The axis of symmetry is that of the jet. The loaded domain is a circle with its radius rapidly growing from zero to the jet radius. The pressure in the loaded domain is non-uniform both in time and space. Three metal alloys (aluminium, copper-nickel and steel) are considered as the body material. The loading of the body surface in all the cases corresponds to the impact of a water jet with the radius 100 μm and the velocity 300 m/s. It has been shown that under such impact a nanometer pit arises on the body surface at the center of the domain of the jet action. The profile of the pit and its maximal depth depend on the body material.

1. Introduction

Study of shock impact of high-speed jets and drops on surfaces of bodies is of interest because of their erosion in the conditions of their contact with liquid of varying pressure [1]. Though the impact of a single jet or drop, as a rule, does not lead to visible changes in the surface, the multiple repeated impacts can cause its damages. For understanding the nature of this destructive action it is important to study features of the impact of a single jet. There is a lot of research devoted to this problem (for example, [2-6]), the majority of which is experimental. Meanwhile, possibilities of experimental study of the changes in the body are very limited. Therefore numerical modeling becomes very important. In numerical modeling one should take into account both the features of loading and the mechanical properties of the body material.

Under impact of a jet the surface of an elastic plastic body is deformed, a pit arises on it, sometimes with a ledge at the edges [5]. If the stress intensity in the body does not reach the yield stress, the surface of the body is restored after removing the load. If the pressure on the surface is large enough then plastic deformations arise in the body and residual damages remain after removing the load.

In the present paper the deformation of surfaces of bodies made of different metal alloys under the load arising during impact of a high-velocity liquid jet with the hemispherical end as well as the influence of the properties of the material of the body on its surface deformation are numerically studied.

2. Problem statement

Shock impact on a body of a high- velocity jet with the hemispherical end striking the body orthogonally to its flat surface is investigated. Figure 1 shows the radial profiles of the pressure $p(r, t)$ (r is the radial coordinate counted from the jet axis, t is time) on the body surface during impact at the moments $t_1 = 0.01\tau$, $t_2 = 0.04\tau$, $t_3 = 0.07\tau$, $t_4 = 0.14\tau$, $t_5 = 0.31\tau$, $t_6 = 0.57\tau$, $t_7 = 2\tau$ where $\tau = 47.4$ ns. The pressure function $p(r, t)$ was determined by a spline approximation of the results computed for the jet impact on the perfectly rigid wall (the computations were executed by T.S. Guseva using the technique [7]). It corresponds to the water jet (ρ=1000 kg/m^3, C=1500 m/s) with the radius R = 100 μm and the velocity V = 300 m/s.

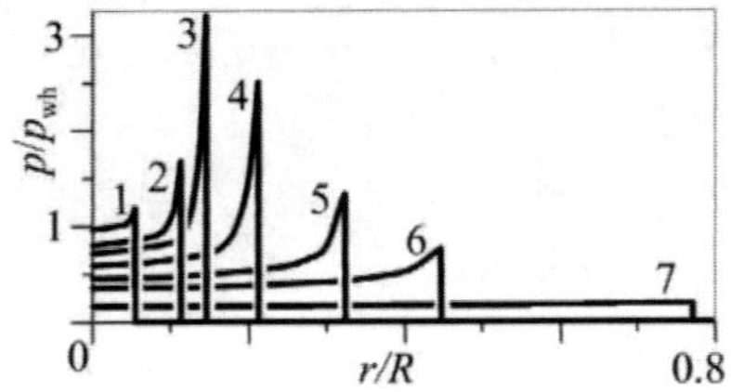

Figure 1. Radial profiles of the pressure on the body surface in seven successive time moments t_1-t_7

The body surface deformation is defined by four independent parameters: the modulus of elasticity E (Young modulus), the Poisson ratio υ, the density ρ and the yield stress Y_0 [8]. In the present work the bodies made of three various metal alloys: aluminium D16T, monel 400 (a copper-nickel alloy), and alloyed steel 30H3MF (table 1) are taken. They are often used for production of elements contacting water or their sheeting.

Table 1. Mechanical characteristics of alloys

Alloy	E (GPa)	υ	ρ (kg/m^3)	Y_0 (MPa)
D16T	71	0.34	2700	325
Monel 400	173	0.3	7850	250
30H3MF	210	0.3	7850	835

The parameters of these metals differ rather strongly, which allows one to separately estimate their influence on the body surface deformation.

The body is considered to be the elastic plastic isotropic semi-space. The deformations and displacements in the body are assumed small.

Dynamics of the body is governed by the equations

$$\rho\frac{\partial u}{\partial t} = \frac{\partial(S_{rr} - P)}{\partial r} + \frac{\partial S_{rz}}{\partial z} + \frac{S_{rr} - S_{\varphi\varphi}}{r}, \qquad \frac{\partial S_{zz}}{\partial t} = -\frac{2}{3}\mu\left(\frac{\partial u}{\partial r} - 2\frac{\partial v}{\partial z} + \frac{u}{r}\right),$$

$$\rho\frac{\partial v}{\partial t} = \frac{\partial S_{rz}}{\partial r} + \frac{\partial(S_{zz} - P)}{\partial z} + \frac{S_{rz}}{r}, \qquad \frac{\partial S_{rr}}{\partial t} = \frac{2}{3}\mu\left(2\frac{\partial u}{\partial r} - \frac{\partial v}{\partial z} - \frac{u}{r}\right), \tag{1}$$

$$\frac{\partial S_{rz}}{\partial t} = \mu\left(\frac{\partial u}{\partial z} + \frac{\partial v}{\partial r}\right), \qquad \frac{\partial S_{\varphi\varphi}}{\partial t} = -\frac{2}{3}\mu\left(\frac{\partial u}{\partial r} + \frac{\partial v}{\partial z} - 2\frac{u}{r}\right), \qquad \frac{\partial P}{\partial t} = -K\left(\frac{\partial u}{\partial r} + \frac{\partial v}{\partial z} + \frac{u}{r}\right)$$

Here z is the axial coordinate; u, v – the velocity components along the axes of the cylindrical coordinates r and z, respectively; S_{rr}, S_{zz}, $S_{\varphi\varphi}$, S_{rz} are the components of the deviator S of the stress tensor σ; $S = \sigma + Pg$, P is the hydrostatic pressure, g is the metric tensor, $\lambda = E\upsilon\ (1+\upsilon)/(1-2\upsilon)$ and $\mu = 0.5E/(1+\upsilon)$ are the Lame parameters, ρ is the undisturbed density of the body material.

In the plastic zones the von Mises condition $\sigma_i = Y_0$, is satisfied where

$$\sigma_i = \frac{1}{\sqrt{2}}\sqrt{(\sigma_z - \sigma_r)^2 + (\sigma_z - \sigma_\varphi)^2 + (\sigma_r - \sigma_\varphi)^2 + 6\tau_{rz}^2}$$

is the stress intensity. If the condition of the plastic yield is violated, i.e., if σ_i exceeds the yield strength in a point, then the deviator of the stress tensor is corrected: $S_{corr} = S \cdot Y_0/\sigma_i$.

The moment of contact between the body surface and the jet is taken as the initial time moment $t = 0$. At this moment the deformations in the radial and circular directions are absent, the static tension state is realized

$$\sigma_z = -p_0,\ \sigma_r = \sigma_\varphi = -\lambda(\lambda + 2\mu)^{-1} p_0,\ \tau_{rz} = 0,\ u = v = 0. \tag{2}$$

On the body surface z = 0 it is assumed that

$$\sigma_z = -p_L(r,t),\ \tau_{rz} = 0. \tag{3}$$

At the infinite distance from the place of the load application the body remains undisturbed.

Problem (1)-(3) is solved numerically by the classical method of S. K. Godunov [9]. The detailed description of the computational procedure and illustration of its application can be found in [10, 11]. It should be noted that in that technique the infinite area of the body is replaced by a finite cylindrical domain with the non-reflecting conditions on its artificial boundaries [12].

3. Deformation of a body surface under impact of a jet

Under the loading specified above the surface layer of the bodies of each of the three considered metals is subjected to elastic plastic deformation. Figure 2 illustrates the dynamics of the surface of the elastic-plastic body under impact of the jet with the hemispherical end. The bodies made of three metal alloys are presented. It should be noted that in this figure the scales in the radial and axial directions largely differ (the picture is strongly stretched along the axis z). One can see that the impact results in appearance of a small deepening (pit) on the body surface. The center of the pit coincides with that of the loading area while its depth depends on the body material. In all the three cases the radial sizes of the pit constantly increase with time, their depth grows by the moment t_6 and then somewhat decreases. By the moment t_3 the ledges are produced on the edge of the pits. By the moment t_4 they increase and then disappear by the moment t_6. The largest pit and the greatest ledge arise on the surface of the body of the aluminium alloy, and the smallest ones on the surface of the body of the alloyed steel. The maximum depth of the pit in the body made of an aluminium alloy is 1.9 times greater than that in the body of monel, and is 4.7 times greater than that in the body of steel.

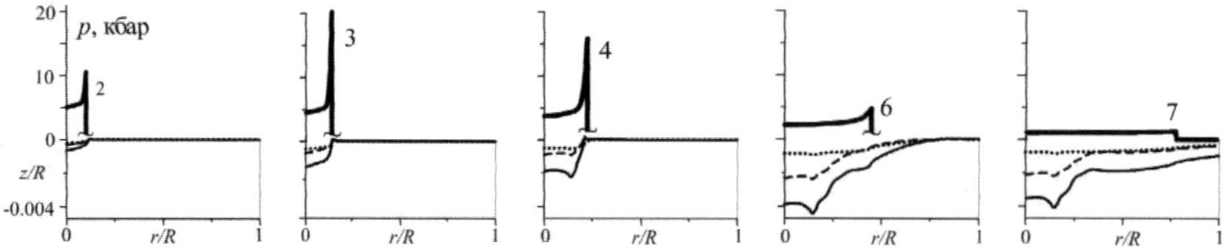

Figure 2. Deformation of the surface of the bodies under impact of the jet with the hemispherical end in five characteristic time moments corresponding to those in figure 1. The solid, dashed and dotted lines are the surfaces of the bodies made of aluminium alloy, monel and alloyed steel, respectively

It is easy to see that the correlation between the distributions of the depth of the pit on the surfaces of the considered bodies corresponds to the correlation between the values of the Young modulus of their materials. At the equal densities of monel and steel the depth of the pit in the body made of steel is 2.5 times smaller than in that made of monel. It is obvious that the lower the yield strength of the body material, the greater the deformation of its surface; however, the pit on the surface of the body of monel is less than that on the surface of the body of aluminium alloy though the yield strength of the first material is nearly 1.5 times greater than that of the second one.

Table 2 presents the values of the maximum and residual depths of the pits obtained in computations.

Table 2. The depth of the pits for various body materials.

Body material	Maximum depth of the pit (micron)	Residual depth of the pit (micron)
D16T	0.442	0.408
Monel 400	0.238	0.220
30H3MF	0.094	0.086

Acknowledgments

The authors are grateful to T.S. Guseva for the computational results used in deriving the approximate load presented in figure 1. This work was supported by the Russian Foundation for Basic Research (grant 14-01-97004)

References

[1] Field J E 1999 *Wear* **233** 1
[2] Voinov O V and Voinov V V 1976 *DAN USSR* **227** No 1 63
[3] Tomita Y and Shima A. 1986 *Journal of Fluid Mechanics* **169** 535
[4] Choi J - K, Jayaprakash A, Kapahi A, Hsiao C-T and Chahine G L 2014 *Journal of Materials Science* **49** 3034
[5] Bowden F P and Brunton J H 1961 *Proceedings of the Royal Society A* **263** 433
[6] Lush P A 1983 *Journal of Fluid Mechanics* **135** 373
[7] Aganin A A and Guseva T S 2014 *Uchenye zapiski Kazanskogo universiteta Seriya fiziko-matematicheskiye nauki* **156** Book 2 55
[8] Fortes-Patella R, Challier G, Reboud J L and Archer A 2013 *Journal of Fluids Engineering* **135** (1) 011303-1
[9] Godunov S K, Zabrodin AV, Ivanov M Y, Krayko A N and Prokopov G P 1976 *Numerical solving of gas dynamics multidimensional problem* (Moscow: Nauka) p 400
[10] Aganin A A, Ilgamov M A and Khismatullina N A 2013 *Uchenye zapiski Kazanskogo universiteta Seriya fiziko-matematicheskiye nauki* **155** Book 2 131
[11] Aganin A A and Khismatullina N A 2014 *Uchenye zapiski Kazanskogo universiteta. Seriya fiziko-matematicheskiye nauk* **156** Book 2 72
[12] Ilgamov M A and Gilmanov A N 2003 *Nonreflecting conditions on the boundary of computational domain* (Moscow: FIZMATLIT) p 240

Journal of Physics: Conference Series **669** (2016) 012032 doi:10.1088/1742-6596/669/1/012032

Generation of low-temperature plasma by low-pressure arcs for synthesis of nitride coatings

O V Krysina[1,2], N N Koval[1,2], I V Lopatin[1], V V Shugurov[1], and S S Kovalsky[1,2]

[1]Institute of High Current Electronics SB RAS, Tomsk, 634055, Russia

[2]National Research Tomsk State University, Tomsk, 634050, Russia

E-mail: krysina_82@mail.ru

Abstract. Experiments were performed to study gas, metal, and mixed metal-gas plasmas. The plasmas were generated with the use of an arc evaporator and a gas-plasma source with a hot filament and hollow cathode that were operated independently or simultaneously. It has been revealed that the arc current of gas-plasma source affects the parameters of the metal-gas plasma and the element concentrations in the coatings. It has been demonstrated that the characteristics of the nitride coatings produced by plasma-assisted vacuum-arc deposition can be controlled by varying the parameters of the arc in the gas-plasma source.

1. Introduction

Physical vapor deposition (PVD) of protective coatings on the surfaces of machine parts, tools, and products is a widely used technique to increase their service life and extend their field of use due to the unique properties of the coatings. In particular, high performance coatings are produced using the well-known binary titanium nitride system that is deposited on cutting tools to form wear-resistant high hardness (H = 20–25 GPa) coatings or on various products to form golden decorative coatings.

The vacuum-arc coating deposition technique based on condensation of the plasma generated by a low-pressure arc with a cathode spot [1] shows versatility, high deposition rate (~10 μm/h), and other advantages, thus being among the most efficient methods of plasma-ion synthesis of functional coatings.

Usually, the nitrogen pressure, discharge current, substrate temperature, bias voltage are the varied parameters by optimization of nitride coatings deposition modes [2–7]. On the other hand, the authors of a great number of publications on plasma-ion deposition of functional coatings point to the use of additional devices (ion or electron sources, plasma filters, etc.) [8–10], or complex compositions of vaporized targets, or working gas mixtures, etc. This makes every work original and enables one to vary the structure and properties of coatings over wide limits. In our case, the coatings were synthesized by the plasma-assisted vacuum-arc deposition technique using the hot filament, hollow cathode plasma generator [11, 12] developed at the Institute of High Current Electronics, SB RAS (Tomsk, Russia) and named PINK (Russian acronym) as an accessory plasma source. The PINK generator is well suited to perform cleaning, activation, and heating of various surfaces, plasma-ion etching to remove layers of different thickness, nitriding, and plasma-ion assistance in depositing various-purpose coatings [13–15].

It is well known that the properties, chemical composition, and structure of a coating produced by vacuum-arc deposition and the pattern of plasma chemical reactions taking place on the substrate are substantially determined by the characteristics of plasma in the arc gap. Therefore, to investigate the parameters of the plasma that is condensed to form a coating is of importance. In the experiment under consideration, plasma-assisted vacuum-arc deposition of coatings was used, and the metal-gas plasma generated by a low-pressure arc was the main object of investigation.

The goal of the experiment was to investigate the generation of gas, metal, and metal-gas plasmas, to measure the plasma parameters, and to elucidate the effect of plasma assistance on the characteristics of the coatings synthesized by the vacuum-arc technique.

2. Equipment and procedures

The experiment aimed to study the low-temperature plasmas of low-pressure arcs and the plasma-assisted synthesis of coatings was carried out using automated plasma-ion facility developed at the Institute of High Current Electronics, SB RAS (Tomsk, Russia). A schematic diagram of the experimental apparatus is shown in Fig. 1. Plasma was produced with two plasma sources: a DI-100 arc evaporator [13, 16] operating based on a cathode spot arc and the PINK gas plasma source operating based on a non-self-sustained arc with a hot filament, hollow cathode [11–13]. The plasma parameters were measured using a cylindrical Langmuir probe (Ø 0.5 mm; $l = 5$ mm) and a unique automated probe system.

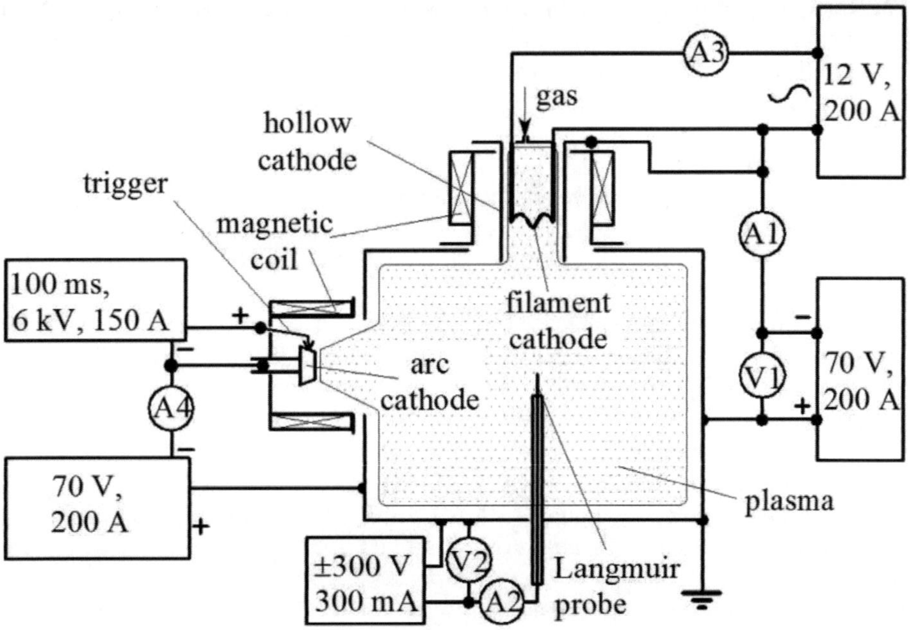

Figure 1. Schematic diagram of the experimental apparatus.

A block diagram of the probe measuring system is shown in Fig. 2. Functionally the system can be divided into two units: a measuring unit (MU) and a probe bias unit (PBU). For measuring the probe current-voltage characteristic, a sawtooth voltage or pulse voltage with variable amplitude is applied from the PBU to the probe. To preclude arcing at the probe surface immersed into plasma, the PBU is equipped with additional protective and output-current limiting circuits.

For measuring the probe voltage, a voltage sensor is connected parallel to the probe circuit. For measuring the current, a current sensor is connected in the probe circuit. Signals from both sensors are transmitted through a buffering and amplifying system to a two-channel analogue-to-digital converter (ADC) with a common input for signal conversion. The system provides a minimum phase shift between the current and voltage probe signals.

The system is controlled with a microcontroller, which provides sync signal processing, a delay between the sync signal and startup of the ADC, data transfer from the ADC to a personal computer (PC) through an optical converter, and sync operation of the MU and PBU.

Journal of Physics: Conference Series **669** (2016) 012032 doi:10.1088/1742-6596/669/1/012032

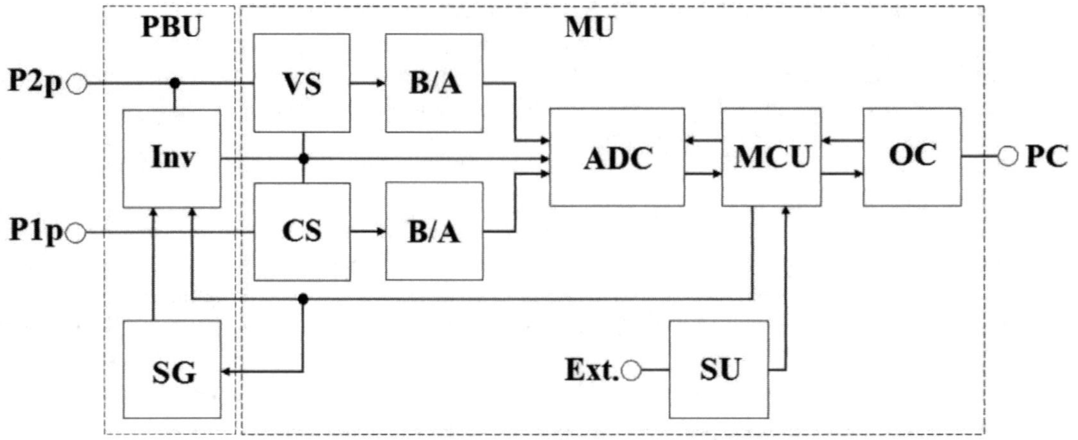

Figure 2. Block diagram of the probe measuring system: P1p – connection for a probe; P2p – connection for a reference electrode (or second probe); VS – voltage sensor; CS – current sensor; B/A – buffers and amplifiers; ADC – analog-to-digital converter; Ext – external synchronizer; PC – port for a personal computer; Inv – power inverter; SG – sawtooth voltage generator; PBU – probe bias unit; MU – measurement unit.

The system can be operated in standby and continuous modes. In the continuous mode, each next point of the current-voltage characteristic is recorded as soon as recording of the previous point is completed. If the PBU, in this case, operates in the pulsed mode the bias voltage is applied for the ADC sampling time.

In the standby mode, the system is triggered by a synchronizer signal. Once the signal is produced, the microcontroller starts up the delay timer (from 5 μs to 10 ms). Within a prescribed delay time, the ADC is triggered. If the PBU operates in the pulsed mode, the voltage to the probe is applied 5 μs before the ADC is triggered. This is required for the transient processes in the probe circuit to be complete before the point of measurement.

When the system is operated in the mode with ion cleaning of the probe surface from contamination and dielectric inclusions, a negative potential (–100 V) with respect to the reference electrode is applied to the probe between measurements. Cleaning of the probe is possible only in the pulsed mode of the PBU.

Measurement data are stored in special PC software for further processing. The software allows adjustment of the automated plasma measuring system and data storage in a format compatible with Excel, Origin, etc.

The system provides the following: a possibility of research with a single probe and double probe; simultaneous measurements of instantaneous voltage and current; input voltages of ±100 V; input currents of ±300 mA; fast automatic discrete adjustment of amplification with a factor of 1, 2, 5, 10; protection of analogue inputs from overvoltages and short circuit; an ADC capacity of 14 bits; control of the number of points in a measurement interval and measurement time; an optical input; an adjustable delay between the measurement point and sync pulse from 5 μs to 10 ms with a step of 1 μs; galvanic decoupling between the measuring unit and PC; a possibility of ion cleaning of the probe at intervals between the measurements.

All measurements were performed with the probe placed at the center of the working chamber, immediately at the cross point of the axes of the plasma sources, 450 mm from the output aperture of the PINK source and 300 mm from the output aperture of the arc evaporator. For each operation mode, the probe current-voltage characteristic was measured from which the floating potential (φ_{fl}) and the plasma potential (φ_{pl}) were evaluated by a graphical method. The electron temperature (T_e) was determined from the probe characteristic plotted on a semilogarithmic scale. The plasma density (n) was evaluated from the electron branch of the probe characteristic.

The vacuum system of the ion-plasma facility, which incorporates a sliding vane rotary pump and a turbomolecular pump, is capable to evacuate a vacuum chamber to a residual pressure of ~10^{-3} Pa. Inert argon and reactive nitrogen were chosen as working gases. The working gas pressure was ~0.3 Pa for all the modes examined. The gas plasma density, controlled by varying the PINK arc current (I_p = 0–100 A), was a variable parameter in the probe measurements of the parameters of gas and metal-gas plasmas. Investigations of metal and metal-gas plasmas were carried out at the evaporator arc current I_d = 50 and 100 A. Commercially pure titanium (Grade 2 titanium alloy: 99.3%Ti) was used for the cathode material. Molybdenum alloy, 12Cr18Ni10Ti steel, and WC-8%Co hard alloy were used as substrates for X-ray analysis, elemental, and mechanical investigations, correspondingly.

The gas plasma density was a variable parameter. Other parameters of the coating deposition modes were fixed (evaporator arc current with a titanium cathode I_d, Ti = 100 A, nitrogen pressure p_{N2} = 0.3 Pa, and substrate was under floating potential φ_{fl}). The substrates were situated at the center of the working chamber like a probe.

It should be noted that the drop fraction is formed at arc deposition of coatings, but in the paper, the influence of gas plasma source on drop fraction was not investigated. That can be find in earlier works [17, 18].

Investigations of the deposited coatings were carried out by micro- and nanoindentation (PMT-3, NHT-S-AX-000X Nano Hardness Tester). The phase composition was investigated by X-ray diffraction analysis (Shimadzu XRD 6000 X-ray diffractometer). The element concentrations in the TiN coatings were measured by the Auger spectroscopy method. The calculation of nitrogen concentration was carried out by deduction of LMM titanium line intensity from that of overlapped LMM and KLL lines. The intensity of LMM titanium line was calculated from ratio of LMM and LMV titanium lines.

3. Results and discussion

In order to reveal how the characteristics of non-self-sustained arc discharge with hot filament and hollow cathode affect the metal-gas plasma parameters the measurements of plasma parameters at (1) independent work of PINK gas-plasma source (gas plasma parameters measurements), at (2) independent work of DI-100 arc evaporator (metal plasma parameters measurements) and at (3) simultaneous work of both plasma sources (metal-gas plasma parameters measurements) were carried out.

3.1. Gas plasma

At first, the gas plasma generated by the PINK plasma source was investigated. The parameters of the argon plasma that was used for pre-cleaning, heating, and activation of the surfaces by plasma-ion etching are presented for different arc operation modes in Fig. 3. An increase in filament current, and, hence, in electron emission current from the filament cathode, increases the PINK arc current and changes the arc operation voltage, leading to an increase in density of the gas plasma.

It has been found that as the arc current was increased from 10 to 90 A, the argon plasma density increased from $0.3 \cdot 10^{16}$ to $5.2 \cdot 10^{16}$ m^{-3}, resulting in an increase in ion current density at the substrate. Under these conditions, the plasma potential relative to the anode was positive and increased from 5.4 to 6.4 V. The electron temperature T_e varied inappreciably throughout the current range, and its average value was ~1.4 eV.

The measured parameters of the nitrogen plasma that was used for nitriding of extended surface layers of the substrates are given in Fig. 4. The average electron temperature of the nitrogen plasma generated by a low-pressure arc was ~1.4 eV, the same as for the argon plasma. The plasma potential also increased as the arc current was increased from 10 to 90 A, but in another range: from 3.3 to 4.2 V. The increase of the PINK arc current in the chosen range at a constant working gas pressure increased the plasma density in the range $(0.2–2.0) \cdot 10^{16}$ m^{-3}.

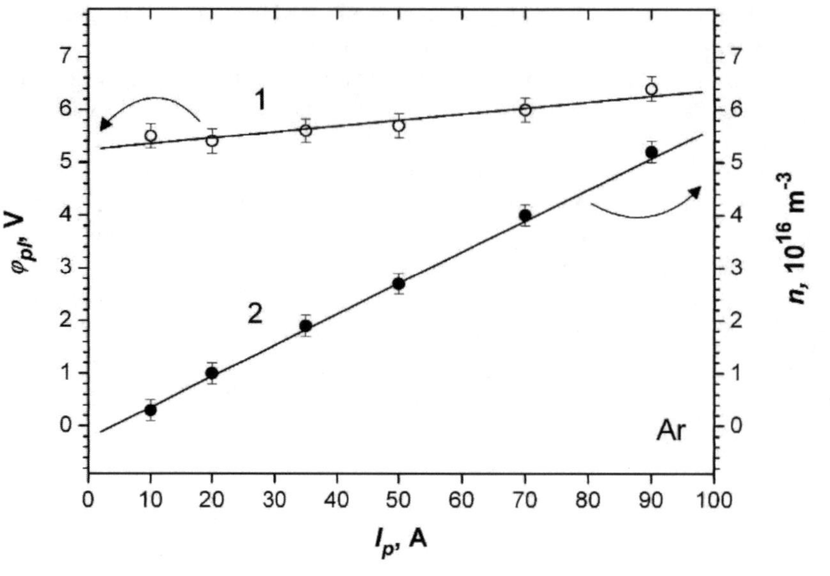

Figure 3. Plasma potential (curve 1) and density (curve 2) versus PINK arc current for argon plasma.

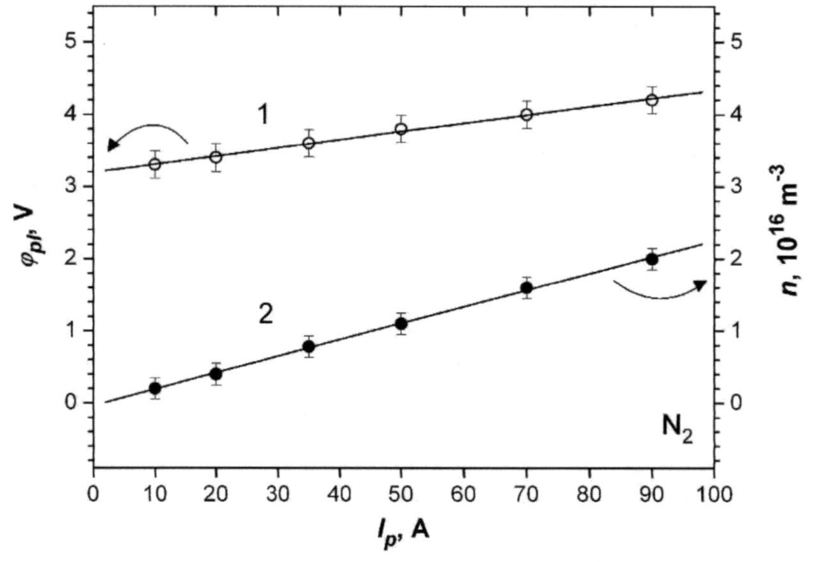

Figure 4. Plasma potential (curve 1) and density (curve 2) versus PINK arc current for nitrogen plasma.

3.2. Metal and metal-gas plasma for synthesis of coatings

The density of the metal plasma produced, not using the PINK source, with argon and nitrogen used as working gases at $I_d = 50$ A (Fig. 5, 6) was $2.6 \cdot 10^{16}$ and $2.2 \cdot 10^{16}$ m^{-3} and its potential was 3.8 and 2.4 V, respectively. The average electron temperature for the arc-generated metal (titanium) plasma was ~0.8 eV.

When the sources of metal and gas plasmas were operated at $I_d = 50$ A with argon used as working gas and the PINK current was increased from 10 to 50 A, the plasma potential increased from 3.8 to 5.5 V (see Fig. 5). In this case, at a fixed current of the arc evaporator, the metal-gas plasma density increased from $2.6 \cdot 10^{16}$ to $4.5 \cdot 10^{16}$ m^{-3}. When nitrogen was used as working gas, the plasma potential increased from 2.4 to 3.4 V (see Fig. 6) and the metal-gas plasma density increased from $2.2 \cdot 10^{16}$ to $3.0 \cdot 10^{16}$ m^{-3}. The average electron temperature for the mixed metal-gas plasma was 0.9–1.0 eV.

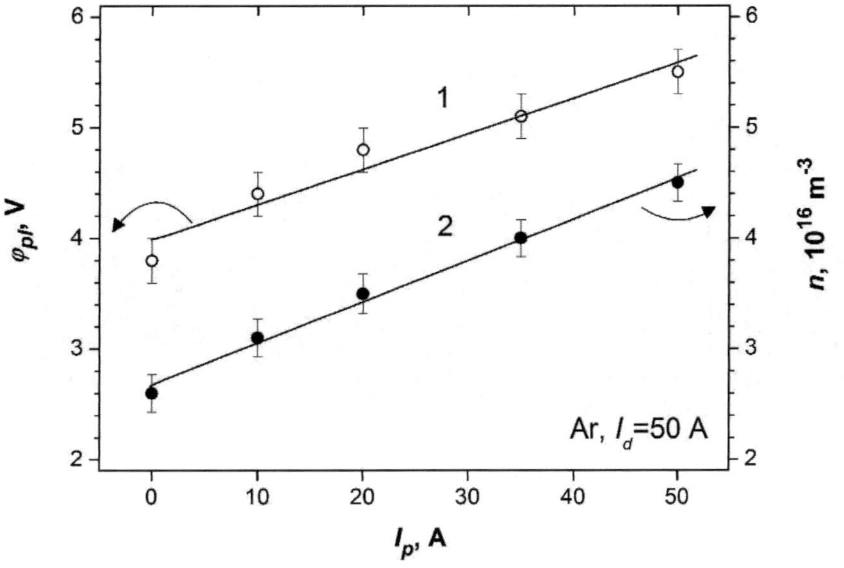

Figure 5. Metal-gas plasma potential (curve 1) and density (curve 2) versus PINK arc current for argon: I_d = 50 A.

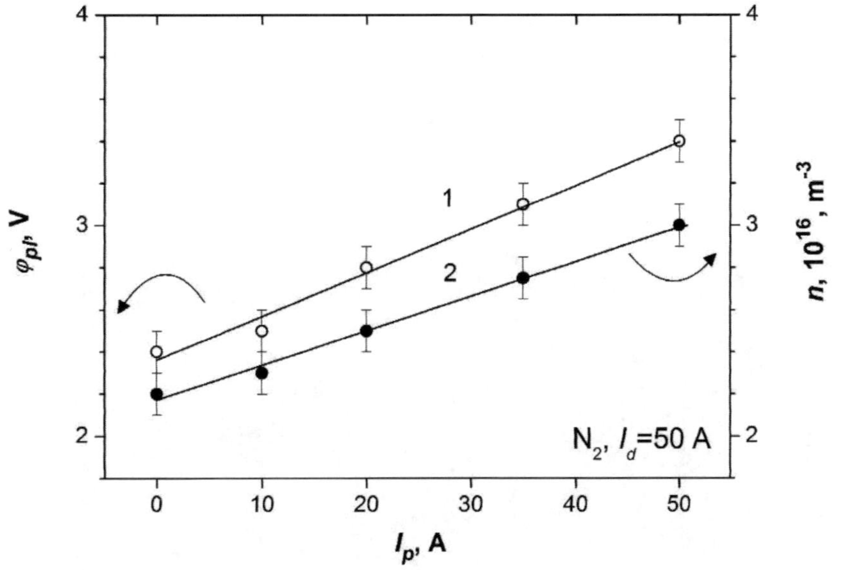

Figure 6. Metal-gas plasma potential (curve 1) and density (curve 2) versus PINK arc current for nitrogen: I_d = 50 A.

It is clear that the density of the metal-gas plasma was approximately equal to the sum of the gas and metal plasma densities measured at the same arc parameters. For the doubled discharge current of the arc evaporator (I_d = 100 A), the plasma potential and density were also observed to vary linearly with the PINK arc current. However, in this case, the density of the mixed plasma was approximately twice that measured at a 50 A current (Fig. 7).

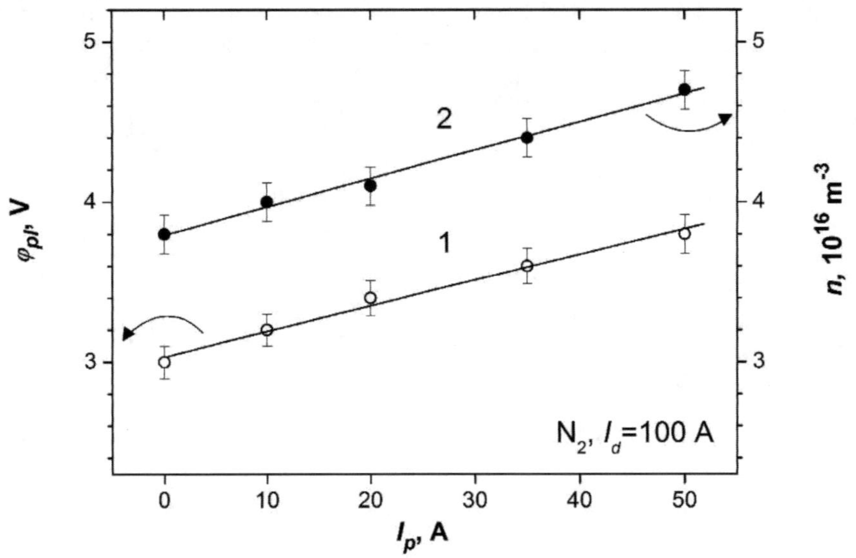

Figure 7. Metal-gas plasma potential (curve 1) and density (curve 2) versus PINK arc current (I_d = 100 A).

3.3. The effect of plasma assistance on the characteristics of the coatings

The experimental data suggested that the concentration of nitrogen charged particles in plasma-ion-assisted vacuum-arc deposition of nitride coatings could be increased by increasing the PINK arc current, thus promoting the formation of nitride compounds on the substrate and in the cathode–substrate gap.

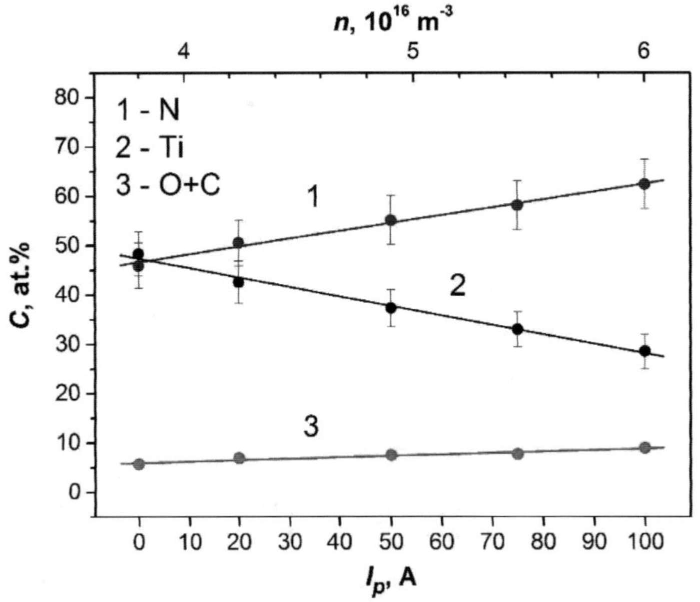

Figure 8. Element concentrations in TiN coatings versus PINK arc current and metal-gas plasma density.

To check this, plasma-assisted deposition of TiN coatings was performed at different PINK arc currents. The elemental composition of the coatings as a function of current is given in Fig. 8. For the TiN coatings deposited (not using the PINK source) from metal plasma in a molecular nitrogen medium at p_{N2} = 0.3 Pa, the titanium-to-nitrogen concentration ratio is, within the limits of experimental error, 1:1. The Ti:N ratio for the coating deposited with plasma assistance at a rather low PINK arc current (≤ 20 A) can also be taken equal to 1:1, within the limits of experimental error. As the discharge current is further increased, and, hence, the fraction of nitrogen charged particles (gas plasma density) increases, the nitrogen content in the TiN coating increases, whereas the titanium content accordingly decreases. When the evaporator arc current and the PINK arc current are in the ratio 1:1, the concentration of titanium is approximately half that of nitrogen.

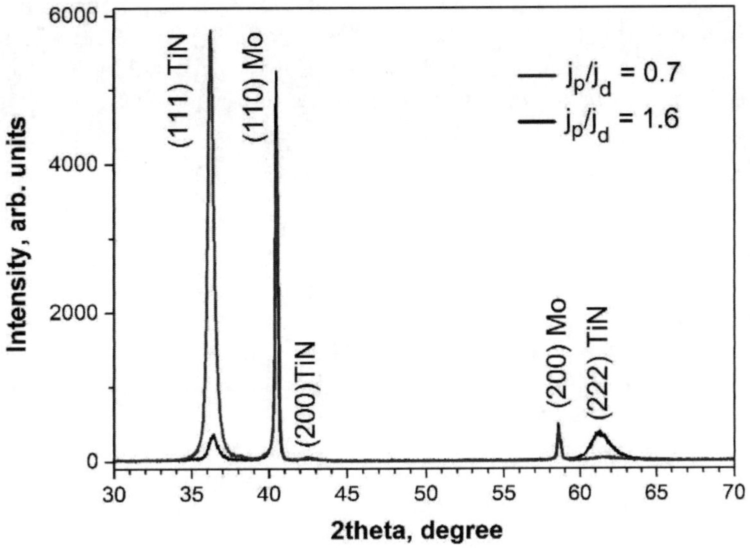

Figure 9. Diffraction patterns of TiN coatings deposited at different ratio j_p/j_d.

It should be noted that the color of coatings changes in dependence of increase of gas plasma source current and accordingly in dependence of nitrogen content from gold to copper-gold (Table 1). It was revealed that hardness of coatings increases from 24 GPa to 35 GPa at increasing nitrogen concentration.

Table 1. Parameters of TiN coatings in dependence of ratio j_p/j_d.

j_p/j_d	H (GPa)	CSR (nm)	color
0.7	24	41	gold
1.2	28	30.5	pink-gold
1.6	35	17	copper-gold

The results of X-ray analysis of TiN coatings deposited at different ratio of current densities of PINK source and arc evaporator (j_p/j_d) are shown in Fig. 9 and Tabl. 1. It is clear, that the coating consists of polycrystalline TiN with a NaCl-type FCC structure at j_p/j_d=0.7. TiN crystallites have have preferred orientation along the plane [111]. In the case the size coherent-scattering region (*CSR*) is 41 nm. At increase of ratio j_p/j_d up to 1.6 the considerable increasing (222)TiN reflex intensity and broadening of TiN peaks are observed. The fact deals with decreasing crystallite size as size of *CSR* decreased to 17 nm.

It should be noted that in all coatings synthesized by the chosen method, oxygen and carbon were detected, which was related to the conditions of evacuation. The total concentration of these elements in the coatings was not above 9 at. %. However, the authors of [19] showed that the presence of a large amount of oxygen (up to 7 at. %) and carbon (up to 4.7 at. %) in the coatings based on TiN causes no superhardness degradation. In addition, the oxygen existence in the TiN coating can be the cause of titanium oxides formation on a coating surface. It was investigated in detail by XPS analysis and shown in [20]. It should be noted that in the present work the oxide or carbide phases were not detected by X-ray analysis.

4. Conclusion

The paper presents the results of an experiment performed to examine gas, metal, and mixed metal-gas plasmas. The plasmas were generated by an arc evaporator operating based on a self-sustained arc with cathode spot and by the PINK plasma source with hot filament and hollow cathode operating based on a non-self-sustained arc. The plasma sources were used independently or simultaneously, depending on the type of desired plasma. The plasma parameters were measured using a cylindrical Langmuir probe and a unique automated probe system described in the paper.

It has been demonstrated that the elemental composition of the titanium nitride coatings produced by plasma-assisted vacuum-arc deposition can be controlled by varying the arc parameters of the PINK hot filament, hollow cathode plasma source. It has been found that the PINK arc current affects the parameters of the metal-gas plasma and the element concentrations, structure and characteristics of the coatings. The effect revealed can be harnessed to synthesize 1) graded coatings where the stoichiometry could be varied varying the arc current of gas-plasma source; 2) the coatings with fine grained structure and improved mechanical properties. This could be a matter of convenience in the coating deposition using automated technological facilities.

Acknowledgments

The work was supported by the grant of Russian Science Foundation (Projects No 14-29-00091).

References

[1] Boxman R L *et al* 1995 *Handbook of Vacuum Arc Science and Technology* (Noyes Publishing, Ridge Park NJ)
[2] Shanyong Zhang 1993 *Journal of Materials Processing Technology* **39** 165
[3] Saoula N *et al* 2009 *J. Plasma Fusion Res. SERIES* **8** 140
[4] Roquiny Ph *et al* 1999 *Surf. Coat. Technol.* **116–119** 278
[5] Sundgren J E *et al* 1986 *J. Vac. Sci. Technol. A 4* **5** 2259
[6] Lousa A *et al* 2007 *Vacuum* **81** 1507
[7] Wei Yongqiang *et al* 2013 *Vacuum* **89** 185
[8] Borisov D P *et al* 1998 *IEEE Trans Plasma Sci* **26** 1680
[9] Ryabchikov A I *et al* 2005 *Vacuum* **78** 445
[10] Ryabchikov A I *et al* 1992 *Review of Scientific Instruments* **63** 2428
[11] Akhmadeev Yu H *et al* 2003 *Laser and Particle Beams* **21** (2) 249

[12] Vintizenko L G *et al* 2001 *Russ. Phys. J.* **44** 927
[13] Koval N N *et al* 2015 *Russian Journal of General Chemistry* **85** (5) 1326
[14] Goncharenko I M *et al* 2003 *Surf. Coat. Technol.* **169–170** 419
[15] Ivanov Yu F *et al* 2012 *Surf. Coat. Technol.* **207** 430
[16] Koval N N *et al* 2012 *Proceedings of the 25th International Symposium on Discharges and Electrical Insulation in Vacuum* (Tomsk, Russia) **2** 537
[17] Koval N N *et al* 2000 *Proceedings of the 19th International Symposium on Discharges and Electrical Insulation in Vacuum* (Tomsk, Russia) **2** 590
[18] Vershinin D S *et al* 2001 *Proceedings of the Fifth Russian-Korean International Symposium on Science and Technology KORUS '01* **1** 328
[19] Korotaev A D *et al* 2007 Physical mesomechanics **10** (3-4) 156
[20] Iordanova I *et al* 2007 *Vacuum* **81** 830

Journal of Physics: Conference Series **669** (2016) 012033 doi:10.1088/1742-6596/669/1/012033

Hybrid plasma system for magnetron deposition of coatings with ion assistance

K V Vavilin[1], E A Kralkina[a, 1], P A Nekludova[1], A. K Petrov[1], A M Nikonov[1], V B Pavlov[b, (1)]

A A Airapetov[2], V V Odinokov[2], G Ya Pavlov[2], V A Sologub[2]

[1] Faculty of physics, Lomonosov MSU, Moscow, 119991, Leninskie Gory 1, build. 2
[2] PJSC "Research institute of precision machine manufacturing", Russia. 124469, Moscow, Zelenograd, Panfilovsky avenue 10

[a] ekralkina@mail.ru , [b] vb_pavlov@mail.ru

Abstract. The results of the study of the plasma hybrid system based on the combined magnetron discharge and high-frequency inductive discharge located in the external magnetic field is presented. Magnetron discharge provides the generation of atoms and ions of the target materials while the flow of accelerated ions used for the ion assistance is provided by the RF inductive discharge. An external magnetic field is used to optimize the power input to the discharge, to increase the ion current density in the realm of substrate and to enhance the area of uniform plasma. The joint operation of magnetron and RF inductive discharge leads to a substantial increase (not equal to the sum of the parameters obtained under separate operation of two hybrid system channels) of the ion current density and intensity of sputtered material spectral lines radiation. Optimal mode of the hybrid plasma system operation provides uniform ion current density on the diameter of at least 150mm at 0.7PA argon pressure. The optimal values of the magnetic fields in the region of the substrate location lie in the range 2-8 mTl, while in the region of the RF input power unit lie in the range 0.5-25 mTl.

1. Introduction

The creation of innovative products in electronics, mechanical engineering, medical technology and other fields is related usually to the use of new composite and multicomponent materials. In general, a multifunctional thin film layers must be deposited by the low cost, environmentally friendly, technically accessible and safe methods. In the last decade a set of methods for the coating of complex shape surfaces has been intensively developed [1,2]. Among them the scientific and technological preconditions have been established for the practical utilization of clean and safe vacuum plasma processes. Ion sputtering of the solid targets in gas discharge plasma is the multipurpose and efficient method of the target material atoms and ions generation. So, it is sputtering vacuum-plasma magnetron and vacuum arc gas discharges systems that are widely used for the film deposition. However, the simple usage of the sputtering systems does not allow to control with necessary accuracy the nanostructure and chemical composition of the deposited layers. This requires the introduction of the assisting ion bombardment of the substrate simultaneously with the film deposition [3,4], i.e. one

Journal of Physics: Conference Series **669** (2016) 012033 doi:10.1088/1742-6596/669/1/012033

needs to develop a hybrid plasma system, where in addition to the magnetron or vacuum arc device the source of the accelerated ions with the flux density matched with the deposition system rate is present. This paper presents the first results of a study of the hybrid plasma system, where in addition to magnetron the RF inductive discharge placed in an external magnetic field is used for generation of accelerated ions flow for ion assistance.

2. Construction of the hybrid plasma system

The design of the hybrid plasma system (HPS) is shown on Figures 1 and 2.

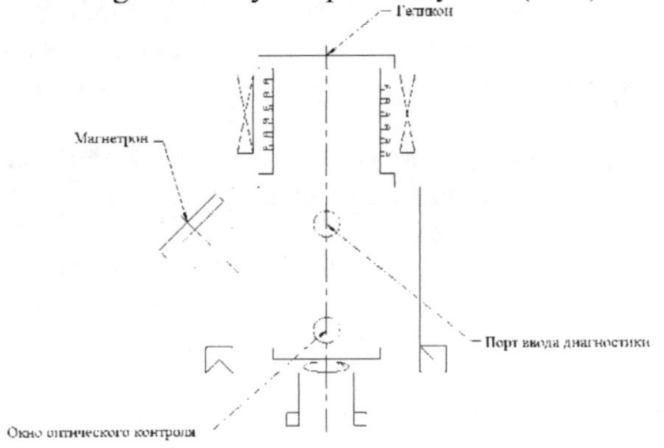

Figure 1. Schematic view of the Hybrid plasma system.

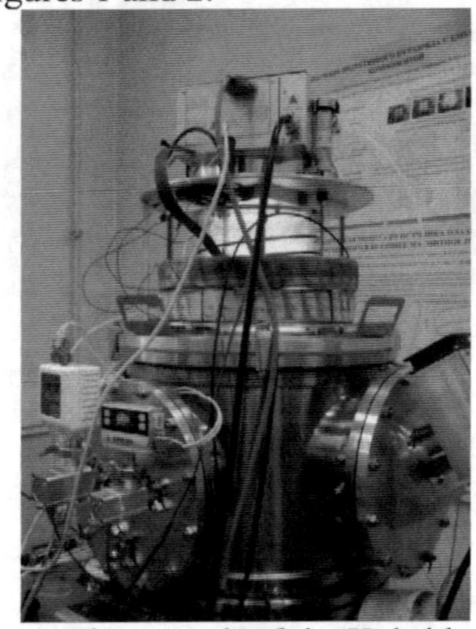

Figure 2. Photograph of the Hybrid plasma system.

The HPS consists of two parts. The main part is a steel cylindrical chamber with a diameter of 500 mm and a height of 350 mm. At the bottom of the chamber the rotatable table for placing the processed samples is assembled. During tests and optimization of the technological regimes of HPS a system of probes for monitoring plasma parameters was mounted on the table. A special port is mounted on the side surface of the chamber for output of the signals from probes to the recording equipment. To carry out spectrometric studies of plasma parameters in the area near the table two windows strictly arranged against each other are assembled. On the side surface of the chamber a magnetron is installed. Below the main chamber an electromagnet, which yields an external magnetic field induction near the table in the range 0 -150Gs is situated.

Above the main chamber quartz cylindrical plasma source with a length of 250 mm and a diameter of 220mm is mounted. On the top the source is closed by glass flange and at the bottom - by the metal flange with an opening providing plasma access to the main plasma chamber. The magnetic field in the upper chamber in the range of 0 -300Gs is provided by electromagnet. To ignite and maintain the discharge the helical antenna is used. The antenna ends are connected via a matching system to the RF power supply with the operating frequency 13.56MHz and output power up to 1000W.

To investigate the plasma uniformity in the region of the substrate the 25 flat probes are mounted on the table. 13 probes are parallel to the symmetry axis of the magnetron (the x axis) and 12 probes are perpendicular thereto (hereinafter referred to axis y). To measure the ion saturation current the potential -60V with respect to the walls of the main chamber is applied to the probes. The plasma radiation via optical fiber is sent to the monochromator MDR – 40 at which output the photomultiplier is located. The signal from the photomultiplier is amplified and registered by computer. Scanning the spectrum is carried out in the range of 400-700nm.

3. Results of the operating regimes of the HPS study

Note that in the above range of external magnetic fields and RF power supply power, the inequality is valid [5]:

$$\omega_{Li} << \omega << \Omega_e << \omega_{Le},$$

where ω_{Li}, ω, Ω_e, ω_{Le} are ion Langmuir, operating, electron cyclotron and electron Langmuir frequency. The range of plasma parameters satisfying the inequality corresponds to an area where resonant excitation of linked helicon and oblique Langmuir waves is possible. Under conditions of resonance the energy input to plasma is optimized. In addition, the penetration of the RF field into the plasma under conditions of resonance provides the possibility to organize extensive area of long and uniform plasma. Below RF inductive discharge placed in an external magnetic field, the induction of which corresponds to the conditions of resonance excitation of helicon and oblique Langmuir waves will be called helicon one

Experimental study of the discharge in the GPS has shown that the imposition of a uniform magnetic field leads to significant changes in the length of the discharge. In the absence of a magnetic field the discharge is concentrated in the top quartz chamber. Increase of the magnetic field strength first leads to a plasma appearance in the upper part of the main steel chamber, then the length of the portion of discharge with the high intensity of radiation in the main chamber begins to grow, and finally, the discharge is closed to the table and form an extended plasma column. Thus, optimization of magnetic field value gives the possibility to increase ion current density near the substrate from 0.5 to 3mA/cm^2 at the RF power generator power 500W. The diameter of the plasma column is approximately equal to the diameter of the top quartz chamber.

Experiments have demnstrated the mutual influence of the magnetron and helicon discharge. Figures 3 and 4 shows the results of radial ion saturation current distribution measurements along the x and y axes, obtained under operation of the helicon discharge and magnetron individually, and under their simultaneous operation.

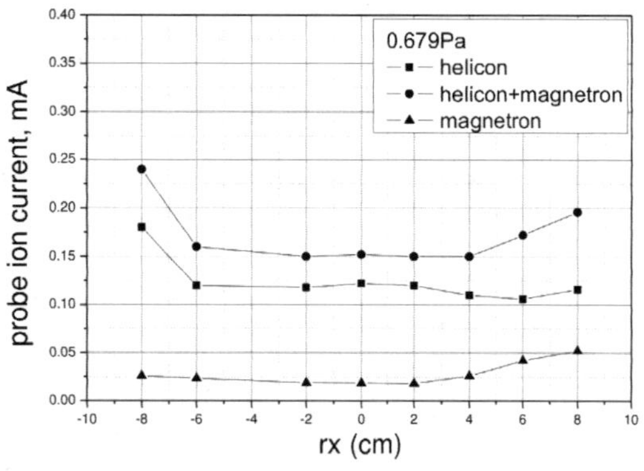

Figure 3. Radial dependence of saturation current along *x* axis

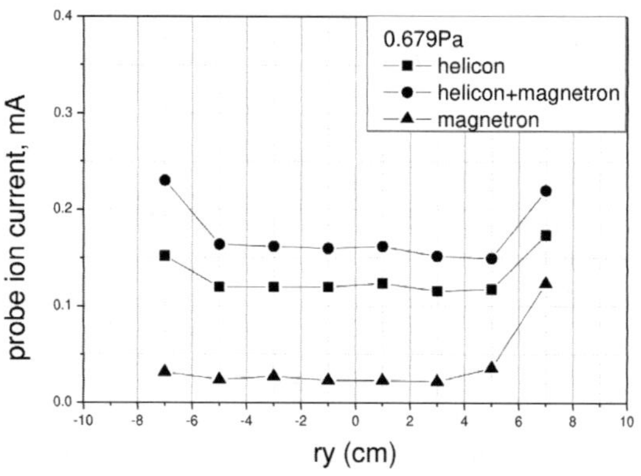

Figure 4. Radial dependence of saturation current along *y* axis

As can be seen, the joint implementation of the two discharges leads to a significant increase in the values of the ionic current. It is higher than the sum of the values measured under separate operation of helicon discharge and magnetron. This conclusion is valid for all the considered experimental conditions. The growth of Ar pressure from 0.3 to 0.7Pa, as experiments have shown, leads to a significant improvement of the radial distribution uniformity of the ion current, however, the absolute value of the ion current falls. A further increase in the pressure to 1 Pa leads to a further drop in the values of the ionic current.

The results presented above indicate the impact of the magnetron discharge on the parameters of the hybrid system. Influence of the helicon plasma source on the magnetron with a titanium target operation can be seen on the basis of the plasma spectral studies. Figure 3 shows a part of the plasma radiation spectrum, where the intensive TiI spectral lines are localized, measured at a pressure 0.3Pa in the case of the separate magnetron operation and joint helicon and magnetron operation.

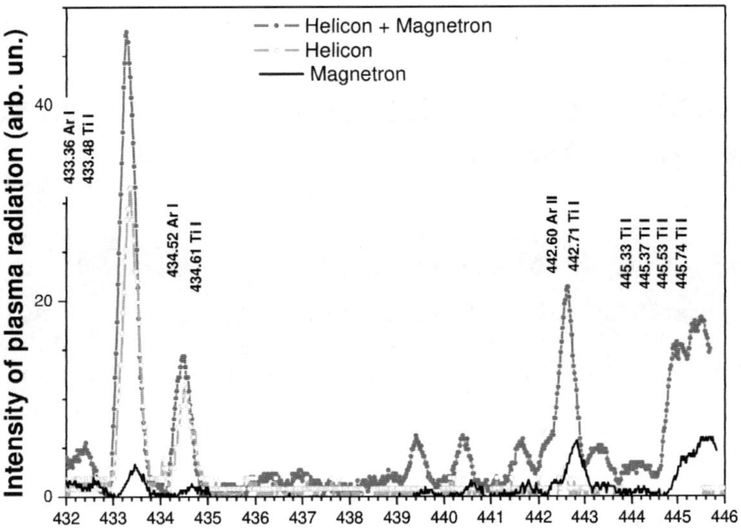

Figure 3. Part of the radiation spectra with high intensity TiI lines.

Experiments have shown that under conditions of separate magnetron operation The TiI spectral lines can be hardly identified due to the small number of Ti atoms in the discharge. However, under joint operation of the devices the intensity of TiI spectral increase, indicating the growth of titanium atoms concentration. In addition it was found that the intensity of the TiI lines in the region of the substrate was slightly less than the intensity in the region of the magnetron location.

Optimal mode of the hybrid plasma system operation has been found to provide uniform ion current density on the diameter of at least 150mm at 0.7PA argon pressure. The optimal values of the magnetic fields in the region of the substrate location lie in the range 2-8 mTl, while in the region of the RF input power unit lie in the range 0.5 – 25 mTl.

References

[1] Donald M. Mattox 2003 *The foundation of vacuum coating technology* N.Y.: Noyes publications.
[2] Zhurin V V 2012 *Industrial ion sources. Broad beam Gridless Ion Source Technology* N.Y.: Wiley-VCH Verlag & Co.
[3] John S. Colligon 2004 Ion-assisted sputter deposition *Phil. Trans. R. Soc. Lond. A* **362**, 103-16.
[4] Winters H et al. 1998 Influence of energy reflected from the target on thin film characteristics *J.Vac.Sci.Technol. A* **11** 657-63.
[5] Alexandrov A.F., Bogdankevich L.S., Rukhadze A.A. *Principles of Plasma Electrodynamics*, Heidelberg: Springer Verlag, 1984.

In situ X-ray diffraction investigation of nitride coatings at high-temperature oxidation

O V Krysina[1,2], N N Koval[1,2], A N Shmakov[3], Z S Vinokurov[4]

[1]Institute of High Current Electronics SB RAS, Tomsk, 634055, Russia

[2]National Research Tomsk State University, Tomsk, 634050, Russia

[3]Boreskov Institute of Catalysis SB RAS, Novosibirsk, 630090, Russia

[4]Novosibirsk State University, Novosibirsk, 630090, Russia

E-mail: krysina_82@mail.ru

Abstract. Structural and phase researches of the multicomponent nanocrystalline coatings synthesized by plasma-assisted vacuum arc method at high-temperature influence by method of X-ray diffraction with the use of synchrotron radiation *in situ* have been carried out. The main features of these coatings are the superhardness (39 – 45 GPa) and nanocrystalline structure (5 – 20 nm). The analysis of results of structural and phase researches, physical and mechanical characteristics after oxidation of multicomponent nitride coatings at high-temperature heating in open air is presented.

1. Introduction

Despite existence of a large amount of materials with various element composition, within technical progress of the modern industry, new materials and coatings with unique service properties are required constantly. Their preparation is one of the actual technical problems today.

It should be noted, that in many cases, the creation of product made of expensive materials is not economically beneficial, and the formation of thin modified layer or a coating with the required physical and mechanical characteristics on their working surface is more preferably. That provides long service life and high service properties of the products.

In this case multicomponent nanocrystalline coatings based on transitional metals with unique characteristics is peculiar such as super- and ultrahardness (40 – 100 GPa) [1, 2], low friction coefficient (≤ 0.2) [3], high degree of elastic recovery (80 – 94%), elastic deformation more than 10%, high tensile strength of 10 - 40 GPa [4], high thermal stability (up to 1700 °C) [5], resistance to oxidation at high temperatures (> 1000 °C) [6, 7] possess great potential opportunities. One of striking examples – multicomponent nitride coatings based on traditional binary TiN system with addition of such elements, as Si, Al, Cu, Cr, etc. in its composition [4].

Multicomponent thin (up to 5 μm) coatings based on titanium nitride with additional elements, such as Ti-Cu-N, Ti-Si-N, Ti-Al-N, Ti-Cr-N [8 – 11] were the objects of researches in this work. These coatings were developed and obtained in Institute of high current electronics SB RAS (Russia, Tomsk) on the modern automated plasma-ion equipment for the vacuum- arc plasma-assisted deposition [11 – 12]. The synthesized coatings possess nanocrystalline structure (d = 5 - 30 nm), high hardness (up to 50

GPa), low friction coefficient (≤ 0.3), high wear resistance (in 2 - 3 times higher, than that of traditional TiN coatings), and etc. It should be noted that such superhard wear-resistant coatings are used for hardening and increase of service life of the cutting and other tools, details and products. And during operation the temperature of a working surface can increases to ≥ 600 °C (it is critical temperature for binary TiN coatings) that in turn can lead to intensive corrosion and destruction, both coatings, and the products surface protected by them. Therefore, the reveal of regularities of nanostructured coatings stability at high-temperature influence gains special relevance, both with practical, and from the fundamental point of view.

The purpose of this work was carrying out *ex situ* researches of resistance of multicomponent coatings based on titanium nitride with various additional elements to oxidation at high-temperature heating in open air by methods of modern materials science (physical and mechanical characteristics, structural and phase composition), and also *in situ* research by methods of X-ray diffraction analysis with the use of synchrotron radiation. And, in particular, that is reveal of critical temperature at which there are an essential structural and phase changes of coatings, and their degradation and destruction.

2. Equipment and procedures

The deposition of thin (3 – 5 μm) multicomponent nanocrystalline coatings based on titanium nitride was carried out on the unique automated vacuum plasma-ion QUINTA installation developed in Institute of high current electronics SB RAS [12] with the following key parameters: dimension of a vacuum chamber: 750 × 600 × 600 mm^3; working pressure: 10^{-1} – 1 Pa; negative bias voltage: up to 1000 V; arc current of gas-plasma source PINK: up to 250 A; arc current of standard arc evaporator: up to 150 A; arc current of gas-plasma PINK-P-0.4M source: up to 150 A; arc current of DI400 arc evaporator: up to 200 A [13 – 15]. The installation is completely automated for the purpose of exact reproducibility of complex technological processes and equipped with five plasma sources – three arc evaporators (sources of metal plasma based on arc discharge with a cathode spot) and two plasma sources based on non-self-sustained arc discharge with filament and hollow cathode (sources of gas-discharge plasma PINK). It allows varying elemental structure of coatings at using few cathodes of different composition or it is essential to reduce the duration of experiments at using few identical cathodes. Besides, existence of gas-discharge plasma sources allows carrying out the process of coating condensation in the mode of plasma assistance. That promotes the formation fine structure (with a size of crystallites up to ~ 1 – 10 nm) dense composite coatings with good adhesion to the substrate.

The composite sintered cathodes of Ti-Cu, Ti-Si, Ti-Al, Ti-Cr systems obtained by the methods of powder metallurgy [16, 17] were used as material of the evaporated cathodes. The specimens made of hard alloy (WC-8%Co) which were polished on diamond paste to roughness R_a = 0.02 μm were used as substrates. The synthesis of multicomponent coatings was carried out in the range of following values of parameters: bias voltage U_b = - (100 ÷ 300) V; working gas pressure p = (0.3 ÷ 0.4) Pa; arc current of evaporator I_d = (50 ÷ 100) A; specimen temperature T = (300 ÷ 400) °C. Thickness of coatings was 3 – 5 μm at coating deposition rate of 1 – 3 μm/h.

It should be noted that at first the coating resistance to oxidation was checked *ex situ* after temperature influence. For this purpose the specimens with coatings were heated in the muffle SNOL-1,6.2,5.1/11-I2 furnace at temperatures of 300, 400, 500 and 800°C in the air atmosphere. The rate of specimen temperature increase was 10 °/min. After endurance of the specimens at a certain temperature within 20 min, these were cooled in furnace volume to room temperature during several hours.

The research of the synthesized coatings before and after thermal influence was carried out by optical microscopy (metallographic microvizor μVizo-MET-221), an optical profilometry (micro-nano-profilometer MNP-1), the scanning electron microscopy and the micro X-ray spectral analysis (scanning electron microscope Philips SEM-515 with the EDAX Genesis XM 260 SEM microanalyzer), the X-ray diffraction analysis (X-ray diffractometer Shimadzu XRD 6000), micro- and nanoindentation (PMT-3 microhardness tester and the Nano Hardness Tester NHT-S-AX-000X). The researches of structural and phase composition were carried out at stations of powder diffraction on channels of the synchrotron radiation (SR) of the VEPP-3 (Siberian Synchrotron and Terahertz Radiation Centre, Budker Institute of Nuclear Physics (BINP), Russia, Novosibirsk) [18] and by the method of transmission electron microscopy (JEOL JEM-2100 F).

In situ experiments on research of coating resistance to oxidation at high-temperature heating in open air were carried out at the "Precision diffractometry and anomalous scattering" located on the 6th channel of synchrotron radiation extraction of the electron store VEPP-3 with the use of the high-temperature X-ray camera of the XRK-900 reactor and the position sensitive parallax-free detector OD-3M350 [19] of production BINP SB RAS. The detector allows to register X-ray pattern in the range of ~ 30 ° (range 2θ = 33.5 ÷ 64.5 ° was chosen here) with discretization ~ 0.01 °, and the record duration of one diffraction pattern is about 3 s. The fixed wavelength of 1.7135 Å was set by single reflection of radiation from a crystal of Ge (111) monochromator, and was determined by a diffraction pattern of α-Al_2O_3 standard (NIST SRM 676). In the required range of angles, the X-ray diffraction pattern of the standard was registered. According that the graduation of the detector was carried out. Further measurements were carried out at a standard position of the detector. The element composition of coatings was investigated by the method of Auger spectroscopy (Auger-electron spectrometer "Schooner-2").

3. Results and discussion

After thermal influence at a certain temperature in open air the condition of specimens was fixed by means of camera and optical microscopy. After thermal influence in air at temperature of ≈ 800 °C all coatings lost the initial color and became dark gray. Besides new formations are revealed on a surface in comparison with an initial surface (Fig. 1). At endurance of the specimens at intermediate temperatures visible changes on a surface were not observed by an optical microscope for all coatings except Ti-Si-N. The oxidation of Ti-Si-N coating surface begins already at temperature of 500 °C. On the other hand, for other coatings the degradation of properties and destruction due to oxidation takes place at higher temperatures. These conclusions are confirmed by the results of surface roughness measurements (Table 1). The values of Ti-Si-N roughness start increasing at T ≈ 500 °C, in the case of TiN and Ti-Cu-N coatings the changes occur only at T ≈ 800 °C.

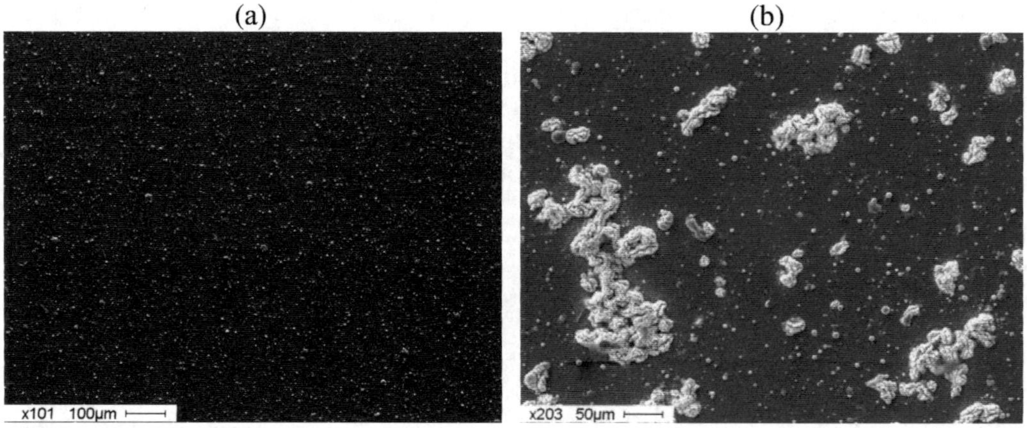

Figure 1. SEM images of specimen with deposited Ti-Cu-N coatings before (a) and after (b) high-temperature oxidation in open air at temperature of 850-850 °C.

Table 1. Microhardness H and roughness R_a of the coatings before and after thermal influence in open air.

T, °C	H, GPa	R_a, μm

	TiN	Ti-Cu-N	TiN	Ti-Cu-N
initial	27.6	44.8	0.4	0.4
300	32.4	43.9	0.4	0.4
400	33.8	46.9	0.5	0.4
500	38.0	48.2	0.5	0.5
800	28.1	20.4	1.0	0.8

In the Table 1 there is hardness of TiN and Ti-Cu-N coatings depending on influence temperature. It is clear that the hardness slightly increases in both cases up to 500 °C, and at achievement of T ≈ 800 °C sharply fall to the substrate hardness. That confirms full coating degradation about this temperature. All above results of researches specify that critical temperature for Ti-Si-N coatings corresponds to temperature ≈ 500 °C; for other coatings – about 800 °C.

The structural researches confirming these assumptions which are carried out by X-ray diffraction analysis are given in Fig. 2. It is clear that besides reflexes from TiN coating and a substrate the set of reflexes, such as TiO_2, WO_3, $CoWO_4$, and some reflexes which identification is complicated because of imposing, appear after thermal influence at temperature ≈ 800 °C on X-ray diffraction pattern.

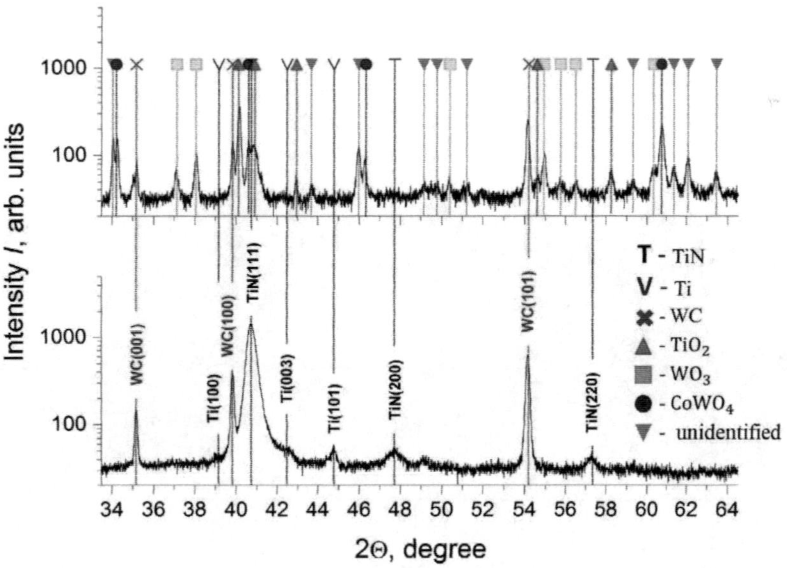

Figure 2. Diffraction pattern of a Ti–Cu–N coating before and after thermal influence in open air up to 850 °C (logarithmic scale).

In situ X-ray diffraction researches of coatings revealed that destruction of nitride phases happens not at a time (Fig. 3, 4; Table 2). The intensity of reflexes from nitride phases at some critical temperature (T_1, usually about 700 °C) starts decreasing, and they completely disappear at higher temperature which is in range of 800-850 °C for coatings based on TiN with various additional elements. Along with the decrease of the intensity of nitride phase reflexes and their disappearance, the emergence of oxides reflexes based on coating elements (for example, TiO_2), and that based on substrate elements is observed (for example, WO_3, $CoWO_4$). And, at achievement of some second critical temperature (T_2) which is higher or it is equal about 800 °C, all reflexes of nitride phases disappear, and only reflexes of oxide compounds and reflexes of substrate (Fig. 4) are observed.

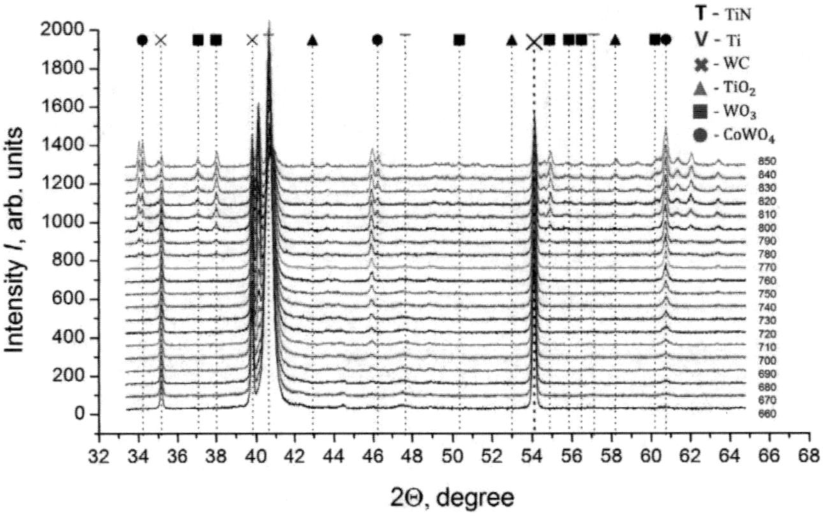

Figure 3. Diffraction pattern of a Ti–Cu–N coating at *in situ* thermal influence in open air up to 850 °C (660-850 °C).

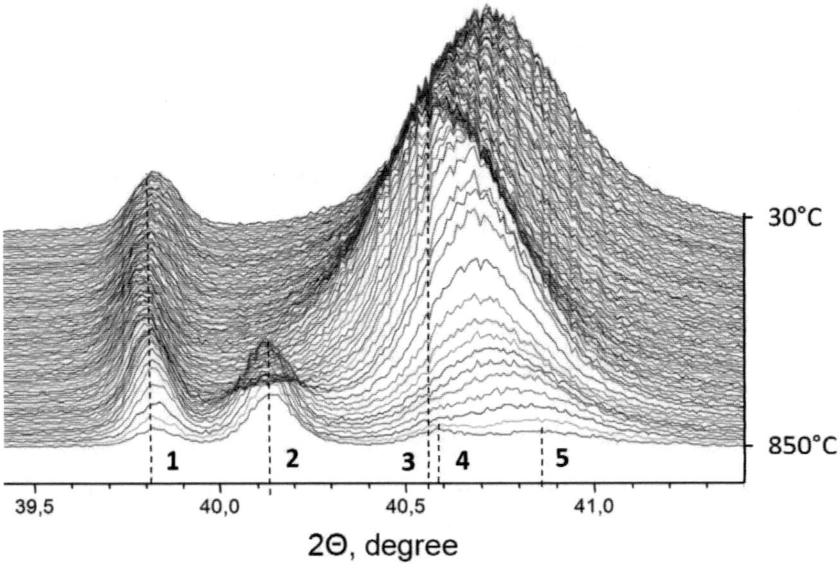

Figure 4. Diffraction pattern (2θ = 39.4 – 41.4 °) of a Ti–Cu–N coating at *in situ* thermal influence in open air up to 850 °C (30 – 850 °C, inverse temperature dependence): 1 – WC; 2 – TiO_2; 3 – TiN; 4 – WO_3; 5 – TiO_2.

The investigations of element composition (by Auger spectroscopy, Fig. 5) of coatings after high-temperature heating in open air completely confirm the conclusions based on the X-ray diffraction *in situ* researches. Namely, at achievement of the second critical temperature (T_2) which for coatings on the basis of TiN with various additional elements is in the range of 800-850 °C (Table 2), there is a full removal of nitrogen from surface layer of coating/substrate system, and the oxygen concentration (≈ 50 at. %) in the layer is caused by formation of oxide compounds based on elements of the destroyed coating and substrate elements.

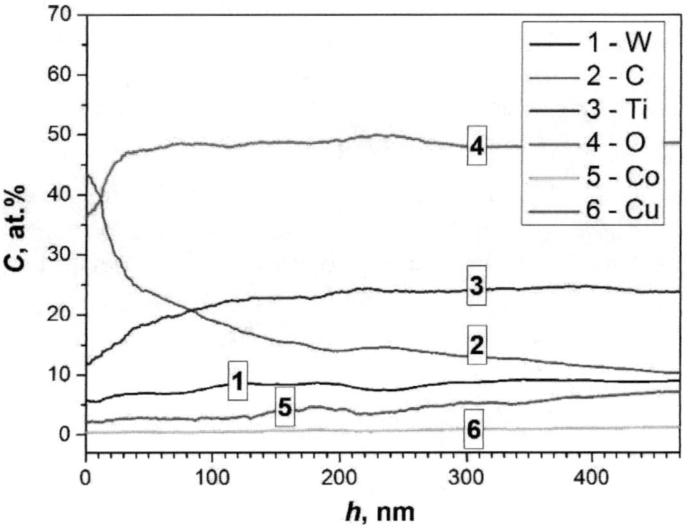

Figure 5. Distribution of element concentration (*C*) in the depth (*h*) of Ti-Cu-N coating deposited on WC-Co substrate after *in situ* high temperature oxidation up to 850 °C (Auger electron spectroscopy)

Table 2. The main quantity changes in phase-structural composition of Ti-Cu-N, Ti-Si-N, Ti-Cr-N, Ti-Al-N coatings.

	Specimen temperature *T*, °C			
	Ti-Cu-N	Ti-Cr-N	Ti-Al-N	Ti-Si-N
Decrease of TiN (111) reflex intensity	750	730	750	760
Disappearance of TiN (111) reflex	840	850	850	820
Disappearance of TiN (200) reflex	740	830	750	730
Disappearance of TiN (220) reflex	310	-	-	850
Appearance of TiO_2 reflexes	720	800	850	720
Appearance of WO_3, $CoWO_4$ reflexes	780	800	820	750

4. Conclusion

The researches and the analysis of the results on thermal oxidation of multicomponent nanocrystalline coatings based on titanium nitride were carried out. *In situ* X-ray diffraction diagnostics of multicomponent coatings at high-temperature heating (up to 850 °C) are revealed that destruction of nitride phases happens not at a time. The intensity of nitride phases reflexes at $T \approx 700$ °C (T_1) starts decreasing, and they completely disappeared at higher temperature (800 – 850 °C). Along with that, the emergence of reflexes from oxide compounds of coating and substrate elements is observed (TiO_2, WO_3, $CoWO_4$, etc.). At achievement of some second critical temperature (T_2) all reflexes from nitride phases disappeared. Thus by the results of the element analysis there is a full removal of nitrogen from surface layer of coating/substrate system, and the concentration of oxygen reaches ≈ 50 at. %.

Acknowledgments
The work was supported in part by the Russian Foundation for Basic Research (Project No. 13-08-98108-r_sibir_a), and with the use of equipment of Siberian Synchrotron and Terahertz Radiation Centre (Budker Institute of Nuclear Physics, Russia, Novosibirsk)

References
[1] Veprek S *et al* 2005 *Thin Solid Films* **476** 1
[2] Musil J 2000 *Surf. Coat. Technol.* **125** 322
[3] Musil J *et al* 2010 *J. Vac. Sci. Technol.* A **28** (2) 244
[4] Cavaleiro A and Hosson J T D 2006 *Nanostructure Coatings* (Springer Science+Business Media, LLC)
[5] Zeman P *et al* 2010 *Thin Solid Films* **519** 306
[6] PalDey S *et al* 2003 *Mater. Sci. Eng.* A **361** 1
[7] Musil J *et al* 2008 *Adv. Appl. Ceram.* **107** 148
[8] Ivanov Yu F *et al* 2012 *Surf. Coat. Technol.* **207** 430
[9] Koval N N *et al* 2007 *Russ. Phys. J.* **50** 146
[10] Krysina O V *et al* 2013 *High Temperature Material Processes* **17** (2–3) 153
[11] Koval N N *et al* 2015 *Russian Journal of General Chemistry* **85** (5) 1326
[12] Kalushevich A A *et al* 2012 *Izvestiya vuzov. Fizika* **55** (12/3) 118
[13] Vintizenko L G *et al* 2001 *Russ. Phys. J.* **44** 927
[14] Shugurov V V *et al* 2010 *Proc. 10th Intern. Conf. on Modification of Materials with Particle Beams and Plasma Flows* (Tomsk, Russia) 49
[15] Koval N N *et al* 2012 *Proceedings of the 25th International Symposium on Discharges and Electrical Insulation in Vacuum* (Tomsk, Russia) **2** 537
[16] Krysina O V *et al* 2013 *High Temperature Material Processes* **17** (2–3) 145
[17] Korosteleva E N *et al* 2009 *Powder Metallurgy and Metal Ceramics* **48** (1) 8
[18] http://ssrc.inp.nsk.su/CKP/beamlines.html
[19] Aulchenko V M *et al* 2009 *Nucl. Instrum. Methods. Phys. Res.* A **603** 76

Radiative heat transfer in plasma of pulsed high pressure caesium discharge

V F Lapshin

Department of Physics, St. Petersburg State Transport University, Moskovsky 9, St. Petersburg, 190031, Russia

E-mail: lapshinvf@mail.ru

Abstract. Two-temperature many component gas dynamic model is used for the analysis of features of radiative heat transfer in pulsed high pressure caesium discharge plasma. It is shown that at a sufficiently high pressure the radial optical thickness of arc column is close to unit (τ_R (λ) ~ 1) in most part of spectrum. In this case radiative heat transfer has not local character. In these conditions the photons which are emitted in any point of plasma volume are absorbed in other point remote from an emission point on considerable distance. As a result, the most part of the electric energy put in the discharge mainly near its axis is almost instantly redistributed on all volume of discharge column. In such discharge radial profiles of temperature are smooth. In case of low pressure, when discharge plasma is optically transparent for own radiation in the most part of a spectrum ($\tau_R(\lambda)$ << 1), the emission of radiation without reabsorption takes place. Radiative heat transfer in plasma has local character and profiles of temperature have considerable gradient.

1. Introduction

Pulsed high pressure caesium (HPC) discharge is of interest as an effective source of visible radiation of high power with a continuous spectrum [1-3]. The investigated discharge is realized in a sapphire tube with an internal radius $R \sim 2\div3$ mm and length ~ 100 mm. Temperature of walls has values $T_W \sim$ 1200-1500 K. In this work, the established mode is considered. In such mode the current pulse of given form $I(t)$ periodically (with a frequency $\nu \sim 1000$ Hz) is passed through the weakly ionized plasma which is cooling down after the previous current pulse or supported by the auxiliary discharge current $I_0 \sim 0{,}1\div1{,}0$ A. The duration of the current pulse has values $t_p \sim 0{,}1\nu^{-1}$ and amplitude $I_{max} \sim$ $10\div100$ A. The amount of caesium per unit length tubes M_a is defined by its saturation pressure at the coldest part of walls. Results of theoretical and experimental study of pulsed HPC discharge are presented in works [4-9].

In the present work radiative heat transfer in plasma of pulsed HPC discharge is considered. The high pressure of plasma and, respectively, high speed of collisional processes in plasma, provide the local thermodynamic equilibrium (LTE) in discharge plasma. It means that Maxwell's distribution for electrons and Saha-Boltzmann equation take place with electron temperature. Let's note that temperatures of electrons and heavy particles can significantly differ. Existence of LTE significantly simplifies consideration of radiation transfer and allows to receive simple expressions for the description of radiative heat transfer in axisymmetric plasma of HPC discharge. For research of HPC discharge two-temperature many component gas dynamic model [5] is used.

2. Basic equations

In the absence of radiation scattering radiative transfer equation for the LTE plasma along the set direction $\vec{\Omega}$ takes the form:

$$\frac{\partial}{\partial s} I_\lambda = k'_\lambda \left(I_{\lambda P} - I_\lambda \right) \tag{1}$$

Here $I_\lambda = I_\lambda(r, \vec{\Omega})$ is the spectral intensity of radiation with wavelength λ, r is the radial coordinate, s is the coordinate along a light beam, k'_λ is the absorption coefficient of caesium plasma [5], $I_{\lambda P} = 2hc^2\lambda^{-5}[\exp(hc/\lambda k_B T_e)-1]^{-1}$ is the Planck's black body intensity, $T_e = T_e(r)$ is the electron temperature, $\vec{\Omega}(\psi, \theta)$ is the unit length vector in the direction of radiation propagation (see figure 1).

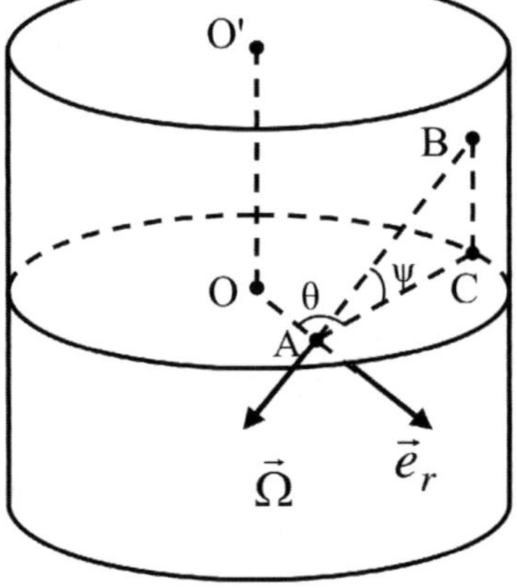

Figure 1. The plasma column geometry.

The integrated solution of the equation (1) has the form

$$I_\lambda(s, \vec{\Omega}) = \int_{s_B}^{s} k'_\lambda(s') I_{\lambda P}(s') \exp\left(-\int_{s'}^{s} k'_\lambda(s'') ds'' \right) ds' . \tag{2}$$

Here s_B is the coordinate of the point B lying on a plasma column surface (see figure 1). To carry out the integration, it is convenient to introduce a new variable $l = (s_A - s')\cos\psi$. It allows to replace the integration along a beam $\vec{\Omega}$ by integration along a projection of this beam to the plane which is perpendicular discharge axes:

$$I_\lambda(r, \psi, \theta) = \int_0^{l_W} k'_\lambda I_{\lambda P} \exp\left(-\int_0^{l} k'_\lambda(l') \frac{dl'}{\cos\psi} \right) \frac{dl}{\cos\psi} . \tag{3}$$

Here integration is carried out along a segment AC in the direction from point A to point C, r = OA, l_W = AC = AB$\cos\psi$, $l_W = l_W(r,\theta) = r\cos\theta + (R^2 - r^2\sin^2\theta)^{1/2}$. Now main spectral characteristics of radiative heat transfer, such as the radial radiation flux F_λ (W/m^3), volume density of energy radiation U_λ (J/m^4) and volume power of energy losses on the radiation W_λ (W/m^4), can be written by means of the equation (3) in a form

$$F_\lambda(r) = \int_{(4\pi)} \left(\vec{\Omega} \cdot \vec{e}_r \right) I_\lambda d\Omega = 4\int_0^{\pi} d\theta \cos\theta \int_0^{\pi/2} d\psi \cos\psi \int_0^{l_W} k'_\lambda I_{\lambda P} \exp\left(-\int_0^{l} k'_\lambda \frac{dl'}{\cos\psi} \right) dl , \tag{4}$$

$$U_\lambda(r)=\frac{1}{c}\int\limits_{(4\pi)} I_\lambda d\Omega=\frac{4}{c}\int\limits_0^\pi d\theta\int\limits_0^{\pi/2}d\psi\int\limits_0^{l_W}k'_\lambda I_{\lambda P}\exp\left(-\int\limits_0^l k'_\lambda\frac{dl'}{\cos\psi}\right)dl\,, \qquad (5)$$

$$W_\lambda(r)=\frac{1}{r}\frac{\partial}{\partial r}(rF_\lambda)=ck'_\lambda(U_{\lambda P}-U_\lambda)\,. \qquad (6)$$

Here $\vec{e}_r$ is the unit length vector in radial direction, $U_{\lambda P}=4\pi I_{\lambda P}/c$ is the equilibrium density of radiation energy. Let's note here that the right part of (6) is the difference between the emission and absorption at a given point. Full losses of energy W_{rad} on radiation can be calculated integrating equation (6) over the area of wavelengths 100 nm - 4000 nm where the radiation contributes significantly to energy balance of the discharge.

To demonstrate features of radiative heat transfer in HPC caesium discharge separate components of W_{rad} corresponding to losses of energy in different parts of a spectrum are calculated:

$$W_{rad}(r)=\int W_\lambda d\lambda=W_1+W_2+W_{nonl}=\int\limits_{\tau_R(\lambda)<0,3}W_\lambda d\lambda+\int\limits_{\tau_R(\lambda)>3}W_\lambda d\lambda+\int\limits_{0,3<\tau_R(\lambda)<3}W_\lambda d\lambda\,. \qquad (7)$$

Here $\tau_R(\lambda)=\int\limits_0^R k'_\lambda dr$ is the radial optical thickness of the plasma discharge column, values W_1, W_2, W_{nonl} designate energy losses on radiation in those parts of a spectrum for which $\tau_R(\lambda)< 0,3$, $\tau_R(\lambda) > 3$ and $0,3 < \tau_R(\lambda) < 3$ respectively. According to the equation (6), the value W_{nonl} we will present in the form:

$$W_{nonl}=W_{nonl}^{(em)}-W_{nonl}^{(abs)}=\int\limits_{0,3<\tau_R(\lambda)<3}ck'_\lambda U_{\lambda P}d\lambda-\int\limits_{0,3<\tau_R(\lambda)<3}ck'_\lambda U_\lambda d\lambda \qquad (8)$$

Here $W_{nonl}^{(em)}$ is the radiation energy emitted from plasma, and $W_{nonl}^{(abs)}$ is the energy absorbed by unit of plasma volume in a unit of time in the spectral range, for which $0,3 < \tau_R(\lambda) < 3$. $W_{nonl}^{(abs)}$ is the energy transferred by radiation to this point of plasma from all other points of plasma volume. This value characterizes not local transfer of energy in discharge plasma. The energy absorbed in the spectral range $0,3 < \tau_R(\lambda) < 3$ by unit of length of arc column in a unit of time, is defined by formulae

$$P_{nonl}^{(abs)}=2\pi\int\limits_0^R rW_{nonl}^{(abs)}dr\,. \qquad (9)$$

The ratio of $P_{nonl}^{(abs)}$ to electric power $P_J=IE$ put per unit length of arc (E is the longitudinal electric field strength in plasma), characterizes a role of not local radiative heat transfer in the discharge.

3. Results of calculations and discussion

The analysis of features of radiative heat transfer in plasma of HPC discharge is carried out for two modes of operation: the mode A (M_a = 0,007 mg/cm, ν = 1350 Hz, t_p = 35 μs, I_{max} = 70 A) and the mode B (M_a = 0,06 mg/cm, ν = 1000 Hz, t_p = 62,5 μs, I_{max} = 120 A). Calculations are carried out for triangular shape of current pulse $I(t)=I_0+(I_{max}-I_0)t/t_p$, $0 < t < t_p$, at R = 2,5 mm, I_0 = 0,2 A. Results of calculations are given in figures 2-6 for t/t_p = 0,5 (it corresponds to the middle of current pulse). As seen in figure 2, in mode A discharge plasma is optically transparent for own radiation ($\tau_R(\lambda) \ll 1$) and in mode B $\tau_R(\lambda) \sim 1$ in large part of spectrum.

Results of calculations of power of Joule heating σE^2 and energy losses because of radiation (W_{rad}), electron heat conductivity (div$\mathbf{q}_e$) and collisions with ions (Q_{ei}) and neutral atoms (Q_{ea}) are shown in

Journal of Physics: Conference Series **669** (2016) 012035 doi:10.1088/1742-6596/669/1/012035

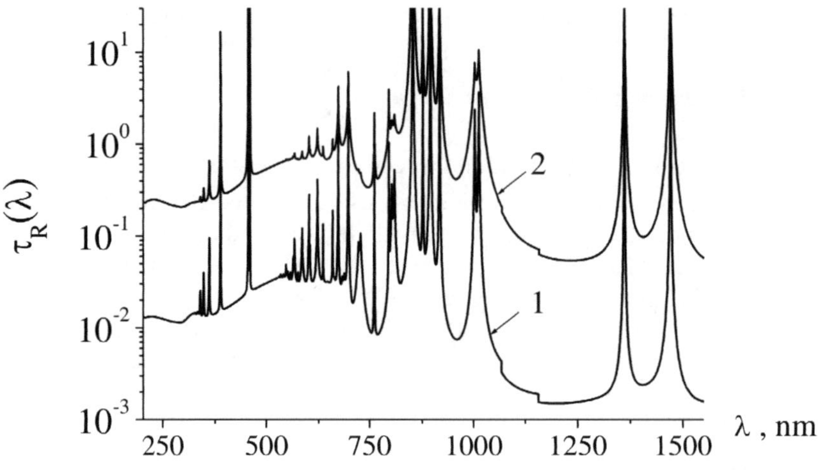

Figure 2. Radial optical thickness of the plasma discharge column $\tau_R(\lambda)$: 1 - the mode A ($p(t)$ = 90 Torr), 2 - the mode B ($p(t)$ = 680 Torr).

figure 3. Here σ is the electrical conductivity of plasma, $Q_{eh} = \frac{3m_e}{m_a}\frac{n_e}{\tau_{eh}} k_B (T_e - T_h)$, τ_{eh}^{-1} is the effective frequency of collisions of electrons with heavy particles, $q_e = -\lambda_e \frac{\partial T_e}{\partial r}$, λ_e is the thermalconductivity of electrons, n_e is the concentration of electrons, m_e and m_a are the mass of electrons and atoms of caesium, T_h is the temperature of heavy particles. As shown in figure 3, electric energy is put in the discharge, mainly, to the hottest central region where electrical conductivity of plasma is maximum. At a high pressure (figure 3b) the most part of the electric energy enclosed in the discharge is radiated by plasma. On the periphery of an arc, in its coldest part, $W_{rad} < 0$.

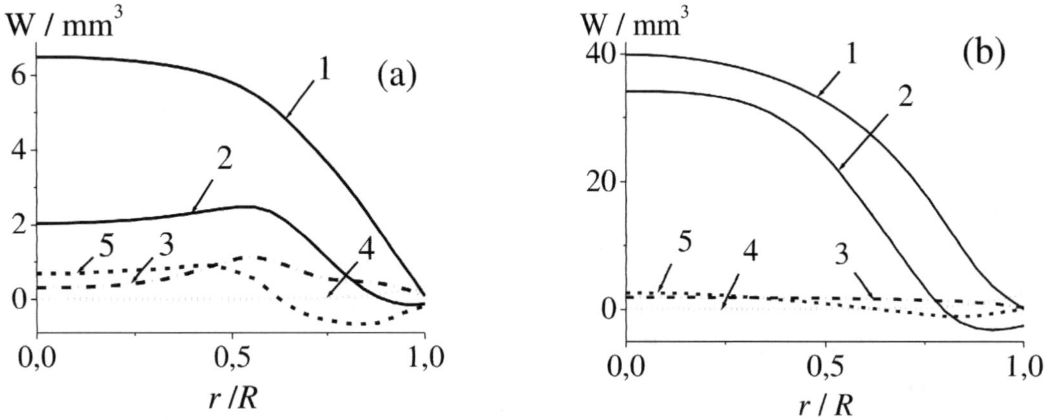

Figure 3. The various mechanisms of energy exchange in plasma: 1 - σE^2, 2 - W_{rad} , 3 - Q_{ei}, 4 -Q_{ea} , 5 - div$\mathbf{q}_e$. Discharge parameters: (a) – mode A, (b) – mode B.

It means that in this area of discharge the plasma heats through absorption of radiation. Let's note that other mechanisms of loss of electron energy play a noticeable role only in cold area near the wall. At the relatively low pressure, when $\tau_R(\lambda) << 1$, loss of energy W_{rad} (see figure 3a) is significantly less. In this case energy losses through collision mechanism and heat conductivity play significantly large role.

Losses of energy on radiation in various parts of a spectrum are compared in figure 4. As is seen from figure 4a, at the relatively low pressure (mode A) the main contribution to radiation losses of energy W_{rad} introduce the quantity W_1 corresponding to emission without reabsorption. In this case heat transfer in plasma has local character. The part of energy absorbed in plasma due to not local heat transfer is small. The ratio $P_{nonl}^{(abs)} / P_J$ is equal 0,17 only. In this case, heating of plasma is non-uniform and temperature profiles have considerable gradients (see figure 5).

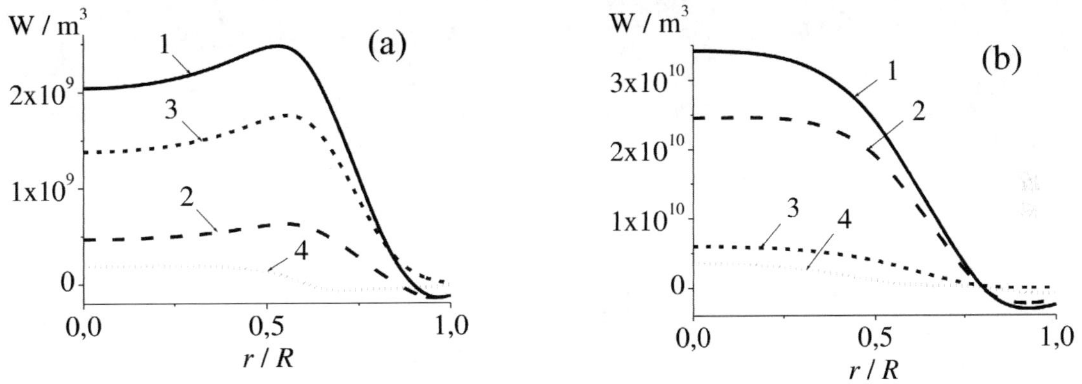

Figure 4. Radial distribution of energy losses to radiation in various parts of a spectrum: 1 $-W_{rad}$, 2 – W_{nonl}, 3 – W_1 , 4 – W_2 . Discharge parameters: (a) – mode A, (b) – mode B.

At a sufficiently high pressure (mode B) radiation losses of energy W_{nonl} in spectral area for which $0,3 < \tau_R(\lambda) < 3$ has a dominant role (see figure 4b). As the length of photon free path in this spectral area is comparable with sizes of the plasma column, radiative heat transfer has in the mode B not local character. In this case, points, in which the photon is radiated and absorbed, may be retired one from the other. In the considered mode the ratio $P_{nonl}^{(abs)} / P_J$ is equal 0,76. Thus, the most part of the electric energy put in the discharge mainly near its axis thanks to not local radiation heat exchange is almost instantly redistributed on all volume of a discharge column. As a result, radial profiles of temperature are smooth. Formation of steep temperature fronts in these conditions is impossible.

The contribution in heating of plasma through reabsorption of radiation from spectral area for which $\tau_R(\lambda) \sim 1$ is shown in figure 6 for the mode B. As illustrated in figure 6, in external, colder part of an arc, reaches the maximum value of $W_{nonl}^{(abs)}$. It means that not local heat transfer plays the same role in heating of electrons here, as well as Joule heating.

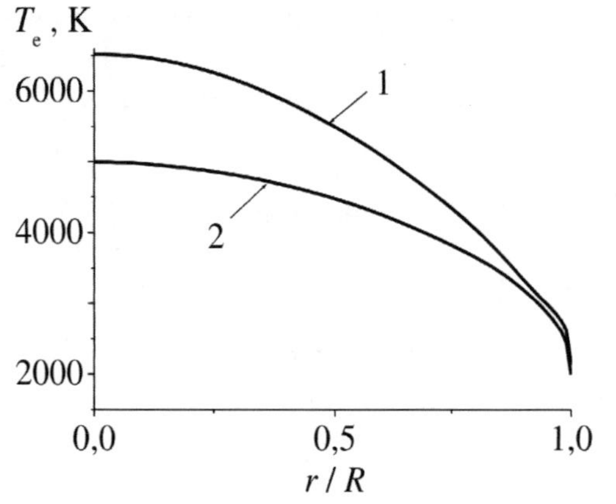

Figure 5. Radial temperature profile $T_e(r)$: 1 – mode A, 2 – mode B.

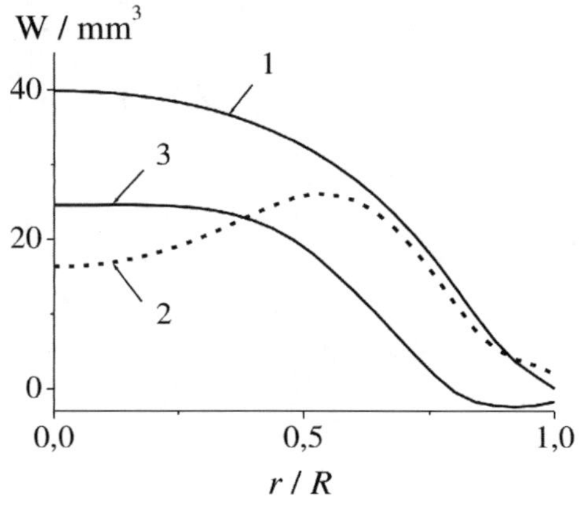

Figure 6. The various mechanisms of plasma heating for mode B: 1 – σE^2, 2 – W_{nonl}, 3 – $W_{nonl}^{(abs)}$.

4. Conclusion

In the present work calculation of radiative heat transfer in plasma pulsed high pressure caesium discharge is carried out. It is shown that at a sufficiently high pressure the radial optical thickness of a arc column $\tau_R(\lambda) \sim 1$ for considerable part of a spectrum. In this case not local radiation heat transfer takes place in the discharge. In these conditions the electric energy put in the discharge is almost instantly redistributed on all volume of plasma. In other cases, when $\tau_R(\lambda) << 1$ or $\tau_R(\lambda) >> 1$, radiation heat exchange has local character.

References

[1] Baksht F G and Lapshin V F 1997 Radiation characteristics of a dense caesium plasma in the visible range *Tech. Phys. Lett.* **23** 954-7

[2] Gu H, Muzeroll M E, Chamberlain J C and Maya J 2001 Pulse modulated high-pressure caesium discharge lamp *Plasma Sources Sci. Technol.* **10** 1-9

[3] Baksht F G and Lapshin V F 2002 The pulse-periodic discharge in caesium as an effective light source *Tech. Phys.* **47** 894-99

[4] Pichler G, Živčec V, Benc R., Mrzljak Ž, Ban T, Skenderović H, Gunther K, Liu J 2003 UV, Visible and IR Spectrum of the Cs High Pressure Lamp *Phys. Scr.* **T105** 98-100

[5] Baksht F G and Lapshin V F 2008 Two-temperature modelling of pulsed high pressure caesium radiative discharge with recombination mechanism of light emission *J. Phys.D: Appl. Phys.* **41** 205201

[6] Baksht F G and Lapshin V F 2009 Formation of the continuous visible emission spectrum of pulsed discharge in cesium vapor *Tech. Phys. Lett.* **35** 1078-1080

[7] Rakić M and Pichler G 2011 Comparison of visible and infrared spectrum of light sources *Opt. Commun.* **284** 2881-2885

[8] Rakić M and Pichler G 2015 Time evolution of the spectrum of the cesium high pressure discharge light source *JQSRT* **151** 169-173

[9] Baksht F G and Lapshin V F 2015 Calculation of the visible radiation fraction in the spectrum of a pulse-periodic discharge in cesium *Tech. Phys. Lett.* **41** 620-623

Acoustic disturbances in a gas with an axial temperature gradient

V M Larionov

Kazan Federal University, 18 Kremlyovskaya str., Kazan 420008, Russian Federation

E-mail: techphys@kpfu.ru

Abstract. Linear analysis acoustic disturbances in one-dimensional gas flow with a longitudinal gradient of the sound speed provided. Known wave equation for the acoustic velocity is used. In the case of linear distribution of the sound speed in the hot part of the flow equation has an exact analytic solution. This allows to define the expression describing the propagation acoustic disturbances in a gas with varying mean temperature and density. The results can be used to calculate the resonance frequencies of the gas oscillations in the laboratory and industrial combustors.

1. Introduction

It was shown [1,2], that the resonant frequency of the oscillation in the gas distribution channels are dependent on the mean temperature, density, and sound speed. It is known [3,4] that the boundaries of combustion stability, the amplitude of steady acoustic oscillations depend significantly on the frequency. Expressions describing acoustic disturbances in the gas at a constant mean temperature, density and speed of sound, known and widely used [5-7]. If the mean temperature of the gas is distributed along the channel, the wave equation becomes non-linear and its analytic solution - complex problem.

The purpose of this work - linear analysis acoustic disturbances propagating in the channel in the case of longitudinal linear distribution of the sound speed.

2. Wave equation and its solution

The wave equation for the acoustic velocity in the one-dimensional flow of an ideal gas with low Mach number is known [8] and is given by:

$$\frac{\partial^2 u'}{\partial t^2} - c^2(x)\frac{\partial^2 u'}{\partial x^2} = 0 \tag{1}$$

where u' - acoustic velocity c - sound speed.

In general, this equation can be solved by approximate or numerical methods. Consider the case where the mean temperature of the gas is reduced so that distribution of the sound speed is linear, i.e.:

$$c(x) = a - bx \tag{2}$$

The solution is sought in the form $u' = F_u(x)\exp(i\omega t)$. After substituting in the expression (1) we obtain equation:

$$\frac{d^2F_u}{dx^2}+\left(\frac{\omega}{c}\right)^2 F_u=0 \tag{3}$$

where ω - cyclic frequency.

We introduce a new variable:

$$y_u=\int\frac{\omega dx}{c}$$

After the conversion equation (3) takes the form:

$$\frac{d^2F_u}{dy_u^2}+\frac{b}{\omega}\cdot\frac{dF_u}{dy_u}+F_u=0$$

The solution is sought in the form $F_u = F^*\exp(\alpha' y)$. Defining values α', obtain:

$$F_u(y_u)=e^{-\frac{by_u}{2\omega}}\left(F_1^*e^{i\beta y_u}+F_2^*e^{-i\beta y_u}\right),$$

$$\beta=\sqrt{1-\left(\frac{b}{2\omega}\right)^2}.$$

Going back to the original variable, believing $2F_1^*=C\exp i\phi$, $2F_2^*=C\exp(-i\phi)$, we have:

$$u'(x,t)=\exp\left(-\frac{b}{2}\int\frac{dx}{c}\right)C\cos\left[\left(\omega\beta\int\frac{dx}{c}\right)+\phi\right]\exp(i\omega t) \tag{4}$$

For distribution (2):

$$\int\frac{dx}{c}=\int\frac{dx}{a-bx}=-\frac{1}{b}\ln(a-bx)+const$$

If the sound speed gradient is absent, acoustic velocity is known [9] and has the form:

$$u'(x,t)=C\cos(\omega x/a+\phi)\exp(i\omega t) \tag{5}$$

Analysis showed that the expression (4) has the form (5), if $b=0$, $\beta=1$, $const=(1/b)\ln a$. Then:

$$u'(x,t)=Ce^{i\omega t}\left(1-\frac{bx}{a}\right)^{1/2}\cos\left[\phi-\frac{\omega\beta}{b}\ln\left(1-\frac{bx}{a}\right)\right] \tag{6}$$

Acoustic pressure can be determined from the linearized continuity equation:

$$p'(x,t)=-\rho_0c^2\int\frac{\partial u'}{\partial x}dt=-\frac{\rho_0c^2}{i\omega}\cdot\frac{dF_u}{dx}e^{i\omega t}$$

where ρ_0 – the mean gas density.

Finally, we have:

$$p'(x,t)=-i\rho_0cCe^{i\omega t}\left(1-\frac{bx}{a}\right)^{1/2}\left\{\frac{b}{2\omega}\cos\left[\phi-\right.\right.$$
$$\left.\left.-\frac{\omega\beta}{b}\ln\left(1-\frac{bx}{a}\right)\right]+\beta\sin\left[\phi-\frac{\omega\beta}{b}\ln\left(1-\frac{bx}{a}\right)\right]\right\}. \tag{7}$$

In excretions (6), (7) phase angle ϕ depends on the boundary conditions at the ends of the channel.

3. Conclusion

The simplest approach - linear distribution of the speed of sound was used. An exact analytical solution of the wave equation for acoustic velocity was obtained. It was possible to find the expressions describing the velocity and pressure perturbations in the gas flow. The results of analysis allow calculating the resonance frequency of the gas in different systems with heat sources such as industrial combustors.

The work is performed according to the Russian Government Program of Competitive Growth of Kazan Federal University.

References

[1] Ermakov R A, Sadykov A F, Galiullin R G, Larionov V M 2011 *Chemical and Petroleum Engineering* **46** 749-751
[2] Larionov V M, Iovleva O V 2003 *Izvestiya Vysshikh Uchebnykh Zavedenij. Aviatsionnaya Tekhnika* (4) 48-51
[3] Beloded O V, Larionov V M 2006 *Russian Aeronautics* **49** 44-48
[4] Iovleva O V, Larionov V M 2007 *Russian Aeronautics* **50** 303-308
[5] Fadeev S A, Kashapov N F, Larionov V M 2014 *IOP Conf. Ser.: Materials Science and Engineering* **69** 012006
[6] Semenova E V, Larionov V M, Kazakova E I 2014 *J. Phys.: Conf. Ser.* **567** 012033
[7] Ermakov R A, Galiullin R G, Larionov V M, Nikolaev A N 2008 *Chemical and Petroleum Engineering* **44** 387-392
[8] Larionov V M, Nazarenko T I 2000 *Izvestiya Vysshikh Uchebnykh Zavedenij. Aviatsionnaya Tekhnika* (4) 69
[9] Larionov V M, Nazarenko T I, Filipov S E 2004 *Izvestiya Vysshikh Uchebnykh Zavedenij. Aviatsionnaya Tekhnika* (1) 32-34

Journal of Physics: Conference Series **669** (2016) 012037 doi:10.1088/1742-6596/669/1/012037

Optimization of hydrocarbon fuels combustion variable composition in thermal power plants

E R Saifullin[1], V M Larionov[1], A V Busarov[2], V V Busarov[2]

[1]Kazan Federal University, 18 Kremlyovskaya str., Kazan 420008, Russian Federation
[2]Limited Liability Company «Center of industrial safety expertise», 90a Julius Fucik str., Kazan 420140, Russian Federation

E-mail: mr.emilsr@gmail.com, techphys@kpfu.ru

Abstract. It is known that associated petroleum gas and refinery waste can be used as fuel in thermal power plants. However, random changes in the composition of such fuels cause instability of the combustion process. This article explores the burning of hydrocarbon fuel in the case of long continuous change of its specific heat of combustion. The results of analysis were used to develop a technique of optimizing the combustion process, ensuring complete combustion of the fuel and its minimum flow.

1. Introduction

It is known [1] that the optimization of the associated petroleum gas (APG below) and wastes of petrochemical plants combustion [2,3] having an unstable structure, that a complex problem. This is due to the fact that changes in the composition and the specific heat of fuel combustion have a random character. This causes instability of the thermal regime of power plants, such as boilers and other devices [4].

The aim of this work – is development of a technique to optimize APG combustion in the boilers in a case of long continuous changes in the specific heat of combustion.

It is assumed that initially the specific heat of combustion of fuel is constant, burning - optimal, that is, the fuel burns completely, and the flow rate of fuel is minimal for a given thermal power production. At some point, the specific heat of combustion of fuel begins to continuously increase or decrease. The dependence of the specific heat of combustion of fuel on the time is represented by a series of small steps. This approach results in [5] can be used.

2. Physical justification of technique

Reducing the output temperature of the coolant at a constant flow of fuel and air means that the reduction of the specific heat of combustion of fuel was began. Since, there was an excess of air is necessary to begin increasing the fuel flow rate, gradually increasing the rate of change. For example, during the first minute of 1% relative to the initial value, the second minute - 2% of the third minute - 3%, etc. The amount of completely burning fuels and emissions of heat increases, which compensates for heat losses due to the decrease of the specific heat of combustion of fuel. When the rate of increase in the fuel flow exceeds the rate of decrease in the specific heat of combustion, growth of output coolant temperature begins. Eventually the outlet temperature of the coolant reaches the initial value. At this point, a complete combustion and optimum stoichiometric air / fuel ratio is achieved. A further increase in fuel flow results in its unburnt [6] and does not change the coolant temperature at the outlet of heat exchanger. It is necessary to stop the fuel flow increase. Because of the continuing decrease in the specific heat of the combustion temperature of the coolant at the outlet of the heat exchanger will start gradually decreasing. As soon as the coolant temperature decrease to a predetermined reference value you need to start again increase the fuel flow. The increase in fuel flow compensates for the reduced rate of heat release. As a result, the coolant temperature is increased again to the initial value. This correction process of the fuel flow periodically repeated during the entire time of reducing the specific heat of combustion of fuel.

If during the period of constant fuel and air flow output value of the coolant temperature does not drop to the reference value, then reduction of the specific heat of combustion completed, and disappeared reason for the decrease in temperature of the coolant at the outlet of the heat exchanger. However, in the future permanence of the coolant temperature at the outlet of the heat exchanger does not provide unambiguous information on the combustion process. This result corresponds to a constant calorific value and the case where it begins to rise.

Increased specific heat of combustion at constant flow of fuel and air leads to incomplete combustion of fuel due to a shortage of air. The increase in heat generation due to the increase of the specific heat of combustion is compensated by its (fuel) underburning, i.e. heat generation and the resulting cooled heat remain constant. As a result, the coolant temperature at the outlet of the heat exchanger is not changed.

It is necessary to reduce the fuel flow to a certain reference value relative the initial value. In the case of constant specific heat of combustion of fuel and constant air flow there is an excess air ($\alpha > 1$). This leads to a decrease in the coolant temperature at the outlet of heat exchanger. In this case, it is necessary return to the previous value of the fuel flow by restoring the optimal combustion process. This control process optimal combustion process it is necessary repeated periodically.

In the second case, after the reduction the fuel flow with increasing calorific value at constant air flow rate of heat increase due to the increase of the specific heat of combustion is compensated by its underburning. The temperature of the coolant at the outlet of the heat exchanger is not changed.

To eliminate the growing underburning it is necessary to start reduction fuel flow, gradually increasing the rate of change. Air flow rate is not changed. Over time, the rate of reduction the fuel flow exceeds the rate of increase of the specific heat of combustion. As a result, the coolant temperature at the outlet of the heat exchanger is reduced because of an excess of air caused by the reduction in the fuel flow [6]. When the temperature of the coolant at the outlet of the heat exchanger drops to a predetermined reference value, it is necessary to stop the reduction of fuel consumption.

Because of the continuing increase in the specific heat of the combustion temperature of the coolant at the outlet of the heat exchanger will rise to the initial value. After that re-appear underburning of fuel and it is necessary to reduce fuel flow rate again, without changing the air flow rate. Further, the periods of constant and reduces fuel consumption follow each other at all times of increase the specific heat.

If during the constant flow of fuel and air temperature at the outlet of the heat exchanger does not reach the initial value, the increase of the specific heat of combustion of fuel is completed. By inertia outlet temperature of the coolant rises to an initial value and remains constant. This means that the specific heat of the combustion was stabilized. Combustion process is stoichiometric.

3. Conclusion

So, the method of optimizing the combustion of APG and other fuels to the long continually changing specific heat of combustion is developed. Her summary of the following:

1. In the case of long continuous decrease of the specific heat of combustion of fuel at a predetermined reference value, it is necessary to start an increase the fuel flow, gradually increasing the rate of change, without changing the air flow rate. If you exceed the rate of increase in fuel consumption over the rate of reducing the specific heat of combustion is the initial value of the coolant temperature, stabilize the fuel flow, and when the temperature of the coolant at the outlet of the heat

exchanger stops declining, while remaining within the accuracy of measurement, complete the optimization of combustion process.

2. The increase in the specific heat of combustion of fuel is determined by periodic control reducing of the fuel flow. When you fixate a long continuous increase of the specific heat of combustion of fuel at a predetermined reference value, it is necessary start to reduce fuel flow rate, gradually increasing the rate of change, without changing the air flow rate. When the temperature of the coolant at the outlet of the heat exchanger is reduced because of an excess of air it is necessary to stop to reduce fuel consumption. If the coolant temperature at the outlet of the heat exchanger reaches the initial value it is necessary to begin to lower fuel flow again. If the coolant temperature does not reach the initial value it is necessary to complete the optimization process.

The work is performed according to the Russian Government Program of Competitive Growth of Kazan Federal University.

References

[1] Røland T H 2010 *Associated petroleum gas in Russia: reasons for non-utilization* (Fridtjof Nansen Institute)

[2] Semenova E V, Larionov V M, Kazakova E I 2014 *J. Phys.: Conf. Ser.* **567** 012033

[3] Akhmadullin A N, Ahmethanov E N, Iovleva O V, Mitrofanov G 2013 *J. Phys.: Conf. Ser.* **479** 012004

[4] Ermakov R A, Sadykov A F, Galiullin R G, Larionov V M 2011 *Chemical and Petroleum Engineering* **46** 749-751

[5] Saifullin E R, Vankov Yu V 2015 *IOP Conf. Ser.: Materials Science and Engineering* **86** 012006

[6] G. F. Knorre, K. M. Aref'ev, A G Blokh 1966 *Russia B. Theory of combustor processes* 1042

Journal of Physics: Conference Series **669** (2016) 012038 doi:10.1088/1742-6596/669/1/012038

External high-frequency control of combustion instability

V M Larionov[1], G A Mitrofanov[1] and A N Kozar[2]

[1]Kazan Federal University, 18 Kremlyovskaya str., Kazan 420008, Russian Federation
[2]Kazan cooperative Institute Russian University of cooperation, 58 Nikolaya Ershova, Kazan 420081, Russian Federation

E-mail: techphyskfu@gmail.com

Abstract. The article presents the results of experimental studies of combustion instability in the pulse combustor. Propane-air mixture is burned in the chamber with the flame holder. It was experimentally found that feeding high-frequency sound vibrations into the combustion chamber causes the suppression of pulsating combustion. The oscillation frequency ranges in 870 to 1400 Hz. This corresponds to 9-12 resonance frequencies of oscillations in the combustor. The physical mechanism of the observed phenomenon consists in changing the conditions of formation and destruction of fuel jets in the vortex zone behind the flame holder.

1. Introduction

It was noted earlier that the instability of combustion in power plants with combustion is accompanied by fluctuations of pressure, temperature, speed, etc. [1, 2]. The oscillation frequency depends on a number of factors, but the geometrical characteristics of the combustion chamber is of central importance. The use of pulsating combustion in practice can substantially increase the efficiency of combustion processes.

In theoretical and practical studies of the pulsation of the burning tasks of suppression and creation of pressure pulsations remain relevant [3, 4].

To suppress vibrations applied in a variety of ways. For example, an external periodic perturbation [5], steam injection [6] and others. The possibility of using any way to suppress the oscillations depends on the cost of its practical implementation. For the method of control pulsating combustion of an external periodic perturbation is required to create perturbations of the electrical power.

The study of the physical mechanisms of oscillations in combustion chambers has led to the conclusion that the excitation of oscillations and their suppression may be carried out using the same methods [7]. Thus, the control of combustion instability is a more general term that combines both tasks.

The purpose of the study was to determine the possibility of practical use of external acoustic perturbations of the high-frequency range to control the instability of combustion.

2. The method of investigation

Experimental investigation of control of combustion instability by the method of external periodic perturbations were carried out in the chamber pulsating combustion. Scheme of the experimental chamber is presented in Figure 1.

Propane-butane mixture 3 enters the combustion chamber through the tube 8 into the flame stabilizer 9. Air enters through the tube 4 in an air receiver 6. Then the flow passes through the

Journal of Physics: Conference Series **669** (2016) 012038 doi:10.1088/1742-6596/669/1/012038

confuser and the diffuser 7 into the combustion chamber. The mixing of the fuel gas with air takes place at the working edge of the flame stabilizer and vortex formations behind him. The axial displacement of the flame stabilizer 9 in the diffuser 8 enables the preparation of sustained resonant combustion in the combustion chamber. The pulsation frequency of burning depends on the length of the pipe resonator 12. The sound vibrations created by the sound generator 1 electrodynamics 2. They entered the combustion chamber through the gas inlet tube 5. The pressure fluctuations in the combustion chamber were recorded with a microphone 10 and a spectral frequency analyzer 11.

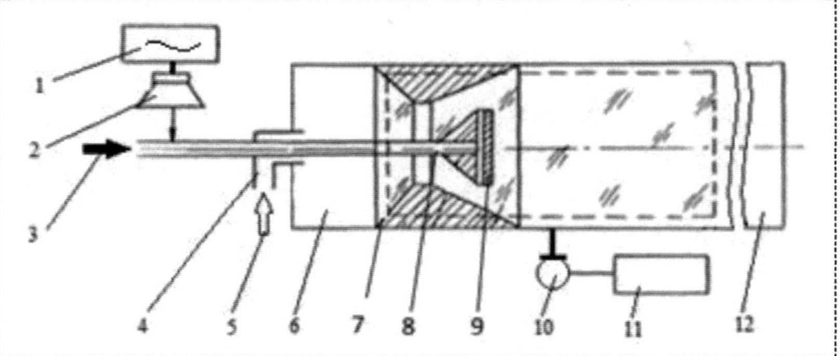

Figure 1. The main elements of the experimental stand.

The result of the influence of external periodic perturbations was assessed by changes in the domains of existence of pulsating combustion in the combustion chamber [7], as well as the change of the sound pressure level of the pulsation of the combustion in the combustion chamber. The boundary of the region of existence of pulsating combustion was built in the coordinate system of the thermal intensity of combustion chamber q (MW/m^2) and the flow rate of air at the working edges of the flame stabilizer v (m/s).

The presence of pulsating combustion in the combustion chamber was determined visually and the spectrum analyzer. The sound pressure level of the pressure oscillations in the combustion chamber depends on the fuel and reaches values of 130 to 150 dB. The research was carried out in the frequency range from 100 to 1400 Hz.

3. The results of the study and discussion

In Figure 2,a is marked with: Lao – source sound pressure level, fa_{IX}, fa_X, fa_{XI} and fa_{XII} – calculation of the resonance mode of self-oscillations in the combustion chamber, L is the sound pressure level in the combustion chamber after exposure. The estimated first resonant mode of the vibrations in the camera for the length of the resonator was equal to 105 Hz. Figure 2,b presents the sound pressure levels of the external sound disturbance for the relevant frequencies L_{ei} measured in the combustion chamber. In Figure 2,c is a graph showing changes of electric power at electrical dynamics spent to generate acoustic oscillations external disturbances.

The graphs are observed intervals reduce the level of sound pressure oscillations in the combustion chamber in the frequency range from 880 up to 980 Hz, from 1030 to 1050 Hz and from 1150 to 1300 Hz. In this case, the resulting sound pressure level reaches levels from 123 to 127 dB.

The sound pressure level of external disturbances in the combustion chamber is changed from 83 to 105 dB. It depends on the frequency of disturbance.

Diagram of electric power on the electrodynamics spent on the excitation of disturbances depends on the disturbance frequency. The minimum value of the electric capacity range from 3 to 5 W in the first and from 5 to 12 W in the third frequency intervals of exposure. In the second the frequency value of electrical power is much higher and ranges from 50 to 75 W.

It should be noted that the study of the reaction of the combustion chamber to external stimuli were carried out in a wide enough range of frequency increments of 10 Hz. The first effects of the damping of the pulsation in the combustion chamber was observed in the frequency range from 30 to 280 Hz. This corresponded to the frequency range from the first to third resonance modes of vibrations of the

Journal of Physics: Conference Series **669** (2016) 012038 doi:10.1088/1742-6596/669/1/012038

camera. The frequency variation from 290 to 870 Hz did not provide the effect of reducing the sound pressure level in the combustion chamber.

The effect of suppressing self-oscillation is observed only in the frequency range of the external perturbations from 880 to 1300 Hz. Thus, we can assume that in the high-frequency impacts the physical mechanism of suppression of self-oscillations differs from the physical mechanism of suppression of low-frequency oscillations under the influence.

A graph of the areas of existence of pulsating combustion is presented in Figure 3.

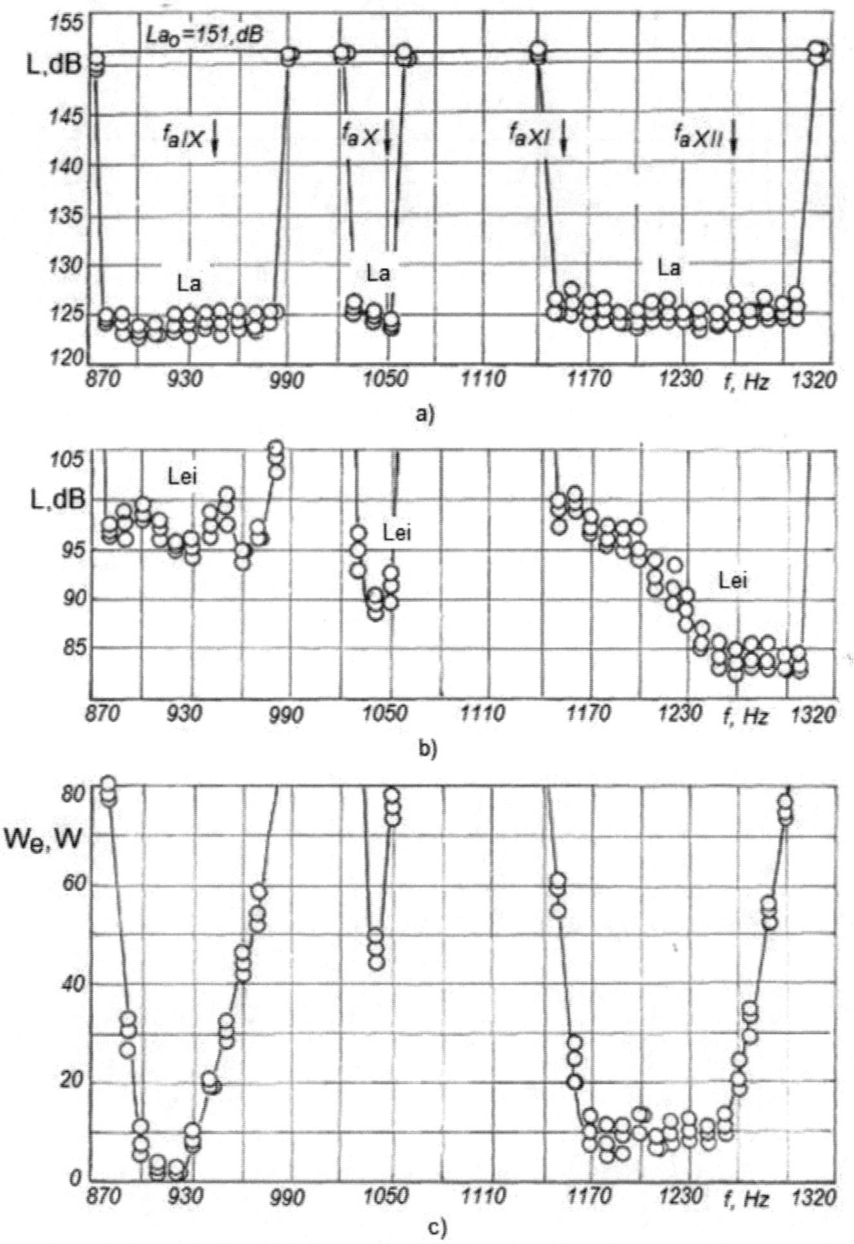

Figure 2. Graphs of sound pressure level in the combustion chamber after exposure (a); the sound pressure level of the external acoustic perturbation, as measured in the combustion chamber (b); diagram of electric power on the electrical dynamics (c).

Unlike the changes of the areas of existence of pulsating combustion with low frequency external influences [7] the graph is compressed in the left margin. Noted feature proves suggested another mechanism of damping of self-oscillations in the combustion chamber during high-frequency effects.

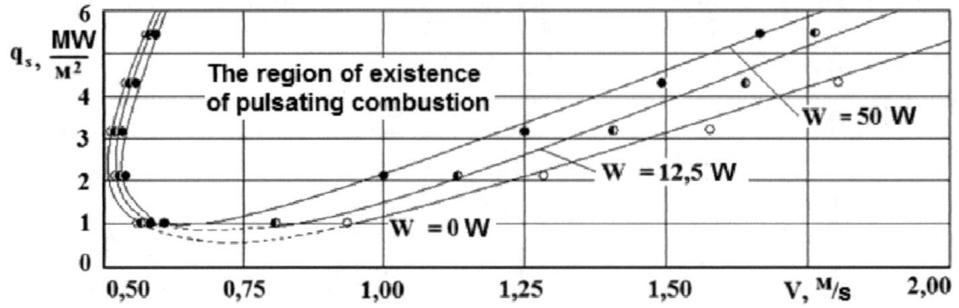

Figure 3. Modify areas of existence of pulsating combustion in high-frequency sound effects.

4. Conclusion

The physical mechanism of the observed phenomenon consists in changing the conditions of formation and destruction of fuel jets in the vortex zone behind the flame holder.

The work is performed according to the Russian Government Program of Competitive Growth of Kazan Federal University.

References

[1] Raushenbah B V 1961 *Vibrazionoe gorenie* Russia, Moscow 500
[2] Semenova E V, Larionov V M, Kazakova E I 2014 *J. Phys.: Conf. Ser.* **567** 012033
[3] Ermakov R A, Sadykov A F, Galiullin R G, Larionov V M 2011 *Chemical and Petroleum Engineering* **46** 749-751
[4] Beloded O V, Larionov V M 2006 *Russian Aeronautics* **49** 44-48
[5] Bykovets A P, Larionov V M, Marchukov E Yu 1992 *Izvestiya Vysshikh Uchebnykh Zavedenij. Aviatsionnaya Tekhnika* (3) 71-74
[6] Larionov V M, Mitrofanov G A, Iovleva O V 2014 *J. Phys.: Conf. Ser.* **567** 012028
[7] Akhmadullin A N, Ahmethanov E N, Iovleva O V, Mitrofanov G A 2013 *J. Phys.: Conf. Ser.* **479** 012004

Journal of Physics: Conference Series **669** (2016) 012039 doi:10.1088/1742-6596/669/1/012039

Features of formation concentration profile in structured materials

T S Pavlenko[1], A A Lisenkov[1] and N A Babinov[1]

[1] Department of electronic instruments and devices, Saint Petersburg State Electrotechnical University them. Ulyanov (Lenin) "LETI", St. Petersburg, 197376, Russia

Abstract. The paper presents analysis of dependence of the penetration depth of the implanted aluminium ions in structured titanium. Ion implantation was performed repetitively pulsed particle beam ion source "MEVVA-V.RU". In the interpretation of the observed patterns of energy accounted for heterogeneous composition of the beam vacuum-arc source, is represented by three components. Within the simulation found that in samples with relatively fine grains (ultrafine samples) largely contribute to diffusion processes, in particular radiation-induced diffusion in comparison with fine-grained samples.

1. Introduction

One of the important directions of modifying the physicochemical properties of structural materials is the method of exposure beams of metallic ions. This method allows the synthesis of ion in the surface layers of metals and alloys and, thus, receive the phase of a transition metal and aluminium, which have high mechanical strength, wear resistance, corrosion resistance [1]. The use of titanium in different structural states is lately increasingly used in various fields of science and technology. To date, a number of studies on the effect of ion irradiation on the structural and phase characteristics and other properties of the metal in a coarse state.

However, implantation nanostructure and fine-grained materials is considerable interest in the prospect of wide application. Microscopic processes that occur in the process of ion implantation of titanium in different structural states, little analyzed and investigated. Therefore, studies of the characteristics of the formation of concentration profiles of implanted ions in titanium with different structural states are relevant.

The starting material of the target for the implantation of aluminium ions selected rod of commercially pure titanium VT1-0. For the formation of nano- and submicrocrystalline (SMS) state titanium billet was subjected to repeated uniaxial pressing with the change to the deformation axis at a rate of 10-2 s-1 with decreasing temperature in the range of 773-673 between the pressing cycle. After the stage, the deformation of the titanium specimens were deformed by rolling in multiple brook rolls at room temperature, then to increase the plasticity of the titanium billet-Gigue. The average amount received by the elements in the structure of the titanium after such processing, was 0.2 microns – SMS state. Titanium samples with different grain sizes were made of titanium rods SMS by annealing time at various temperatures. Implantation of titanium materials with aluminium ions was carried out on vacuum-arc ion source "MEVVA-V.RU" in frequency pulsed mode [1]. Terms titanium implant specimens are shown in table 1. Obtained by layering Auger electron spectroscopy concentration profiles of aluminum in titanium samples are shown in figure 1 [1].

2. Model and Results

Ion source "MEVVA-V.RU" maintains the uniformity of the beam and is characterized by a decrease in the flow of ionic contamination.

Table 1. Modes of irradiation of titanium with different structural states of aluminium ion source "MEVVA-V.RU"

Modes	Grain size (microns)	Type of structure	The dose of irradiation (ion/cm^2)	Time implantation (h)	The target temperature (K)	The accelerating voltage (kV)
1	0,3	Ultrafine-grained (UFG)	$1\cdot10^{18}$	5,25	623	50
2	1,5	Fine-grained (FG)				

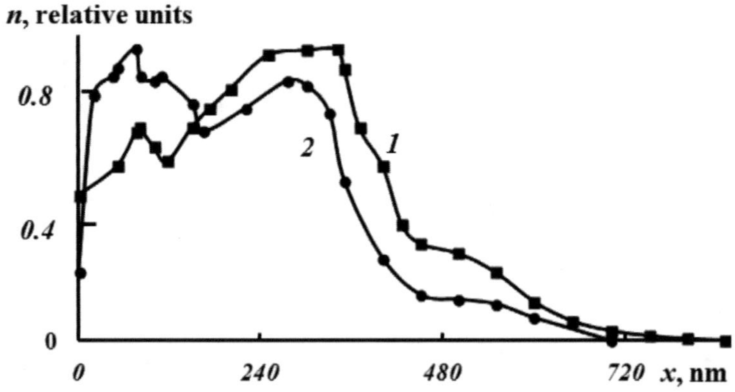

Figure 1. Experimental concentration profiles of aluminum ions in titanium modes: 1 – mode 1; 2 – mode 2

However, the proportion of the aluminium ions in the beam is 85% (balance nitrogen impurity (9.5%), oxygen (4.5%) and hydrogen (1%)), which affects the mass transfer processes taking place during implantation. When using ion sources of vacuum arc ion beam energy is heterogeneous and contains 38% Al +, 51% Al2 +, Al3 + 38% [2]. Consequently, the composition of the energy beam is represented by three components with energies of 50, 100, 150 keV.

To describe the concentration profiles for materials with grain sizes used in any two-stage model. It is assumed that in the first minute implantation (for a small dose of implanted ions and low concentration of generated defects) impurity advantageously distributed statistically (the profile is described using functions Pearson fourth type and sputtering surface of the target). In subsequent times the formation of the profiles contributing to the processes of radiation-stimulated diffusion and heat. The total energy of the inhomogeneous profile for the beam (in this case the three components aluminium) based additive process can be represented as:

$$n(x)=\sum_{i=1}^{3}F_i n_p(x,E_i)+F_i n_d(x,E_i)+F_i n_\tau(x,E_i), \qquad (1)$$

where F_i – the dose of implanted ions, corresponding to the energy E_i; $n_p(x, E_i)$ – describes the statistical distribution of the original; $n_d(x,E_i)$ – takes into account the contribution of radiation-stimulated diffusion; $n_\tau(x,E_i)$ – is the contribution of thermal diffusion.

The following expressions for the $n_p(x, E)$ and $n_d(x,E_i)$, used in the simulation are shown in [3]. In modeling the values of the moments of distributions of ions for each energy components were taken from the program SRIM-2013. The contribution of thermal diffusion was estimated based on their expression:

$$n_{\tau}(x,t)=\int_{0}^{\infty} n_{p}(\xi)G(x,\xi,t)d\xi+\int_{t_0}^{t}\int_{0}^{\infty} f(\xi,t)G(x,\xi,t-\tau)d\tau d\xi, \qquad (2)$$

where $n_p(x)$ profile formed at time t_0; t – the total time of implantation, the function $G(x,\ \xi,\ t)$ for the Green's function of the boundary value problem [3]. The contribution of diffusion processes is taken into account in the framework of radiation-stimulated diffusion [3] for polyenergetic beam.

The share of deposits of statistical and diffusion processes was determined by modeling on the basis of a better agreement between theory and experiment is shown in table 2. The resulting concentration profiles taking into account the contributions of all the mechanisms are shown in figure 2, together with the experimental. Model parameters are given in table 3.

Table 2. Mechanisms of formation of concentration profiles in the implantation of titanium ions aluminum

Modes	Mechanisms of formation profiles, share	
	statistical	diffusion
1	0,33	0,67
2	0,40	0,60

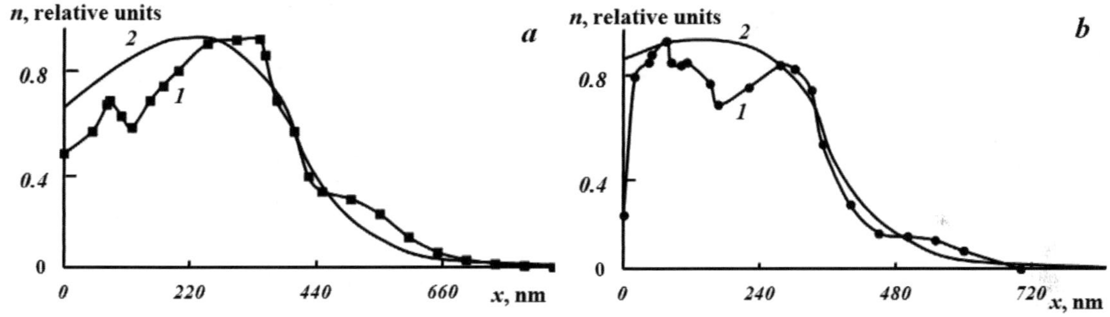

Figure 2. Experimental (curve 1) and theoretical (curve 2) profiles of aluminum ions in titanium modes: a – mode 1, b – mode 2

Table 3. Model parameters to describe the concentration profile of aluminum in titanium

Parameters	Energy (keV)		
	50	100	150
The asymmetry factor S_k	0,15	-0,10	-0,26
The sputtering coefficient S (at./ion)	1,3	1,8	2,0
Gain diffusion due to vacancies d	40	42	44
The diffusion coefficient defects D_d (cm^2/s)	$1{,}8\cdot10^{-14}$	$2{,}0\cdot10^{-14}$	$2{,}1\cdot10^{-14}$
The effective diffusion coefficient of the grain boundaryD_{eff} (cm^2/s)	$2{,}2\cdot10^{-13}$	$2{,}3\cdot10^{-13}$	$2{,}5\cdot10^{-13}$
The diffusion length of defects L_d (cm^2/s)	$3{,}0\cdot10^{-6}$	$3{,}1\cdot10^{-6}$	$3{,}2\cdot10^{-6}$

The modeling found that the share of deposits of diffusion (including thermal diffusion) processes in the formation of the profiles in the UFG titanium slightly higher (about 7-10%) than in the FG titanium, which is associated which is associated with an increase in the grain boundary and the corresponding diffusion processes.

3. Conclusion

The regularities of mass transfer for the UMP and the Ministry of Health of titanium implant on the vacuum-arc ion source "MEVVA-V.RU". In the interpretation of the observed regularities taken into account heterogeneous composition of the beam, represented by three components. Studies have shown that the formation of concentration profiles in titanium occurs mainly by two mechanisms: an impurity originally distributed statistically, but with increasing doses of implanted particles, the concentration of the ion beam generated by structural defects and spraying the surface, begin to affect a variety of diffusion processes.

It is found that in samples with relatively fine grains (UFG titanium) in the formation of the depth profiles of the target to a greater extent contribute to diffusion processes, in particular radiation-induced diffusion, as compared with the FG samples. The proposed two-stage model allows to obtain qualitative agreement between theory and experiment and identify advantageous mechanisms of profiles at various stages of implantation.

References

[1] Kurshina I A, Popova N A, Kalashnikov M P et al. 2011 *News of higher educational institutions. Physics.* 118.
[2] Nikolaev A G, Oks E M, Yushkov G Y 1998 *Technical Physics* **68** 43.
[3] Vahnii T V, Vershinin G A, Sharkeev Y P et al. 2010 *Surface. X-ray, neutron and synchrotron the Exploration.* **4** 99.

Journal of Physics: Conference Series **669** (2016) 012040 doi:10.1088/1742-6596/669/1/012040

The distribution of excess charges in the diffusion flame of hydrocarbons

S M Reshetnikov, I A Zyryanov, A P Pozolotin and A G Budin

Physics Department, Vyatka State University, 610000, Kirov, Moskowskaya Street, 36, Russia

b185@mail.ru

Abstract. The paper discusses the experimental investigation results of the potentials distribution (probing) and the excess charges in diffusion flames of propane and polymers - PMMA (polymethyl methacrylate) and divinyl rubber. The study of the excess charges distribution was made by analysis of flame deformation in the presence of an external lateral electrostatic field. The existence and the accumulation of excess charges in the flame are generally explained on the basis of consideration of the interaction processes of condensed phase particles and charges (electrons) generated during combustion in the presence of a dusty plasma. However, this mechanism does not give the opportunity to consider the sign and magnitude of the excess charges distribution in the flame, depending on the products composition and the value of the oxidant excess. It is unclear how oppositely charged immiscible streams of flame are formed and coexist.

The study of the chemo-ionization phenomenon - the formation of charged particles during the combustion, is critical and fundamental problem. Since the combustion process is accompanied by electrical phenomena, it is very tempting to use electrical phenomena to control the combustion process (ionization). Therefore that problem of consideration of mechanism and using electrical properties during combustion is actual. The complexity of solving the chemo-ionization problem is caused by the fact that before the ionization an electron is a quantum object, and after ionization it becomes a classic one. Equally urgent is the problem of existence and value of the local excess charges in various locations of the flame. The existence of its own electric field in the flame torch was discovered long ago by Michael Faraday. However, until now the mechanism of appearance and the accumulation of excess charges in the combustion zone is unclear. The process of charges accumulating in the reaction zone is explained by the ability of dispersed particles of dusty plasma to absorb and desorb with different intensity ions and electrons.

Dusty plasma is an ionized gas containing charged particles of condensed matter [1]. The condensed phase is formed in the combustion zone during the conversion of hydrocarbon fuel. The growth rate of the particles is comparable to the rate of chemical reactions [2]. The sizes of particles produced during the conversion of propane, are between tens of angstroms to tens of micrometers. The presence of dispersed particles of different sizes requires taking into account the characteristics of their interaction with the charges.

The accumulation of charge on the condensed phase particle is investigated in the study [3]. It is shown that the dispersed particle can accumulate on its surface charge, reaching 10^4 elementary, creating around itself the potential of tens millivolts. In the study [4] the dynamics of charging carbon black particles in the flame over time is analyzed. It is shown that at the initial stage of carbon black growth (10^{-8}-10^{-3}s since the formation) mostly negative charge is accumulated on the particles. Negative charge accumulating process is caused by the interaction of carbon black particles and electrons having high mobility [4]. Further, due to the interaction of negative carbon black particles and positive ions, concentrations of negative and positive carbon black particles are equalized. Through 10^{-1}s since the formation carbon black particles begin to accumulate positive charge [4].

The authors investigated the potential distribution and the excess charge in the diffusion flames of hydrocarbon fuels, propane and polymers [5, 6]. Photo of propane flame with marked equipotential lines are shown in Fig. 1 and 2. Diffusion flames were organized on the burner consisting of two coaxial pipes: first with propane, diameter 8 mm at the center, and second, outside - with air, diameter 20 mm [5]. Fuel consumption varies between 0,5 - 2,7 cm^3/s, oxidant - 13-16 cm^3/s, which allows to vary the oxidized excess ratio α.

The position of the front in diffusion flame is realized in accordance with the solution of Burke-Schumann equation [8], when the ratio of oxidizer/fuel is equal stoichiometry, and the potential is zero. In the flames with a lack of oxidant (α < 1) in the vicinity of the combustion front a region of negative potential (Figure 1) is found, bounded by the surface of zero potential. In the central part of the flame upstream region of positive potential is detected. The maximum negative potential is 120 mV, positive 80 mV. In the field of negative potential localization is located the area of maximum temperature. The temperature there is 1150 K.

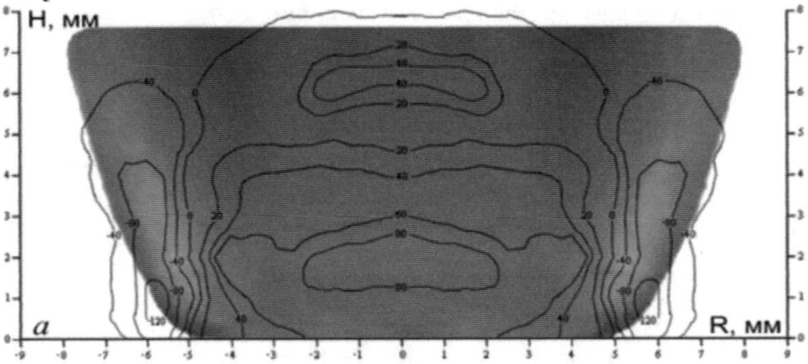

Figure 1. The potential distribution in the flame of a propane when α <1

The potential distribution in flames when α > 1 is presented in Figure 2. In the bottom part of the flame a negative potential region is detected, the maximum value of which is detected at R = ± 4 and H = 1 mm. Upstream in the flame a positive potential region is found. The maximum value of the positive potential is achieved at a height H = 4 mm. The values of potentials in absolute value are 150 mV. The area of maximum temperature is located in the upper part of the flame in the area of positive potential. The temperature there is 1200 K.

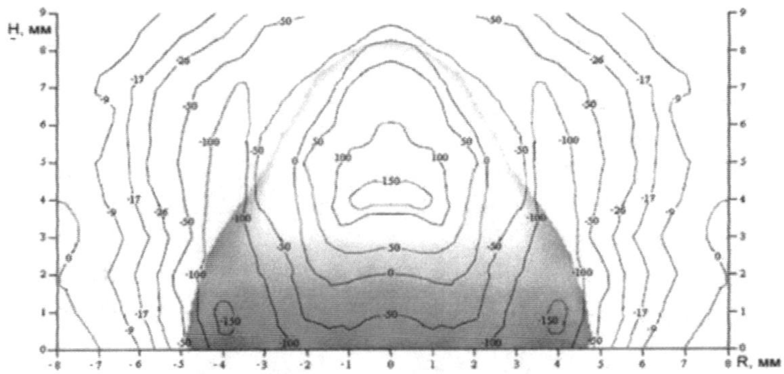

Figure 2. The potential distribution in the flame of the propane with excess of oxidant

According to the kind of the distribution of excess charge in flame polymers can be divided into two groups. In the first group of polymers (for example PMMA) flame is deflected to a negative electrode (Figure 3a). The flames of the second group of polymers (for example divinyl rubber) in the lateral electric field are divided into two torches, that is shown in Figure 3b. Separation of the flame in the electric field indicates presence in flame excess positive and negative charges which do not interact with each other.

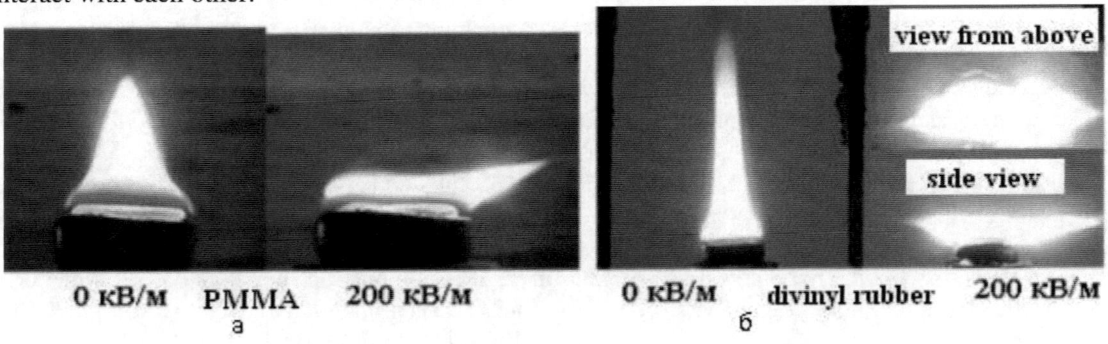

Figure 3. Flames of polymers in the lateral electric field

Accumulation of positive excess charge in the flame is usually associated with the difference in mobility of ions and electrons, for example, [7]. Electrons having higher mobility, leave the reaction zone, and the flame becomes positively charged. This approach does not explain the accumulation of negative charge in the flame. The accumulation of excess charge both positive and negative in a flame can be explained on the basis of the interaction of dispersed particles with ions and electrons.

The charge accumulated in the dispersed particles is determined by the time of its location in the combustion zone [4]. Formation of carbon black occurs in the front of the reaction, further particles entrained by gas flow. In authors experiments, carried out with the propane flame gas velocity is about 15 mm/s. In the initial stages of life in front of burning carbon black particles become negatively charged. During the characteristic time 10^{-1}s particles have time to go from the front of the combustion a distance about of 1.5 mm to form a region of negative potential, which have good agreement with the experimental results (Fig. 1). When removed for a distance more than 1.5 mm from the front of the combustion, charge of the dispersed particles is changed to positive. Thus, with a deficiency of oxidant, the flame front along its entire surface is negatively charged. The negative charge of the carbon black particles during the motion upstream is replaced by positive, which results in forming a region of excess positive charges upstream of the combustion front, (Fig. 1).

In the flames with $\alpha > 1$ the formation of carbon black begins in the end of the burner in the bottom part of flame. Here an excess negative charge is found (Fig. 2). Further, at movement upwards flow there is a growth and recharging the condensed particles. The excess charge in the flame is formed by the competition of the charges of the particles formed in the front and in the bottom of the flame. Thus, neutrally charged region is created in the flame (Fig. 2, h = 2 mm). When moving upstream the positive charge accumulated on the particles formed in the lower part of the flame becomes much larger than charge of particles formed at a given height in the front of the reaction. This leads to formation of positively charged area.

The distribution of electric potential in the area of the combustion of propane and polymers has been experimentally investigated. It was found that if $\alpha < 1$ in the flame front there is detected a negative potential and there is realized the maximum temperature. In the diffusion flames when $\alpha > 1$ a negative potential is detected in the bottom of the torch. At the top of the flame the area of maximum temperature is detected, which is located in the area of positive potential. An explanation of the results obtained on the basis of the process of interaction carbon black particles and charges in flame.
Experimentally found separation of polymers flames in an electric field, which indicates the existence of excess charges with a different sign in flame, which do not interact with each other.

References

[1] Fortov V E, Hrapak A G, Hrapak S A, Molotkov V I, Petrov O F, Dusty Plasma, *Moscow, Successes Of Physical Sciences,* 2004, V.147, No5, P.495-544.

[2] Mansurov Z A, Carbon black formation in combustion processes, *Novosibirsk, Physics Of Combustion And Explosion,* 2005, V.41, No 6, P.137-156.

[3] Dautov G U, Sabitov Sh R, Fairushin I I, Research Potential Distribution And Concentration Of Electrons In Dusty Plasma, *Kazan, Messenger of Kazan National ResearchTechnikal University named after A N Tupolev,* 2007, No 1, P.29-32.

[4] Savelev A M, Starik A M, Interaction Features Of Ions And Electrons With Nano-Particles In The Plasma Formed During Combustion Of Hydrocarbon Fuels, *Saint Petersburg, Journal of Technical Physics,* 2006, V.76, No 4, P.53-60.

[5] Reshetnikov S M, Bobrov A S, Zyryanov I A, Effect Of The Electric Field On The Structure Of A Diffusion Flame With Various Coefficient Of Excess Oxidant, *Kazan, Proceedings Of The Universities. Aviation Technics,* 2010, No 2, P.59-62.

[6] Reshetnikov S M, Zyryanov I A, Pozolotin A P, Combustion Features Of Polymers In Electrostatic Field, *Taganrog, Proceedings Of SFU. Technical Science,* 2013, No 8, P.30-36.

[7] Stepanov E M, Dyachkov B G, Flame Ionization And The Electric Field, *Metallurgy,* 1968, 312P.

[8] Burke S P, Schumann T E W, Diffusion Flames, *Combustion Symposium Industrial And Engineering Chemistry,* 1928, V.20. No 10 P.998–1004.

Journal of Physics: Conference Series **669** (2016) 012041 doi:10.1088/1742-6596/669/1/012041

Mechanism of influence water vapor on combustion characteristics of propane-air mixture

V M Larionov[1], G A Mitrofanov[1], A V Sachovskii[1] and N K Kozar[2]

[1]Kazan Federal University, 18 Kremlyovskaya str., Kazan 420008, Russian Federation
[2]Kazan cooperative Institute Russian University of cooperation, 58 Nikolaya Ershova, Kazan 420081, Russian Federation

E-mail: techphyskfu@gmail.com

Abstract. The article discusses the results of an experimental study of the effect of water vapor at the flame temperature. Propane-butane mixture with air is burning on a modified Bunsen burner. Steam temperature was varied from 180 to 260 degrees. Combustion parameters changed by steam temperature and its proportion in the mixture with the fuel. The fuel-air mixture is burned in the excess air ratio of 0.1. It has been established that the injection of steam changes the characteristics of combustion fuel-air mixture and increase the combustion temperature. The concentration of CO in the combustion products is substantially reduced. Raising the temperature in the combustion zone is associated with increased enthalpy of the fuel by the added steam enthalpy. Reducing the concentration of CO is caused by decrease in the average temperature in the combustion zone by applying steam. Concentration of active hydrogen radicals and oxygen increases in the combustion zone. That has a positive effect on the process of combustion.

1. Introduction

The combustion efficiency can be improved in various ways. These methods are known as methods for combustion control [1].

The great interest in the practice of creating have methods based on the use of various types of fuel additives. In recent years, researchers are paying attention to water-steam additives. Experiments indicate that the presence of water vapor in the gaseous fuel increases the environmental safety of power plants. Furthermore, water vapor have a catalytic effect on the combustion process. This may involve the afterburning of carbon monoxide.

Partial reduction of the temperature in the combustion zone has an effect on nitric oxide. Thus there is a reduction in the concentration of harmful substances [2]. Furthermore, the supply of steam into the combustion zone substantially changes characteristics of the pulsation of the combustion zone [3]. In gas turbine engines marked effect of water injection into the combustion zone for combustion efficiency and emissions of NOx, CO, and others [4].

2. The method of investigation

Combustion of propane-butane mixture with air was carried out on a modified Bunsen burner [2]. Burner of the experimental setup is shown in Figure 1. Propane butane mixture fed into the receiver through the pipeline. Air fed by the air pump. Consumption of air and fuel gas measured by flow

Journal of Physics: Conference Series **669** (2016) 012041 doi:10.1088/1742-6596/669/1/012041

meters Bronkhorst. Water from the tank was fed to the electric heater. It produces a sequence of evaporation of water and heating the resulting steam to a predetermined temperature. The heated steam is fed into the receiver, where it is mixed with the gas-air mixture, and then the resulting mixture was fed to a modified Bunsen burner.

Before the experiments were carried out with the burning of the study process of producing superheated steam. Electric heater had three stages. This provided an opportunity to produce steam temperatures up to 700°C. The steam consumption was calculated on the average volumetric flow rate of water. In addition, the investigated regimes of stable combustion of the fuel mixture (Figure 1), assessed the possibility of visualizing the flame on the burner with a thermal imager Dali LT7 (Figure 2).

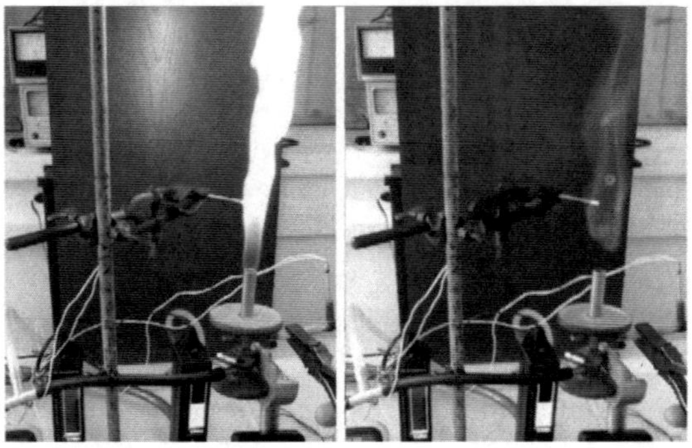

Figure 1. Combustion of the fuel mixture.

In addition, for measuring the average temperature in the combustion zone mica used. The temperature of the heating plate mica flame recorded imager.

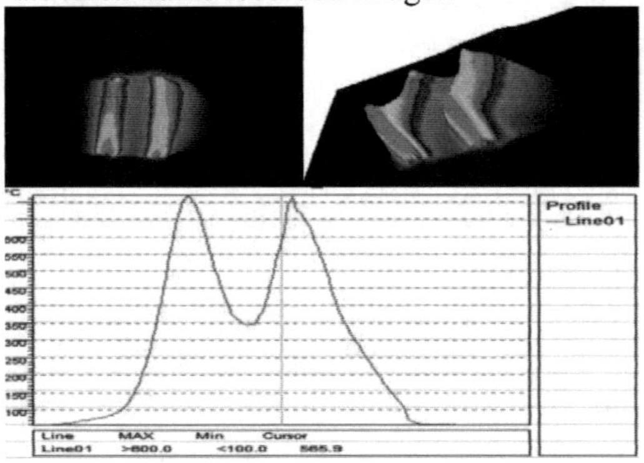

Figure 2. Visualization of the flame on the burner with a thermal imager Dali LT7.

3. The results of the study and discussion

Admission heated or superheated steam into the combustion zone substantially modifies the combustion process. Combustion goes from laminar to turbulent. The sound field produced by an open flame is changed. There are changes of luminosity and colour of the flame. The colour of the flame turns blue (Figure 1). The lack of yellow colour indicates that the concentration of free carbon in the flame during the steam supply drops sharply. The luminosity of the flame changes due to temperature reduction during the combustion zone feed steam.

Increasing the temperature raises the temperature of the steam in the combustion zone. Changes in the relative amount of water vapor in the composition of gaseous fuel also leads to a corresponding change in the wall temperature.

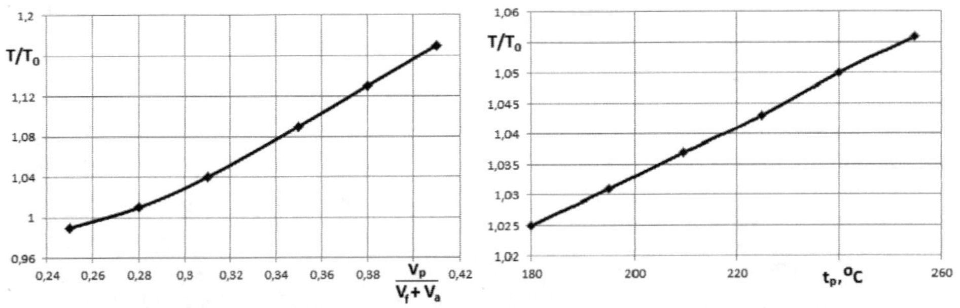

Figure 3. Graphs of the relative temperature of the wall of the quartz tube.

Figure 3 shows graphs of the relative temperature of the wall of the quartz tube T/T_o, depending on the relative amount of water vapor in the composition of gaseous fuel $V_p/(V_a+V_f)$ and depending on the temperature of the steam t_p. T and T_o, - measuring (steam) and initial (no steam) tube wall temperature, V_p, V_f, V_a, - the volume of water vapor, the gaseous fuel and air. Figure 4 is a graph of change in enthalpy of water vapor depending on the temperature [5]. Graphs of the enthalpy of steam and temperature changes similar wall. This suggests that the increase in enthalpy of the fuel leads to a proportional change in the temperature in the combustion zone.

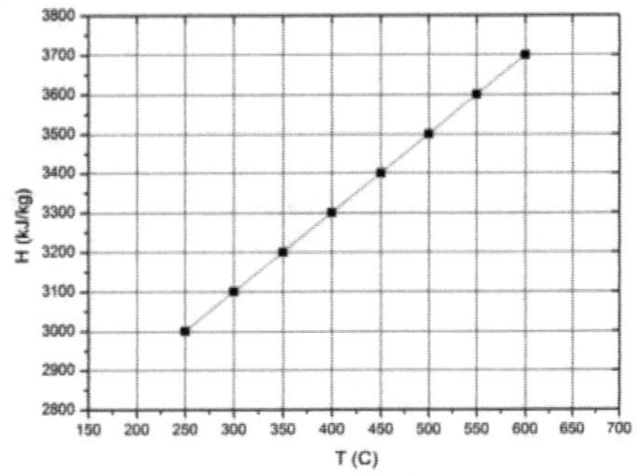

Figure 4. Graph of change in enthalpy of water vapor.

Increasing the temperature of the combustion zone increases the rate of chemical reactions. Reducing the concentration of CO is caused by decrease in the average temperature in the combustion zone by feeding water vapor. Furthermore, when the steam in the combustion zone increases the concentration of active hydrogen radicals and oxygen, which has a positive effect on the process of combustion.

4. Conclusion

Thus, the use of superheated steam in the composition of the gaseous fuel increases the thermal and environmental efficiency of combustion. Changing the combustion gas-air mixture is caused by an increase in temperature of the combustion zone, which is due to the summation of the enthalpy of superheated steam enthalpy of fuel.

The work is performed according to the Russian Government Program of Competitive Growth of Kazan Federal University.

References

[1] Akhmadullin A N, Ahmethanov E N, Iovleva O V, Mitrofanov G A 2013 *J. Phys.: Conf. Ser.* **479** 012004
[2] Larionov V M, Mitrofanov G A, Iovleva O V 2014 *J. Phys.: Conf. Ser.* **567** 012028
[3] Bykovets A P, Larionov V M, Marchukov E Yu 1992 *Izvestiya Vysshikh Uchebnykh Zavedenij. Aviatsionnaya Tekhnika* (3) 71-74
[4] Mingazov B G, Muhametgaliev T H 1992 *Aviatsionnaya i racetno-kosmicheskajaTekhnika.*

Vestnic Samarscogo gosudarstvennogo aerocosmichescogo unuversiteta **5** 204-207
[5] Revcin S L 1980 *Thermal properties of water and steam* Moscow p 424

Journal of Physics: Conference Series **669** (2016) 012042 doi:10.1088/1742-6596/669/1/012042

Interaction of electric and acoustic vibrations in combustion

V M Larionov, K G Sadikov, G A Mitrofanov

Kazan Federal University, 18 Kremlyovskaya str., Kazan 420008, Russian Federation

E-mail: techphys@kpfu.ru

Abstract. The results of experimental studies of the interaction of electrical discharges in the acoustic oscillations in the combustion of isobutane are presented in the article. Electric discharges were created using a pulsed high voltage source at specified intervals. The purpose of the study was to determine the feasibility of using electrical pulse action to control combustion. The study was conducted on the specifically designed pattern of the combustion chamber with a swirl burner in the frequency range from 100 Hz to 1400 Hz. The study found that the method of periodic pulsed electrical influences can be used to control the combustion in the combustion chamber model. There is a steady increase in the amplitude of the oscillations in the combustion chamber model. Effects due to the mechanism of interaction of acoustic waves and oscillations heat release from the combustion zone. Estimated physical mechanism is a periodic change in the concentration of ions in the interaction of the combustion zone with the electric field of high potential. The proposed control method is advantageous in terms of energy costs.

1. Introduction

Combustion in the combustion power plants is accompanied by instability [1]. Increasing the thermal stress causes an increase in the working process of pressure fluctuations, temperature, speed, etc. These phenomena emerge and develop spontaneously. The oscillation frequency depends on the geometry of the combustion chamber. The oscillation amplitude depends on the degree of consistency ripples of heat in the combustion zone with the pressure fluctuations, gas mass flow rate, vortex shedding and other non-stationary processes [2]. Using the positive effects of the pulsating combustion can improve heat generation and ecological characteristics.

Combustion control can be carried out by different methods [3]. Among the known methods there is a group of external periodic perturbation methods is allocated. External periodic perturbations may be sound, electrical and others. These actions are aimed at changing the characteristics of heat.

The purpose of the pilot study is to determine the possibility of using the method of external electrical impulse actions to control the combustion in the combustion chamber model.

2. The method of investigation

General view of the experimental setup is shown in Figure 1.

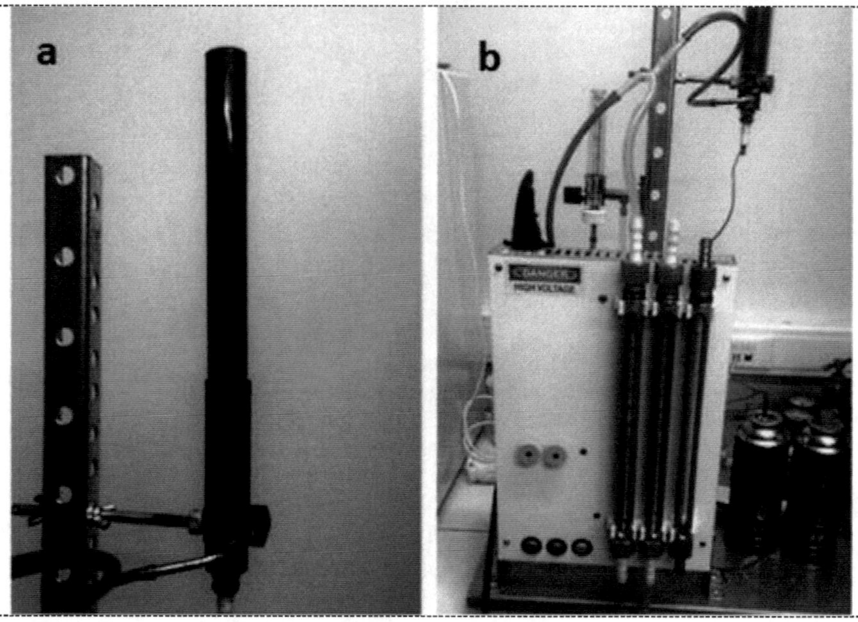

Figure 1. Appearance of the experimental setup: a- the combustion chamber, and b - measuring equipment

The combustion chamber of the experimental setup (figure 1 a) is a cylindrical tube. Its inner diameter is 25 mm, the total length of the resonance part 249 mm. Burner device was manufactured from heat-resistant stainless steel.

At a distance of 55 mm from the lower end of the burner have three tangentially disposed nozzles (Figure 2a).

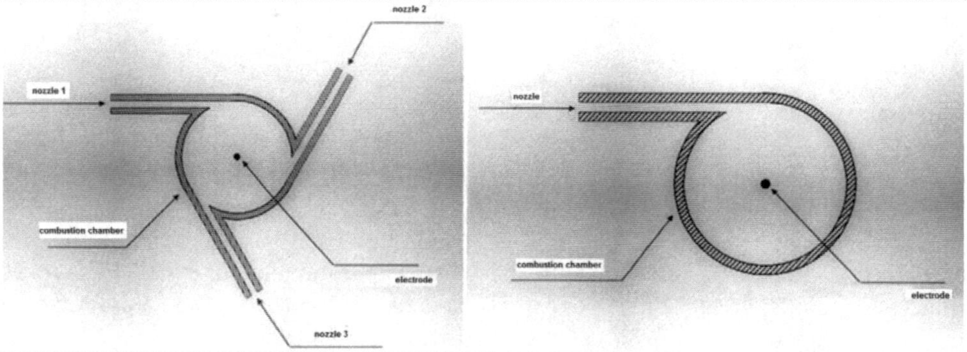

Figure 2. Ways tangential feed gas mixture into the combustion chamber

In addition, for measuring the average temperature in the combustion zone mica used. The temperature of the heating plate mica flame recorded imager.

At the lower end of the burner has a ceramic sleeve for attaching the internal high voltage electrode. The inner electrode is designed as a rod made of refractory steel. The diameter of the electrode was - 1.0 mm. Stem length varied from 50 to 250 mm. The optimal length is 75 - 90 mm.

The combustion zone is formed around the center electrode. The positive potential was applied to the tube body. The negative potential was applied to the rod axis. The fuel gas mixed with air fed into the combustion chamber tangentially at the bottom of a core electrode. Gas mixture swirled counterclockwise. The ignition of the fuel mixture occurs when submitting the first electric pulse

In experimental studies, the supply of gas-air mixture is made on the circuit with a tangential channel (Figure 2, b). The internal diameters of the gas nozzles were 3 mm.

Scheme fuel gas through three nozzles has been more successful. Working gas-air mixture is evenly distributed within the combustion chamber.

The fuel gas is pre-prepared mixture of isobutane and air in stoichiometric ratio. Premixing was carried out immediately prior to feeding the fuel gas to the nozzles.

Costs gases (isobutane and air) were measured by flow meters brand RMF-063G and REF-554000, respectively. Flow Air RMF-063G has full scale of 0.63 m3 / h (10,51 / min). During the experiments, the air flow rate ranged from 7.35 to 8.4 l / min. Flowmeter for isobutane REF-554000 had a full scale of 5 liters / min. During experiments isobutane flow rate ranged from 0.83 to 1.38 l / min.

The sound pressure level pressure oscillations in the combustion chamber are measured by means of an integral sound pressure analyzer IBP-1. Weighted sound pressure level inside the combustion chamber was equal to 140 dBell. Natural frequency of the combustion chamber at the stoichiometric flow rate was equal to 400 Hz.

3. The results of the study and discussion

The experiments were performed in the following sequence.

Initially, we conducted research on the impact of the proposed method of the combustion chamber with a single nozzle. After ignition of the fuel gas to the central axis of the combustion chamber electrode pulse voltage was applied. The amplitude of the sound vibrations in combustion without pulse exposure was equal to 60 dB. Then it was applied pulse voltage. The pulse rate changed. When the pulse frequency was closed to 400 Hz and the excess air ratio exceeding the stoichiometry were increased acoustic oscillations in the combustion chamber. This was appreciable by change of sound pressure level and change of the character of combustion. Thus estimation excess air ratio is in the range from 1.05 to 1.35. Sound pressure level at the measurement point is 120 dB.

The actual amount of air required for complete combustion of 1 kg of fuel in fact different from design. In the practice of burning fuel not all the amount used. Therefore it does not play a part in the combustion reaction resulting in poor mixing of air and fuel. Part of the air does not have time to enter into contact with the carbon in the fuel and leaves flues in the free state. Therefore, the ratio of the amount of air really and theoretical values of the excess air in the combustion chamber differ. The calculation of the excess air ratio by the formula: $\alpha_r = V_r / V_t$, where V_r - the actual volume of air supplied to the furnace at 1 kg fuel; V_t - theoretical volume of air.

Figure 3 is a graph of the theoretical combustion temperature of the air excess factor. Colour allocates the working range of the excess air in an experimental investigation of for the combustion chamber with a one nozzle.

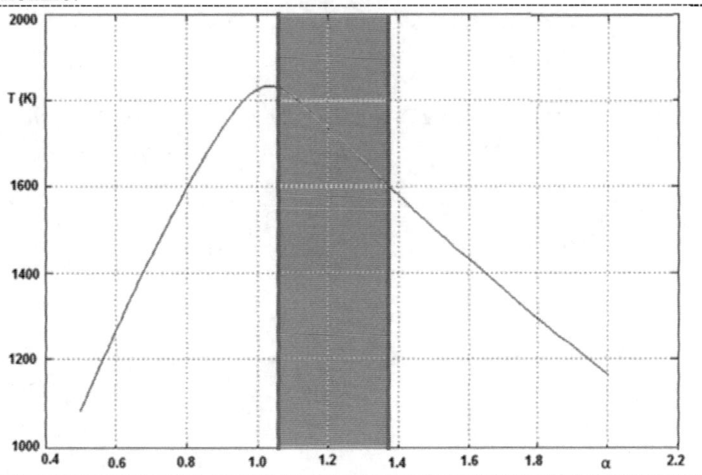

Figure 3. Dependence of the theoretical combustion temperature of the excess air ratio

Then, experiments were conducted to the combustion chamber with three gas inlet nozzles.

In this case, after the ignition of the fuel mixture arose to acoustic oscillations at a resonant frequency of the combustion chamber. This was caused by a known mechanism of excitation of oscillations in the vortex burners [1]

The original sound pressure level is 140 dB. The electrical effects at frequencies close to the resonant frequency of the first half-wave resonator modes of oscillation were increased. Amplification was not significant. As the frequency of external influence to the resonance frequency of the first effect was observed capture frequency. This phenomenon is observed in the range of 2-3 Hz of frequency resonance. Interesting to note that when you change the frequency of the external electric effects of the resonance in the range of 3 Hz frequency change occurred recorded acoustic oscillations. These changes were visible instruments and listening.

The observed phenomenon may explain the mechanism of asynchronous interaction of external periodic perturbations and the natural frequencies of the combustion chamber [3].

The physical nature of this phenomenon is in the interaction of the electric field with the charged particles CHO + [5]. These particles are formed by chemical ionization of fuel gas by oxidation by atmospheric oxygen. It is known that the concentration of ions in the flame can be 4-6 orders of magnitude higher than the concentration that would be observed in a purely thermal mechanism of ionization, and flame can actually behave like weakly ionized plasma. However, the temperature of the flame is not sufficient to ensure that the components of the mixture can be ionized by collisions between the molecules is [4]. The concentration of ions and their charge is dependent on the excess.

4. Conclusion

Method of external electrical disturbances can be used to control the combustion in the combustion chambers of power plants. Depending on the frequency of the external electrical influences has been increasing oscillations in the combustion chamber and at the resonant frequencies close to it. This is due to the mechanisms of synchronous and asynchronous interaction. Estimated physical mechanism is a periodic change in the concentration of ions in the interaction of the combustion zone with the electric field of high potential.

The work is performed according to the Russian Government Program of Competitive Growth of Kazan Federal University

References

[1] Larionov V M, Nazarenko T I, Filipov S E 2004 *Izvestiya Vysshikh Uchebnykh Zavedenij. Aviatsionnaya Tekhnika* (1) 32-34

[2] Semenova E V, Larionov V M, Kazakova E I 2014 *J. Phys.: Conf. Ser.* **567** 012033

[3] Akhmadullin A N, Ahmethanov E N, Iovleva O V, Mitrofanov G A 2013 *J. Phys.: Conf. Ser.* **479** 012004

[4] Starikovskiy A, Aleksandrov N 2013 *Progress in Energy and Combustion Science* **39** 61-110

[5] Fialkov A B 1997 *Progress in Energy and Combustion Science* **23** 399-528

Thermal effect of hydrocarbon fuels combustion after a sudden change in the specific calorific value

E R Saifullin[1], V M Larionov[1], A V Busarov[2], V V Busarov[2]

[1]Kazan Federal University, 18 Kremlyovskaya str., Kazan 420008, Russian Federation
[2]Limited Liability Company «Center of industrial safety expertise», 90a Julius Fucik str., Kazan 420140, Russian Federation

E-mail: mr.emilsr@gmail.com, techphys@kpfu.ru

Abstract. Using associated gas and waste oil refineries in thermal power plants, a complex problem due to the variability in fuel composition. This article explores the burning of hydrocarbon fuel in the case of an abrupt change in its specific combustion heat. Results of the analysis allowed developing a technique of stabilizing the rate of heat release, ensuring complete combustion of the fuel and its minimum flow.

1. Introduction

Currently, one of the acute and unresolved problems in the energy, oil and gas sector in the Russian Federation and in other countries is the problem of utilization of associated petroleum gas (APG below) and wastes of petrochemical plants. The most advanced and affordable method of disposal is burning it mainly in flares or in power plants, such as boiler units and gas turbine installations [1].

APG and wastes of petrochemical production include hydrocarbons and can be used as a "secondary fuels" [2], replacing traditional fuels, such as natural gas [3]. These secondary fuels have a non-permanent structure and therefore a non-constant specific heat of combustion. These deviations have a negative impact on the combustion process and on the heat removal from the hot products of combustion, reduce the efficiency of the combustion process, for example - there is a chemical incomplete combustion, reduces the efficiency of thermal power plants, causes damage to the environment. The task of optimization of the combustion of these fuels is to ensure complete combustion and thermal capacity of a given power plant in the event of changes in fuel composition. Available methods [4], [5] of optimization of combustion are not sufficiently effective, requires continuous analysis of the fuel composition, combustion products and the development of sophisticated control systems.

The aim of this work - the development of a simplified technique of optimization of hydrocarbon fuels combustion in the case of sudden change of specific heat of combustion.

2. The physical model of the optimization process

Mathematical analysis of the combustion process in the case of sudden change of specific heat of combustion of hydrocarbon fuels was carried out in [6]. It was determined the conditions when the fuel burns completely and its flow rate is minimal for the production of a predetermined thermal power:

1. After the reduction of the specific heat of combustion of fuel at a small relative value, to restore the initial thermal conditions is necessary to increase the fuel consumption by the same relative value, without changing the air flow.

2. After raising the specific heat of combustion of the fuel at a small relative value, necessary to reduce the fuel flow rate to the same relative value without changing the air flow.

These results have the following physical explanation. It is assumed that the initial specific heat of combustion of fuel is constant, burning - optimal, that is, the fuel burns completely, and the fuel flow rate is minimal for the production of a predetermined thermal energy.

Assume that in a burns fuel in a short time, there is an abrupt increase in the proportion of heavy fractions. Further, the fuel composition does not change.

This case corresponds to incomplete combustion of fuel, as available air is not enough for its complete combustion. Then the reduction of fuel consumption at a constant flow of air increases the completeness of combustion of fuel and combustion temperature [7]. At the same time, due to the decrease of fuel flow rate the amount of thermal energy contained in the combustion products is decreasing. As a result, the transmitted to the coolant heat flux through the wall of the heat exchanger is not changed, that is reflected in the constant temperature of the coolant at the outlet of the heat exchanger. Next steps is to periodically reduce fuel flow rate. Reducing of the fuel flow rate is completed when the temperature of the coolant at the outlet of the heat exchanger is reduced by an amount greater than the accuracy of the measurement of this parameter. This means that the amount of fuel burned decreased so that there was an excess of air, leading to a decrease in the coolant temperature at the outlet of the heat exchanger, as shown in Figure 1. In this case is necessary to increase the fuel flow rate to the previous value. The optimization process of combustion is considered complete, as reached the minimum fuel flow rate, combustion of which provides a predetermined amount of heat flow imparted to the coolant in the heat exchanger.

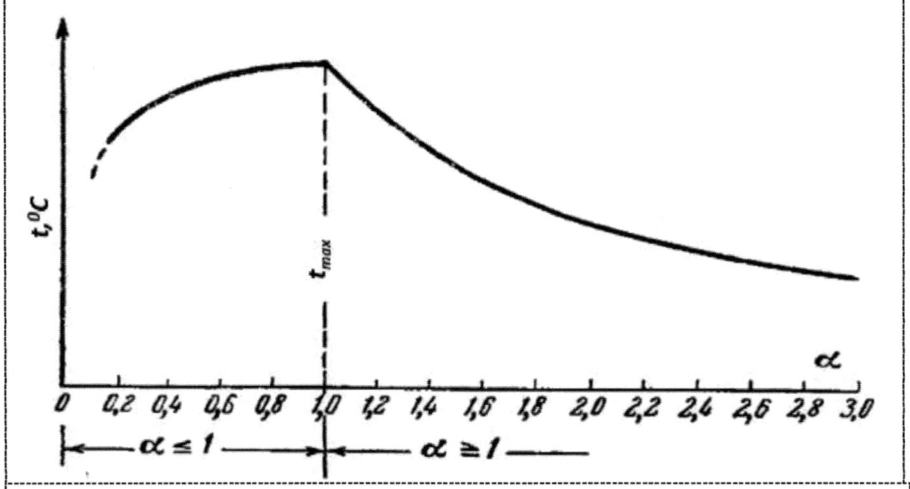

Figure 1. Character of the curve theoretical combustion temperature of fuel in case of shortage and excess air

Consider another case where in a burned fuel in a short time, a rapid increase in the proportion of light fractions. Further, the composition of the fuel is not changed.

In this case, after reduction of the fuel flow rate at a constant air flow rate outlet coolant temperature becomes lower than a predetermined value. Differently from the first case, the fuel burns completely, because the excess air higher than theoretically necessary for complete combustion ($\alpha > 1$), which can be seen from Figure 1. Therefore, necessary to reduce the air flow. If the reduction of air flow leads to the fact that the outlet temperature of the coolant becomes larger than a predetermined value by more than the measurement accuracy of this parameter is necessary to complete gradual reducing of the air flow rate. After stabilization the air flow is necessary to begin a discrete reduction in the fuel supply. When after a reduction of the fuel flow rate cannot increase the temperature of the coolant at the outlet of the heat exchanger to a predetermined value, is necessary to return the flows of fuel and air to the previous values. This means that there has been the case with the stoichiometric air/fuel ratio ($\alpha = 1$), providing complete combustion of fuel. Further, for some time oversee the outlet coolant temperature. If the temperature remains constant, the optimization of combustion considered completed.

3. Conclusion

The physical model and methods of optimization of hydrocarbon fuel combustion in the case of sudden change in the specific calorific value is developed. In the case of an abrupt change in fuel composition, causes an increase in the specific heat of combustion is necessary to conduct periodic reduction of fuel flow rate. When the temperature of the coolant at the outlet of the heat exchanger is reduced by greater than the accuracy of measurement of this parameter value, it's needed to return the fuel flow rate to the previous value, and complete optimization of the combustion process. In the case of an abrupt change in fuel composition, causes a decrease in the specific heat of combustion it's needed to produce a periodic reduction of fuel and air flow rates to raise the temperature of the coolant at the outlet of the heat exchanger. When next reduction of air flow rate don't increases the temperature of the coolant at the outlet of the heat exchanger to a predetermined value, the air flow rate is returned to the previous value, and complete optimization of the combustion process.

The work is performed according to the Russian Government Program of Competitive Growth of Kazan Federal University.

References

[1] Gur'yanov A I, Evdokimov O A, Piralishvili Sh A, Veretennikov S V, Kirichenko R E, Ievlev D G 2015 *Aircraft and rocket engine design and development* **58** 205–209
[2] Semenova E V, Larionov V M, Kazakova E I 2014 *J. Phys.: Conf. Ser.* **567** 012033
[3] Iovleva O V, Larionov V M 2012 *Russian Aeronautics* **55** 388-391
[4] B. A. Shtrambrand, A. F. Feodos'evich, RU Patent No. 2 300 705, (2005)
[5] Kazarinov L S, Shnaider D A, Kopcev L A, Kolesnikova O V, Kinash A V, Sedel'nikov S V 2011 RU Patent No. 2 425 290
[6] Saifullin E R, Vankov Yu V 2015 *IOP Conf. Ser.: Materials Science and Engineering* **86** 012006
[7] G. F. Knorre, K. M. Aref'ev, A G Blokh 1966 *Russia B. Theory of combustor processes* 1042

Journal of Physics: Conference Series **669** (2016) 012044 doi:10.1088/1742-6596/669/1/012044

Modeling the dielectric barrier micro-discharge in argon at atmospheric pressure

A I Saifutdinov[1,2], A A Saifutdinova[1] and N F Kashapov[1]
[1]Kazan Federal University, Institute of Physics, Kazan 420008, Kazan, Russia
[2]St. Petersburg State University, Faculty of Physics, 198504, St. Petersburg, Russia

E-mail: as.uav@bk.ru

Abstract. Dielectric barrier micro-discharge in argon at atmospheric pressure is simulated. Spatial and temporal discharge parameters are presented.

1. Introduction

The non-thermal plasma of gas discharges at high pressure for many years is of great interest to scientists, as an object for the study of fundamental plasma phenomena [1], and from the point of view of the many practical applications. Discharge plasma combines a low temperature of heavy particles and high temperature of electrons.

Various types of bits in which non-equilibrium plasma is supported at the present time are the subject of study by many research groups. The main reasons for this fact are the low cost of equipment for creation of these discharges, and ease of their use. This causes a wide range of applications of such systems – various radiation sources, the generation of ozone, plasma displays, surface treatment, treatment of industrial and exhaust (or destruction of harmful pollutants in the atmosphere), sterilization of medical instruments and products, disinfection of food products and packaging and plasma in biomedicine.

Typically, DC discharges under these conditions are not sustainable [for example, see 1]. Therefore, currently being developed various methods to control the parameters of the plasma DC discharges [2]. Another way to obtain a stable discharge in atmospheric pressure is to change the conditions of burning, in particular the transition to a non-stationary discharge - a barrier, pulsed, high-frequency, microwave discharges [1, 3-6] and others. One of the most common types of these discharges is a dielectric barrier discharge (DBD), in which one or both of the metal electrodes are covered with a dielectric layer. Dielectrics accumulate charges on the surface during the current pulse. These currents in turn create a self-induced field in the gas gap, which suppresses the discharge until the current density exceeds a threshold value.

Despite the low cost of equipment for creation of sources DBD discharges now there is still a problem for their stable operation and optimization. Therefore, at the present time good instrument to predict the parameters of DBD is numerical simulation. Various methods of modeling discharges were considered earlier by the authors in [2,7-8].

Journal of Physics: Conference Series **669** (2016) 012044 doi:10.1088/1742-6596/669/1/012044

2. Model description

The model included balance equations for the densities of charged (electrons, ions) and the excited particles n_k, the electron energy density n_ε, and the Poisson equation for the potential φ. The stationary state was determined by solving the time-dependent equations. The fluxes of charged (electrons, ions) $\mathbf{\Gamma}_k$ and electron energy flux $\mathbf{Q}$ were given in the drift-diffusion form. The complete system of balance equations is

$$\frac{\partial n_k}{\partial t} + \nabla \cdot \mathbf{\Gamma}_k = \sum_j c_{k,j} R_j, \tag{1}$$

$$\frac{3}{2}\frac{\partial n_\varepsilon}{\partial t} + \nabla \cdot \mathbf{Q} = -q_e \mathbf{\Gamma}_e \cdot \mathbf{E} + \frac{3}{2}\delta \nu_{ea} n_e (T_e - T) + \sum_j \Delta\varepsilon_j R_j, \tag{2}$$

$$\mathbf{\Gamma}_k = -D_k \nabla n_k + z_k \mu_k \mathbf{E} n_k, \quad \mathbf{Q} = -D_\varepsilon \nabla n_\varepsilon - \mu_\varepsilon \mathbf{E} n_\varepsilon. \tag{3}$$

$$\nabla \varepsilon\varepsilon_0 \nabla \varphi = -q_e \left(\sum_{k=1}^{N} Z_k n_k \right), \tag{4}$$

Where q_e – is the electron charge, and ε_0 – is the permittivity of free space, Z_k – is a charge particle grade *k*, $\mathbf{E}$ – is the electric field intensity, $\delta = 2m_e / M_a$ – is the ratio of the electron mass m_e to the mass of an atom of gas M_a, ν_{ea} – is the rate of elastic electron-atom collisions, $T_e = (2/3) n_\varepsilon / n_e$ – is the electron temperature.

The right-hand side of equation (1) is defined by reactions in the discharge. Each reaction gives a positive contribution to the function of the source, if it is formed corresponding to the type of particle and the negative contribution, if that kind of particle disappears. In this expression *j* – is the index of the reaction (of 32 plasma-chemical reactions taken into account), $c_{k,j}$ – is the number of particles of type *k* resulting in a reaction of the type *j* (both positive and negative). Reaction rate $c_{k,j}$ is determined by constants corresponding processes $k_j(T_e)$ and they are proportional to the product of the concentrations of the reactants: $R_j = k_j(T_e) \cdot \prod_{k=1}^{N} n_k$ where $N = 2$ for the reactions between the two types of particles, $N = 3$ – for the reactions between three varieties of particles.

The first term on the right side of the first equation (2) corresponds to the Joule heat in the electric field of the electrons and neutral particles of gas, respectively. The second term in (2) describes the energy exchange in elastic collisions of electrons with neutral gas particles. The third term in (2) describes the inelastic energy losses of the electron gas, which $\Delta\varepsilon_j$ – is the energy lost (or acquired if $\Delta\varepsilon_r < 0$) the electrons in this reaction, and R_j – is the reaction rate, which is determined by a constant corresponding inelastic process $R_j = k_j(T_e) \cdot n_e \cdot n_n$, where n_n – is the sort of neutral particles.

Discharge model is a hybrid, because for electron coefficient of mobility μ_e, μ_ε and diffusion D_e, D_ε, as well as some constants processes with their participation, calculated by the convolution of the EDF $f(w, T_e)$, resulting from the decision of the local Boltzmann equation with a cross section of the corresponding process as follows

$$k_j(T_e) = \sqrt{2q_e / m} \int_0^\infty f(w, T_e) \sigma(w) w dw. \tag{5}$$

Here, the index j constants of processes proceeds only by the reactions 11 [9]. When we found the EDF in the kinetic equation takes into account the heating of electrons in the longitudinal electric field and change their energy due to the elastic, inelastic collisions and electron-electron collisions.

The calculations were made for argon using a kit comprising 32 plasma-chemical reactions [9], and taking into account the three effective excited atomic level Ar^*, Ar_r^*, Ar^{**} and two excimer level Ar_2^*, Ar_2^{**}, three kinds of ions Ar^+, Ar_2^+, Ar_3^+. Ion mobility μ_i was taken from [9], and the diffusion coefficients D_i were calculated by the Einstein relation [9].

Boundary conditions were set at the cathode $(z=0)$ and the anode $(z=L)$ to the density of charged particles, the electron energy density and potential:

$$\mathbf{n}\cdot\mathbf{\Gamma_e}\big|_0 = v_{e\,th} n_e/2 - \gamma\,\mathbf{n}\cdot\mathbf{\Gamma}_k,\ \ \mathbf{n}\cdot\mathbf{\Gamma_e}\big|_L = v_{th\,e} n_e/2, \tag{6}$$

$$\mathbf{n}\cdot\mathbf{Q}\big|_0 = \mathbf{n}\cdot\mathbf{Q}\big|_L = 5/6 v_{th\,e} n_e, \tag{7}$$

$$\mathbf{n}\cdot\mathbf{\Gamma}_k\big|_0 = v_{i\,th} n_k,\ \ \mathbf{n}\cdot\mathbf{\Gamma}_k\big|_L = v_{i\,th} n_k + \mu_k\,\mathbf{n}\cdot\mathbf{E}, \tag{8}$$

$$\varphi(0) = 0,\ \varphi(L) = V_{rf},\ V_{rf} = V_0 \sin(wt), \tag{9}$$

where $\gamma = 0.1$ – is the secondary emission coefficient, $v_{e\,th} = \sqrt{8k_B T_e/\pi m_e}$ and $v_{i\,th} = \sqrt{8k_B T_e/\pi M}$ – are the electron and ion thermal velocity, respectively, V_0 – is the voltage amplitude, $w = 2\pi\nu$ – is cyclic frequency, ν – is frequency.

Surface charge accumulation is added to the dielectric surfaces that are adjacent to the gap where the plasma forms by way of the following boundary condition:

$$\mathbf{n}\cdot(\mathbf{D_1} - \mathbf{D_2}) = \rho_s, \tag{10}$$

where ρ_s – is the surface charge density, which is computed by solving the following distributed ODE on the surfaces:

$$\frac{d\rho_s}{dt} = \mathbf{n}\cdot\mathbf{J}_i - \mathbf{n}\cdot\mathbf{J}_e, \tag{11}$$

where $\mathbf{n}\cdot\mathbf{J}_i$ – is the normal component of the total ion current density at the wall, and $\mathbf{n}\cdot\mathbf{J}_e$ – is the normal component of the total electron current density at the wall.

3. Results

Modeling of barrier micro-discharge parameters was carried out in the framework of a one-dimensional geometry of the discharge Fig.1. The magnitude of the discharge gap, the distance between the inner surfaces of the dielectrics L = 2 mm, the thickness of the dielectric barriers d = 1 mm and relative permittivity coefficient $\varepsilon = 10$.

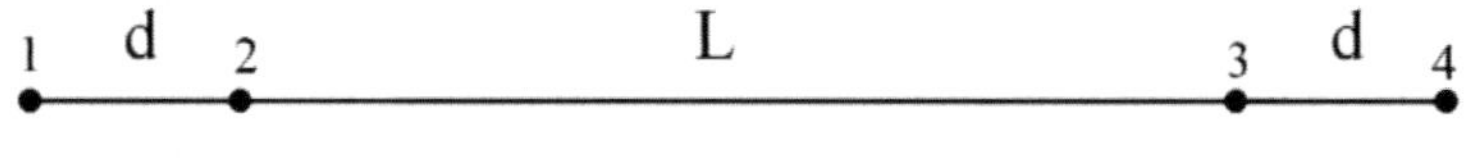

Fig. 1. The scheme of one-dimensional geometry of the DBD. 1 is left-electrode; 1-2, 3-4 are dielectrics; 2-3 is the discharge gap; 4 is right electrode.

The left electrode is grounded and the other is energized V_{rf}. In our calculation we assumed: V_0=2kV, ν = 50 kHz. Atmospheric pressure is expected.

As a result of numerical experiments for the conditions we observed a pulse current to half the period applied to the discharge voltage. It occurs with increasing the potential drop across the gap to reach the breakdown voltage, and is accompanied by a sharp decrease in the voltage drop across the barriers. It is due to the neutralization of the surface charge of the dielectric barriers to the flow of current (Fig. 2). In Fig. 3 shows the space-time dependence of the electron concentration in the discharge gap. It is seen that when the t = 40 mks character of the distribution parameters in time begins to repeat. The maximum concentration of charged particles observed in moments of current pulses.

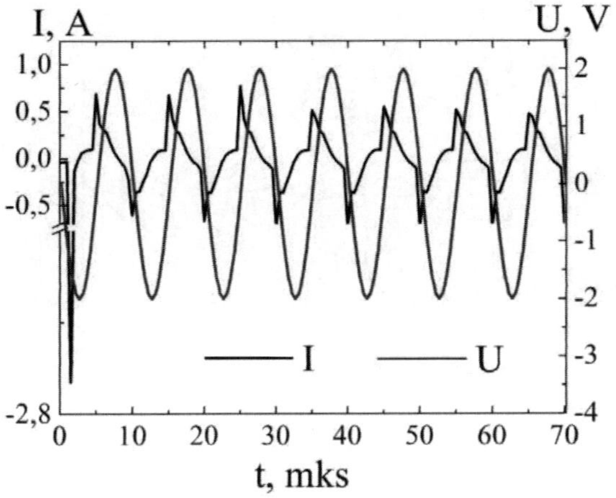

Fig. 2. Time dependence of the voltage across the electrodes and the discharge current.

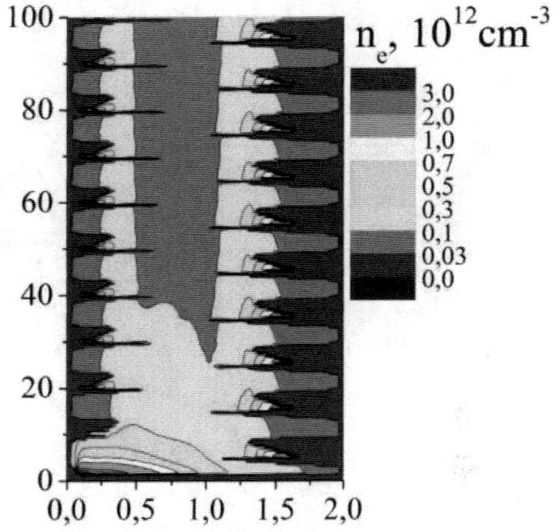

Fig.3. The space-time dependence of the electron density in the discharge gap.

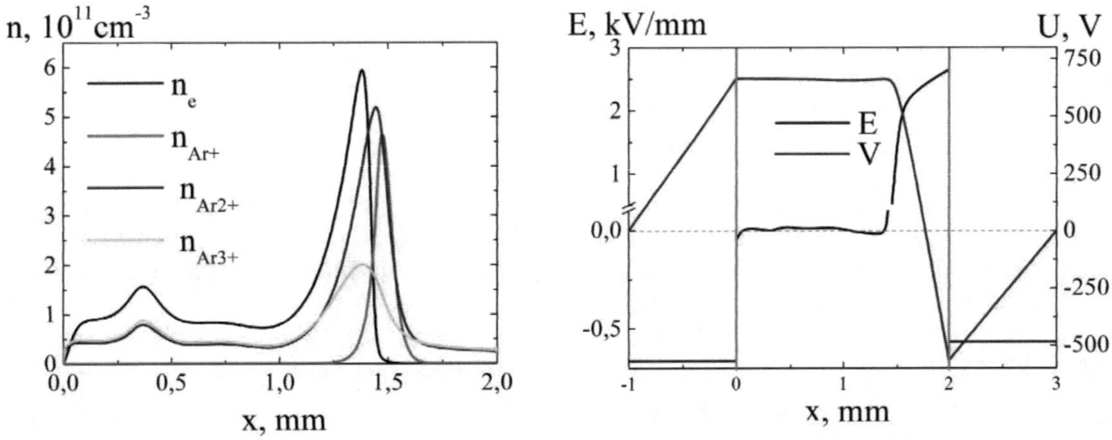

a) *b)*

Fig. 4. *a)* The density distribution of charged particles in the discharge gap, *b)* the electric field and potential distributions at the time t = 45 mks.

The pattern of distribution of basic parameters of the barrier micro-discharge at times when the current reaches the maximum value at the different polarity of the applied voltage are shown in Fig. 4-5. It can be seen that there is some similarity distributions with distributions in DC glow discharge. In both cases, there are layers of a positive charge near dielectric. A large space charge occurs near the negative electrode. Following the positive charge layer there is a maximum concentration of charged particles with a minimum of the electric field (the region of negative glow NG). It is worth noting that while there is an interesting phenomenon – a reversal of the sign of the electric field. The main ions in the high density of charged particles (NG) are Ar^+ and Ar_2^+.

The following is the second point of the treatment field (transition region between the dark space (FDS) and positive column (PC). It is followed by a region with a uniform distribution of plasma parameters (PC) with a characteristic small maximum concentration of the charged particles and a minimum of electric field. In this lengthy plasma region Ar_2^+ and Ar_3^+ are the main ions. In the transition from a uniform distribution of the discharge parameters in a PC to a small space charge of the positive electrode is observed reversal of the sign of electric field.

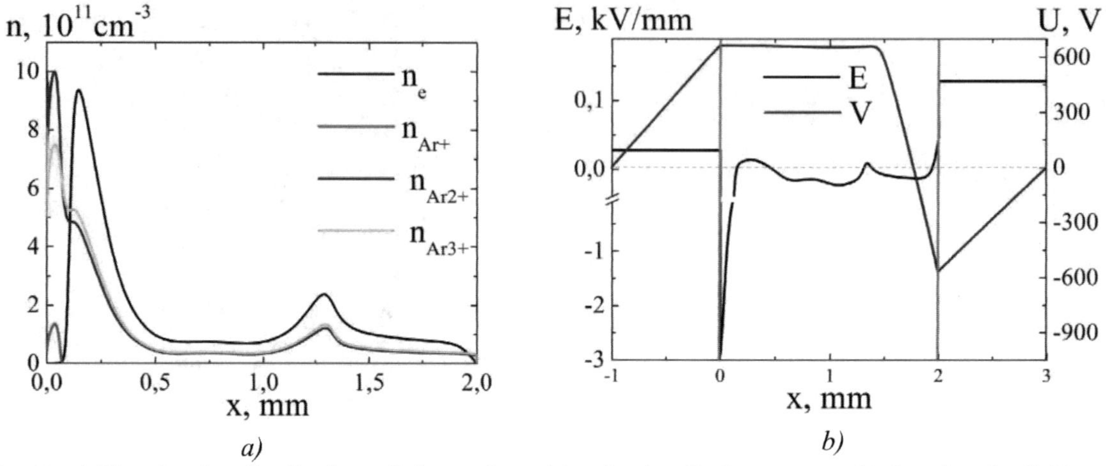

Fig. 5. *a)* The density distribution of charged particles in the discharge gap, *b)* the electric field and potential distributions at the time $t = 50$ mks.

4. Conclusions

Thus, in the simulation result was obtained spatial-temporal distribution pattern parameters of dielectric barrier micro-discharge at atmospheric pressure. It has been shown that there is a current pulse at half period the voltage supplied to the discharge for the conditions of the discharge and the account number of plasma chemical reactions occurring in the discharge. Graphs the distribution of the main DBD parameters are presented in moments of current pulses with different polarities. It is shown that the distribution pattern similar to the distributions in DC glow discharge. It was found a reversal of the sign of the electric field. The described model of allows us to describe the main parameters of DBD plasma at atmospheric pressure and it is a useful tool in predicting their basic properties under various external conditions.

The work is performed according to the Russian Government Program of Competitive Growth of Kazan Federal University.

References

[1] Raizer Y P 1992 *Gas Discharge Physics* (Dolgoprudny: "Intellekt") p 734
[2] Saifutdinov A I, Fadeev S A, Saifutdinova A A, Kashapov N F, 2015, *JEPT letters* **102** 637-642
[3] Benard N, Mizuno A, Moreau E, 2009, *J. Phys* D **42** 235204
[4] Opaits DF, Shneider M N, Miles R B, Likhanskii A V, Macheret S O, 2008, *Phys. Plasma* **15** 073505
[5] Li Q, Zhu X M, Li J T, Pu Y K, 2010, *J. Appl. Phys.* **107**, 043304
[6] Xia G, Chen Z, Saifutdinov A I, Eliseev S, Hu Y, and Kudryavtsev A A, 2014, IEEE Transaction on plasma science, **42** 10 2768 - 2769
[7] Saifutdinov A I, Timerkaev B A, and Zalyaliev B R, 2014, *J. Phys.: Conf. Ser.* **567** 012031
[8] Saifutdinov A I, Timerkaev B A, 2012, J. Eng. Phys. Thermophys. **85** 1202 .
[9] Bogdanov E A, Kudryavtsev A A, Chirtsov A S, 2011, *Technical Physics.* **56** 1 55-60

Journal of Physics: Conference Series **669** (2016) 012045 doi:10.1088/1742-6596/669/1/012045

Dynamics contraction of DC glow discharge in argon

A I Saifutdinov[1,2], A A Saifutdinova[1], N F Kashapov[1] and S A Fadeev[1]
[1]Kazan Federal University, Institute of Physics, Kazan 420008, Kazan, Russia
[2]St. Petersburg State University, Faculty of Physics, 198504, St. Petersburg, Russia

E-mail: as.uav@bk.ru

Abstract. The article presents the results of modeling the dynamics contraction of DC glow discharge. The proposed model can be an important tool in the calculation of devices based on glow discharge plasma at high (up to atmospheric) pressure. Furthermore, the model can be used.

1. Introduction

As is well known, it is difficult to obtain a stable glow discharge at relatively high pressures and currents. Contraction of the positive column is formed under these conditions. This phenomenon in atomic and molecular gases and their mixtures, which are working mixture gas lasers, occurs in plane channels and cylindrical discharge tubes. On the one hand, constriction of the discharge phenomenon is negative, which leads to failure of lasing, which limits energy contribution to the active medium. On the other hand, the detachment of the plasma from the wall of discharge tube is helpful in solving problems of spectral analysis of gas mixtures. In addition, based on the contracted discharge plasma developed inertial sensors that allow for the displacement of the plasma column to produce an electrical signal proportional to the acceleration.

However, in those and other cases to analyze and predict the behavior of devices and systems in which there is a DC glow discharge constriction is requires the involvement of both experimental and theoretical methods.

2. Model description

As is well known [1, 2, etc.], The main reasons for the contraction of the positive column are: a) A non-homogeneous heating of the gas with a maximum temperature on the tube axis, which leads (in view of constant pressure) to displace the neutral particles to the periphery. This leads to an increase in the reduced electric field E_z/N in the axial part of the column. Since the ionization rate very strongly (exponentially) depends on the reduced field, it is already a slight increase in this parameter in the axis leads to a strong contraction of the ionization zone; b) The nonlinear dependence of the ionization rate on the electron density (the degree of ionization n_e/N). Maxwellization of electron distribution function (EDF) due to collisions between the electrons leads to an exponential growth in the rate of ionization with increasing n_e/N. Therefore, a decrease in the degree of ionization in the periphery also leads to delays in the area of charged particles; c) The basic process of electron loss occurs due to volume recombination.

It is worth noting that in the plasma glow discharge elastic scattering of electrons is isotropic. It plays main role in the formation of the EDF. The source of the anisotropy is the external field. Therefore f_0-f_1 decomposition [1] 2schsch holds for EDF. Since in a contracted positive column EDF is Maxwellian due to electron-electron collisions, then a chain of kinetic equations for f_0-f_1 considering Maxwell's EDF is easy to obtain a fluid approximation [2], which with sufficient accuracy allows to describe the constricted DC glow discharge from the cathode to the anode.

In this paper was considered the glow discharge at high pressure in an inert gas – argon at conditions when the state of discharge is constricted. Numerical experiments were based on the drift-diffusion model, which was formulated in [3-8].

The model is based on balance equations for the densities of charged (electrons, ions) and the excited particles, the electron energy density, heat equation, which takes into account the heating of the heavy particles in the plasma and the Poisson equation for the potential φ. The stationary state was determined by solving the time-dependent equations. Detailed description of the model and method of the numerical solutions have been considered earlier in [9]. The calculations were made for argon using a kit comprising 23 plasma-chemical reactions [9], and taking into account the two effective excited atomic level Ar^*, Ar_r^*, and one excimer level Ar_2^*, two kinds of ions Ar^+, Ar_2^+.

3. Numerical experiments and results

Numerical experiments were carried out for the traditional cylindrical configuration of the discharge tube radius R=1 cm and an inter-electrode distance L=4 cm. The gas pressure was 50 Torr. The voltage at the source was set to $\varepsilon_i = 2\,\text{kV}$, ballast varied from 10 to 100 kOhms.

Was obtained spatio-temporal pattern of distribution of plasma parameters of constricted glow discharge in argon from the cathode to the anode when $\Omega = 20$ kOhm. Thus in Fig. 1-3 shows the main parameters of the constricted glow discharge - the electron density, the temperature of the heavy particles and the plasma potential distribution at the times t = 1.4 mks, t= 1.8 mks, t = 3.0 mks, t = 5.0 mks and t = 100 mks.

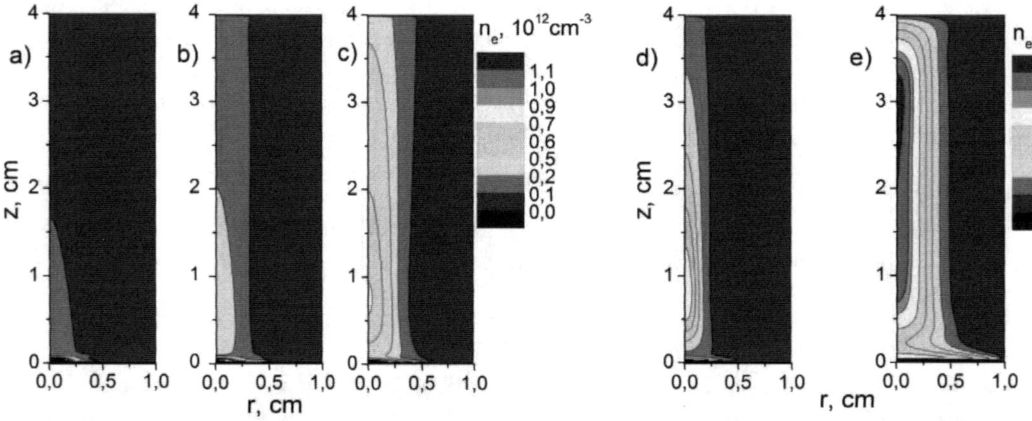

Figure 1. The spatial distribution of the electron density in DC glow discharge at the moments: a) t=1.4 mks, b) t=1.8 mks, c) t=3.0 mks, d) t=5.0 mks, e) t=100 mks

Spatio-temporal pattern formation of the structure of the contracted glow discharge similar to the diffuse form and extends the 3-5 mks:

1) loss of electrons to an anode, in this period significantly increased current density;
2) the process of ionization in the volume of the gas-discharge gap is characterized by the fall of the current density;
3) ionization in the discharge gap, accumulation of ion density near the cathode and the transition to a self-sustaining form of discharge.

The difference lies in the fact that during all the time of formation of the discharge plasma region contracted to a thin cord. The "head" string is distributed at a rate of about 10^6 sm/s. Thus it is seen that heating gas which is observed in the vicinity of the cathode in the first 1 mks, unlike diffuse form discharge begins to spread along the axis of the discharge chamber. A warm front at the same time moving at speed $7 \cdot 10^6$ sm/s. After the "head" cord reaches the counter electrode, there is a sharp drop in the voltage between the electrodes and the substantial increase of the current.

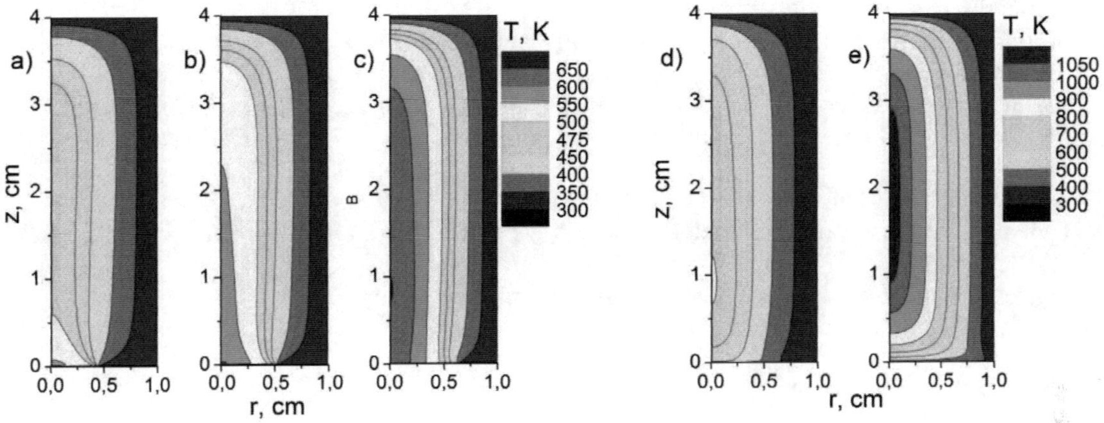

Figure 2. Spatial distribution of heavy particles temperature in DC glow discharge at the moments времени: a) t=1.4 mks, b) t=1.8 mks, c) t=3.0 mks, d) t=5.0 mks, e) t=100 mks

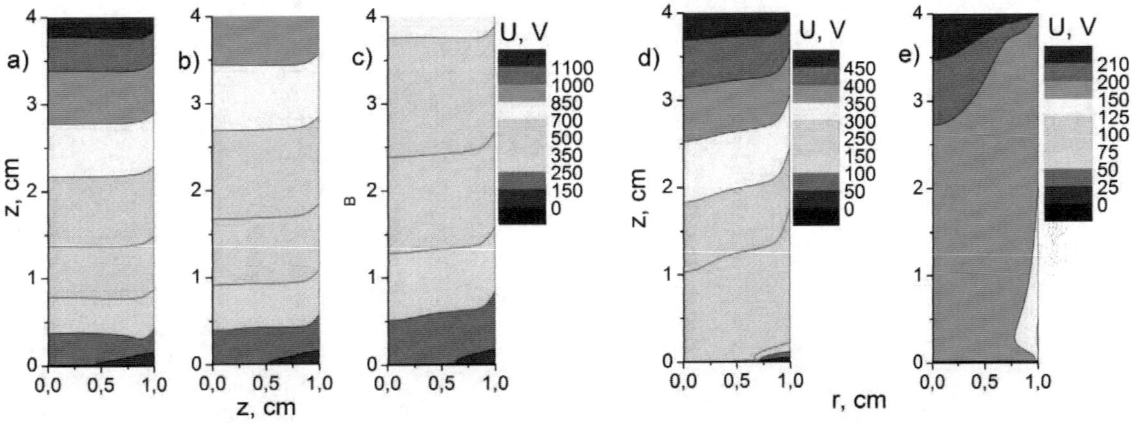

Figure 3. Spatial distribution of electric potential in DC glow discharge at the moments a) t=1.4 mks, b) t=1.8 mks, c) t=3.0 mks, d) t=5.0 mks, e) t=100 mks

4. Conclusions

Thus, the article presents the results of modeling the dynamics contraction of DC glow discharge. The proposed model can be an important tool in the calculation of devices based on glow discharge plasma at high (up to atmospheric) pressure. Furthermore, the model can be used to test different methods for control DC glow discharge the parameters [9, 11, 12], including to improve their stability.

Acknowledgments
The work is performed according to the Russian Government Program of Competitive Growth of Kazan Federal University. The reported study was funded by RFBR, according to the research project No. 16-38-60187 mol_a_dk.

References

[1] Raizer Y P 1992 *Gas Discharge Physics* (Moscow, Science) p 536

[2] Kudryavtsev A A, Smirnov A S and Tsendin L D 2010 *Physics of Glow Discharges* (St. Petersburg, Lan) p 512

[3] Fiala A, Pitchford L C and Boeuf J P 1994 *Phys. Rev. E* **49** 5607

[4] Donko Z, Hartmann P and Kutasi K 2006 *Plasma Sources Sci. Tech.* **15** 178

[5] Bogdanov E A, Demidov V I, Kudryavtsev A A and Saifutdinov A I 2015 *Phys. Plasmas* **22** 024501

[6] Saifutdinov A I and Timerkaev B A 2012 *J. Eng. Phys. Thermophys.* **85** 1202

[7] Surzhikov S and Shang J 2014 *Plasma Sources Sci. Tech.* **23** 054017

[8] Saifutdinov A I, Timerkaev B A and Zalyaliev B R 2014 *J. Phys.: Conf. Series* **567** 012031

[9] Saifutdinov A I, Fadeev S A, Saifutdinova A A and Kashapov N F 2015 *JETP Lett.* **102** 637–642

[10] Shkurenkov I A, Mankelevich Yu A and Rakhimova T V 2009 *Phys. Rev. E* **79** 046406

[11] Fadeev S A, Kashapov N F and Larionov V M 2014 *IOP Conf. Ser.: Mater. Sci. Eng.* **69** 012006

[12] Fadeev S A and Kashapov N F 2013 *J. Phys.: Conf. Series* **479** 012009

Addition agents effects on hydrocarbon fuels burning

V M Larionov, G A Mitrofanov, A V Sakhovskii

Kazan Federal University, 18 Kremlyovskaya str., Kazan 420008, Russian Federation

E-mail: techphys@kpfu.ru

Abstract. Literature review on addition agents effects on hydrocarbon fuels burning has been conducted. The impact results in flame pattern and burning velocity change, energy efficiency increase, environmentally harmful NO_x and CO emission reduction and damping of self-oscillations in flow. An assumption about water molecules dissociation phenomenon existing in a number of practical applications and being neglected in most explanations for physical-chemical processes taking place in case of injection of water/steam into combustion zone has been noted. The hypothesis about necessity of water dissociation account has been proposed. It can be useful for low temperature combustion process control and NO_x emission reduction.

1. Introduction

Hydrocarbon fuels combustion technologies are widely spread and rapidly developing recently. Development is specified by energy market competition and toughening environmental requirements. Using of different fuel additives significantly affecting combustion processes is a way to satisfy it.

Water and steam injection is challenging technology for power systems based on gas-turbine engines [1]. This method is promising for low and average power engineering where use of high performance combined-cycle plants is unreasonable and implies long payback period.

Injection takes place into different air-gas channel areas: compressor inlet, combustion chamber or turbine cooling system. Increasing water/steam temperature by gaining burnt gases heat by means of recovery boiler is reasonable.

2. Review and discussion

Numerous computational simulations and experimental measurements devoted to addition agents effect on hydrocarbon fuels burning were conducted. Water, steam, hydrogen, ammonia etc. diluting fluid and gaseous fuels (black oil, kerosene, methane and different gas mixtures) were investigated [2-7]. Injection was implemented separately as well as being premixed with fuel (fine-dispersed emulsified fuels). Experimental setups complexity varies from simple atmospheric burners up to full-scale sophisticated combustors.

It was determined that addition agents could increase thermal efficiency and power, improve environmental aspect decreasing significantly NO_x and CO emissions and decrease combustion instabilities.

Accepted explanation for these effects are based on additive heat release corresponding components calorific values in case of efficiency improving injection and decreased temperatures of lean flame in areas of NO_x formation in case of emission improving injection. However, these models are not proper for explaining hydrocarbon and carbon monoxide CO burning as well as NO_x

Journal of Physics: Conference Series **669** (2016) 012046 doi:10.1088/1742-6596/669/1/012046

neutralization that take place at low (<900K) temperatures in presence of water steam without any catalyst.

Kalashnikov A.P. [8] described method for limiting NO_x emission by mixing CO_2 with hot combustion products containing water steam and passing through finely porous carbonaceous substance.

It can be explained by following assumption for most combustion processes taking place in presence of steam. Molecules of water dissociate into OH^- and H^+ ions that react with NO_x. Its oxidation state gets decreased to the lowest (NO).

$$NO_3 + H^+ \rightarrow NO_2 + OH^-$$
$$NO_3 + OH^- \rightarrow NO_2 + HO_2$$
$$NO_2 + O^- \rightarrow NO + O_2$$
$$NO_2 + H^+ \rightarrow NO + OH^-$$

Then nitrogen monoxide gets neutralized with CH and NH radicals:

$$NO + CH \rightarrow CO + NH$$
$$NO + NH \rightarrow N_2 + OH^-$$

Intensive NO_x neutralization requires significant inflow of CH that is the final substance of hydrocarbons decomposition that takes place with OH^-, O^+ and H^- participation. These radicals can be contained in fuel components or appear due to water dissociation.

$$CH_4 + OH^- \rightarrow CH_3 + H_2O$$
$$CH_4 + H^+ \rightarrow CH_3 + H_2$$
$$CH_4 + O^- \rightarrow CH_3 + OH^-$$
$$CH_3 + OH^- \rightarrow CH_2 + H_2O$$
$$CH_2 + H^+ \rightarrow CH + H_2$$
$$CH_2 + O^- \rightarrow CH + OH^-$$

Burning of generated hydrogen yields fuel calorific value increase.

Water dissociation can be caused not only by thermal effect but also by water molecules vibrational excitation in case of different impacts – rapid fuel and oxidizer molecular-level mixing process, solid catalyst presence, electromagnetic pulse impact, etc. In such cases energy consumed by dissociation is significantly less than energy released from hydrogen burning.

Technologies for neutralization of NO_x contained in post-combustion gas that use ammonia (NH_3) are linked to forementioned processes. Due to Muzio and Arand experimental data [9] shows that NO_x emission decreases with temperature increase, reaches its minimum in the range of 1200-1250K and then tends to raise again (Figure 1).

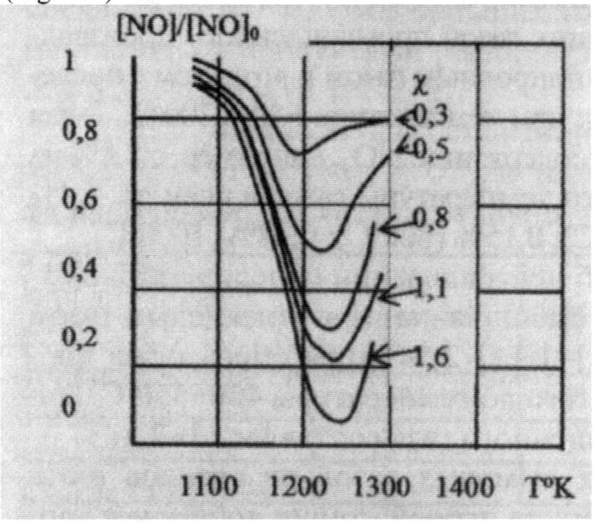

Figure 1. Effect NH_3 injection on NO_x reduction, $\chi = [NH_3]/[NO]$

According to usual burning theory main reactions rate coefficients are in monotonic dependence of temperature and give no explanation about pronounced minimum of NO_x concentration. Thus, it happens due to other reasons. Reaction mechanism proposed by Miller and Bowman [10] considers NO_x reduction with help of self-generated OH^- radicals and NH_3 molecules:

$$NH_3 + OH^- \leftrightarrow NH_2 + H_2O$$
$$NH_2 + NO \leftrightarrow N_2 + H_2O$$

Gaining certain temperature value major part of ammonia gets dissociated and NO_x reduction begins to go down.

3. Conclusion

Effects of water and steam injection on hydrocarbon fuels combustion including methane-air burning are well investigated but effects of probable water dissociation usually are not taken into consideration. In spite of separate researches there is no clear ultimate model taking account of dissociation yet. Relying on available experimental results analysis an assumption about ability to manage combustion processes via water dissociation at different temperatures has been proposed.

The work is performed according to the Russian Government Program of Competitive Growth of Kazan Federal University.

References

[1] Pavlova A A 2011 *Prospects for the forcing of gas turbine and steam power plants* (National research university)
[2] Boyce M P, Vyas Y K and Trevillion W L 1978 *Proc. of the 13th Intersociety Energy Conversion Engineering Conference* **1** 860-865
[3] Touchton G L 1985 *Energeticheskie mashinyi* **107** 118-125
[4] Murotani T, Hattori K, Sato E, Chryssakis C, Babajimopoulos A and Assanis D N 2007 (SAE World Congress)
[5] Boushaki T, Dhué Y, Selle L, Ferret B, Poinsot T 2012 *International Journal of Hydrogen Energy* **37** 9412-9422
[6] Larionov V M, Mitrofanov G A, Iovleva O V 2014 *J. Phys.: Conf. Ser.* **567** 012028
[7] Bykovets A P, Larionov V M, Marchukov E Yu 1992 *Izvestiya Vysshikh Uchebnykh Zavedenij. Aviatsionnaya Tekhnika* (3) 71-74
[8] Kalashnikov A P 1973 Byull. Izobr. SSSR №387725
[9] Muzio L J, Arand J K 1978 *17th Symp. (Int) on Combustion*
[10] Miller J A and Bowman C T 1989 *Prog. Energy Combustion Science* **15** 287-298

Journal of Physics: Conference Series **669** (2016) 012047 doi:10.1088/1742-6596/669/1/012047

Self-exited gas oscillations in Helmholtz resonator type combustor

V M Larionov, E R Saifullin, E V Semenova

Kazan Federal University, 18 Kremlyovskaya str., Kazan 420008, Russian Federation

E-mail: techphys@kpfu.ru

Abstract. When the length of the resonance tube is much less and order of the acoustic wavelength is studied. Experimental set-up with a multiport burner is considered. The "energy balance" method is used to calculate the frequency and amplitude of the gas oscillations. The results of measurements and calculations show that increasing the length of the resonance tube causes the excitation oscillation with a first lower frequency. Further, there are no oscillations. When the length of the resonance tube becomes the order of the wavelength, the second resonant frequency is excited. The calculation results are in good agreement with the experimental data.

1. Introduction

Researches of the combustion-driven oscillations in Helmholtz resonator type combustors have been provided in works [1,2]. If the length of the combustion chamber and the resonance tube is much smaller than the acoustic wavelength we have classic Helmholtz resonator, consisting of acoustic volume (the combustion chamber) and the "throat" (short narrow pipe). This oscillator has a single resonant frequency [3]. When the length of tube is the order of one or more wavelengths we have resonance tube with acoustic volume at the inlet and open at the outlet. Such a system has a number of resonance frequencies [4-6].

The aim of this work is research of combustion-driven oscillations in Helmholtz resonator type combustor with a long resonance tube.

2. Influence of length of the resonance tube at the frequency and amplitude combustion-driven oscillations

Schematic diagram of the combustion device is shown in Figure 1.

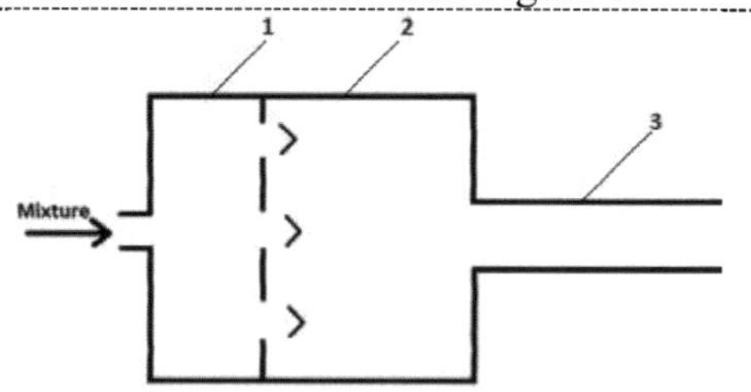

Figure 1. Schematic diagram of the combustion device. 1 – multiport burner, 2 - combustion chamber, 3 - resonance tube

Experimental combustion chamber with a multiport burner and measurement technique were the same as in [7]. Combustion chamber volume was constant $V = 2 \cdot 10^{-5}\ m^3$. The resonant tube has sections. The length of each section of 0.01 m, a diameter of 0.015 m. Stoichiometric propane-air mixture is used. We were obtained according to the frequency and amplitude of pressure fluctuations in the combustion chamber of the length of tube. Oscillations are excited when the tube length reaches 0.05 m and terminated in the $l = 0.1$ m (Figure 2, curve 1).

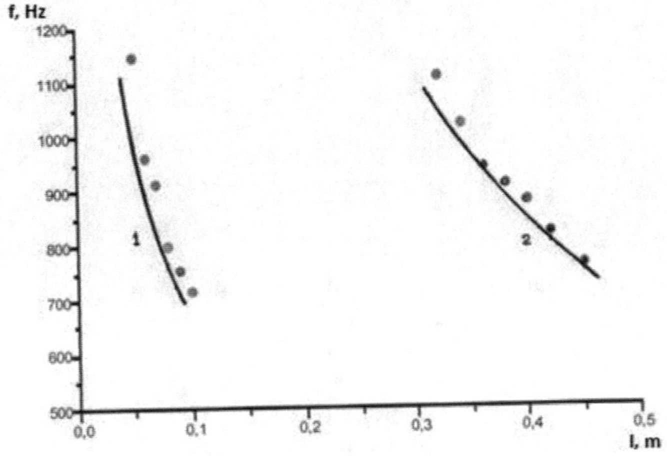

Figure 2. Dependence of the oscillation frequency on the length of the pipe. 1 - first frequency; 2 - second frequency; points - experiment line - calculation

In the interval 0.1 m < l < 0.32 m pulsating combustion is not observed. When the tube length is 0.32 m, the self-excitation of oscillations occurs again, and the combustion pulsation exists as long as the length of the tube is greater than 0.45 m (Figure 2, curve 2). Both curves correspond to the same range of the oscillation frequency. In the interval 0.05 m < l < 0.1 m. Criterion of self-exited oscillations is the same as in the Helmholtz resonator [7].

$$\pi < \omega\tau_u < 2\pi.$$

If the mixture has a constant composition, the delay time of combustion τ_u not changed. Increasing the length of tube causes a decrease in the first frequency ω_1 and the excitation of oscillations is observed, since the phase shift $\omega_1\tau_u$ changed from 2π to π.

Further, for the first frequency should be $\omega_1\tau_u < \pi$. When the length of the resonance tube changes from 0.1 m to 0.32 m we have condition: $\omega_2\tau_u > 2\pi$. Therefore, the conditions of self-excitation of gas oscillations for both frequencies are not performed, and there is no pulsating combustion.

When the length of tube varies from 0.32 m to 0.45 m oscillations with the second frequency are exited, since the phase shift $\omega_2\tau_u$ decreases from 2π to π. If the tube length is more than 0.45 m, gas oscillations with the first and second frequencies are not excited.

Dependence of sound pressure level in the combustion chamber on the length of the resonance tube (Figure 3) have maximum in the middle.

The amplitude of the oscillations reaches its maximum value when the phase shift has an optimum value. In this case, the acoustic power generated by the combustion zone and transmitted to gas oscillations - the largest [8].

"Energy balance" method and techniques developed in [9,10] have been used. The calculation results are shown in Figure 2 and Figure 3. The comparison shows that the calculation results are in good agreement with the experimental data.

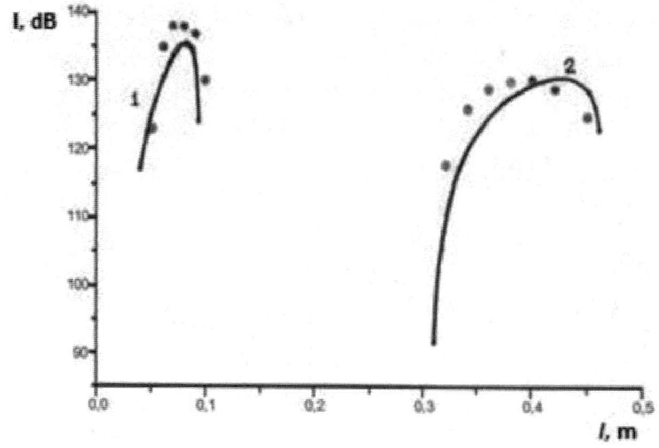

Figure 3. Dependence of the sound pressure level on the length of the resonance tube

3. Conclusion

The study of self-exited oscillations in the Helmholtz resonator type combustor was provided. "Energy balance" method was used. The results of measurements and calculations show that the increase in the length of the resonance tube causes the excitation oscillations with the first lower frequency. Further, there are no oscillations. When the length of the resonance tube becomes the order of the wavelength, oscillations with the second resonant frequency are excited. The calculation results are in good agreement with the experimental data.

The work is performed according to the Russian Government Program of Competitive Growth of Kazan Federal University.

References

[1] Beloded O V, Larionov V M 2006 *Russian Aeronautics* **49** 44-48
[2] Iovleva O V, Larionov V M 2007 *Russian Aeronautics* **50** 303-308
[3] Iovleva O V, Larionov V M 2012 *Russian Aeronautics* **55** 388-391
[4] Ermakov R A, Galiullin R G, Larionov V M, Nikolaev A N 2008 *Chemical and Petroleum Engineering* **44** 387-392
[5] Ermakov R A, Sadykov A F, Galiullin R G, Larionov V M 2011 *Chemical and Petroleum Engineering* **46** 749-751
[6] Bykovets A P, Larionov V M, Marchukov E Yu 1992 *Izvestiya Vysshikh Uchebnykh Zavedenij. Aviatsionnaya Tekhnika* (3) 71-74
[7] Larionov V M, Iovleva O V 2003 *Izvestiya Vysshikh Uchebnykh Zavedenij. Aviatsionnaya Tekhnika* (4) 48-51
[8] Larionov V M, Nazarenko T I 2000 *Izvestiya Vysshikh Uchebnykh Zavedenij. Aviatsionnaya Tekhnika* (4) 69
[9] Yallina E V, Larionov V M, Iovleva O V 2013 *J. Phys.: Conf. Ser.* **479** 012017
[10] Semenova E V, Larionov V M, Kazakova E I 2014 *J. Phys.: Conf. Ser.* **567** 012033

Journal of Physics: Conference Series **669** (2016) 012048 doi:10.1088/1742-6596/669/1/012048

Experimental setup for plasma treatment of disperse materials in the arc plasma jet

K Sadikov, I Fayrushin, A Shamsutdinov and N Kashapov
[1]Kazan Federal University, Institute of Physics, Kazan 420008, Kazan, Russia

E-mail: fairushin_ilnaz@mail.ru

Abstract. The experimental setup for the plasma treatment of particulate materials was created. Spend the processing of composite powder of chromium oxide with aluminium oxide.

1. Introduction

Thermophysical properties of plasma and the treated particulate material, the ratio of their mass spending, how is organized the process of mixing the dispersed phase with thermal plasma flow, that is the scheme of the reactor, as well as the structure and the parameters of the plasma stream formed by plasma installation, - all these parameters have a direct influence on the effectiveness of treatment of a dispersed material in a plasma device [1, 2]. From the chosen technological process, its nuances and principle, depend on the composition of the plasma gas and the nature of the processed material, the weight ratio, as well as the scheme of the reactor, in which the cellular process plasma will processing. In contrast, the parameters and structure of the plasma flow are more free and controlled parameters. Efficiency itself heterogeneous plasma processes, is ultimately determined by the heat exchange between the plasma and the dispersion particles of the processed material [3-5].

2. Experimental setup

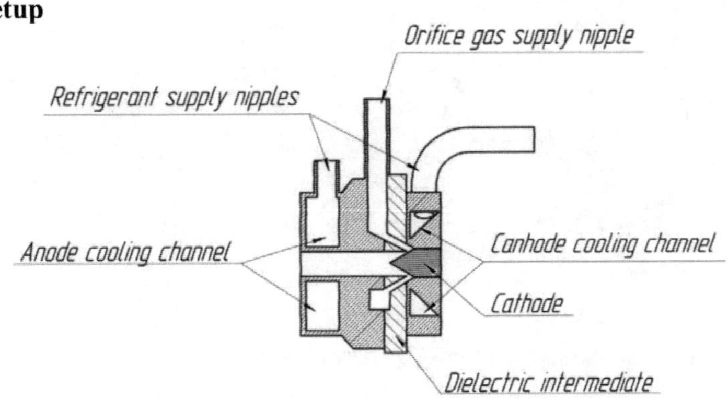

Figure 1. Schematic circuit of arc-plasmatron.

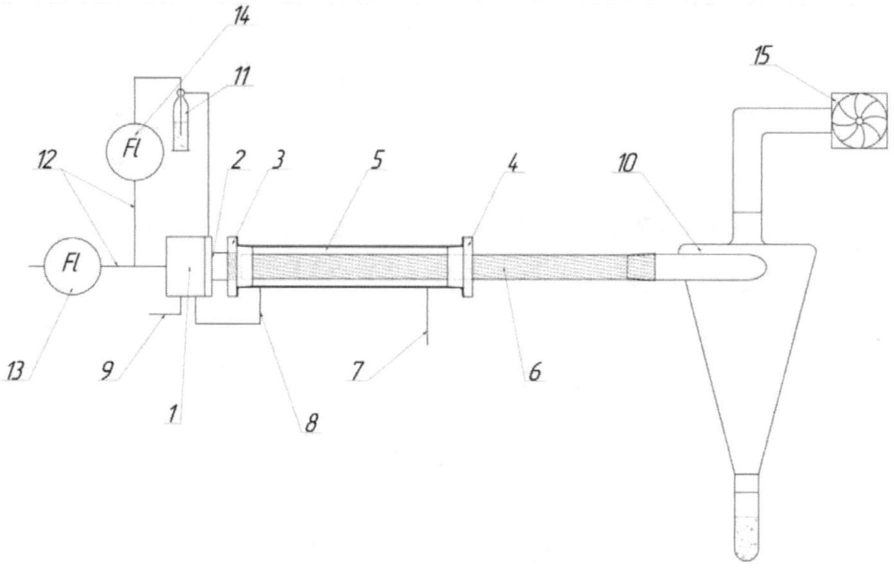

Figure 2. Schematic circuit of experimental setup.

Experimental setup consists of arc-plasmatron (1 and Fig.1) with self-setting arc, fastened with inch thread steel nozzle adapter (2), two flanges (3,4) for glass refrigerator (5) and a heat-resistant quartz tube (6), which is the flow of thermal dusty plasma. Through the fridge through the pipes (7, 8) circulates the water. The plasma torch has water cooled too (8, 9). The quartz tube is connected with a cyclone collector (10), intended for collecting the treated plasma powder (dust phase). As a plasma-forming gas is helium. The arc in the plasma torch is ignited by means of electric pulse oscillator. Power to the plasma torch is supplied by the Larionov's Scheme from the three-phase network (380 V). The pressure of plasma gas at the inlet to the plasmatron is 1.5 from the weather. The powder supply is carried out using the Drexel bottle (11), connected in parallel with the gas supply system to the plasma torch (12). For flow measurement gas is connected through the rotameters on the common entrance (13) and separately for gas passing through the Drexel bottle (14). The outlet gas from the cyclone dust collector is carried out using forced ventilation systems (15).

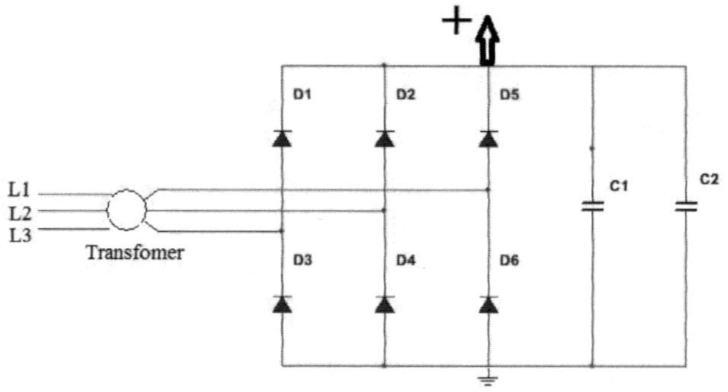

Figure 3. Schematic circuit of experimental setup's power supply.

3. Results

Catalytic tests of plasma treated samples showed that with the increase of plasma processing time in the dehydrogenation activity passes through a maximum (at 55 seconds A = 40.0%). Further increase plazmoobrabotki time leads to a decrease in catalytic activity due to sintering of the active component in the three-dimensional crystals of the less active α-Cr_2O_3.

time, sec	energy, W	Activity, %
45	1000	37,0
45	800	33,4
55	1000	40,0
65	1000	33,0
zero point (heat-treated sample)		43,1

4. Conclusions

Thus, the plasma treatment boehmite support coated with chromic anhydride at 45-55 seconds accompanied with the formation of a catalytically active phase CrO_3 Cr_2O_3, however, the plasma energy is not enough to boehmite alumina, resulting in a smaller surface and thus plasma treatment dehydrogenation activity of the catalysts. With increasing duration of plasma treatment there is no complete phase transition γ-AlOOH $\rightarrow$ γ-Al_2O_3, however, observed catalyst deactivation due to agglomeration of X-ray amorphous to crystalline Cr_2O_3 α-$Cr2O_3$.

Acknowledgments
The work is performed according to the Russian Government Program of Competitive Growth of Kazan Federal University.

References

[1] Dautov G Yu, Sazonov M I 1971 *Journal of Applied Mechanics and Technical Physics.* **4** 86
[2] An'shakov A S, Dautov G Yu, Timoshevskii A N 1973 *Journal of Applied Mechanics and Technical Physics.* **5** 867
[3] Kornienko E E, Lapushkina E J, Kuzmin, V I, Vaschenko S P, Gulyaev I P, Kartaev E V, Sergachev D S, Kashapov N, Sharifullin S, Fayrushin I 2014 *Journal of Physics: Conf. Series* **567** 012010
[4] Fayrushin I , Dautov I, Shamsutdinov A 2014 *Journal of Physics: Conf. Series* **567** 012007
[5] Fayrushin I, Kashapov N, Dautov I 2014 *Journal of Physics: Conf. Series* **567** 012009

Journal of Physics: Conference Series **669** (2016) 012049 doi:10.1088/1742-6596/669/1/012049

Surface hardening of cutting elements agricultural machinery vibro arc plasma

S N Sharifullin[1], N R Adigamov[2], N N Adigamov[2], R Y Solovev[3], K S Arakcheeva[3]
Kazan Federal University[1]
Kazan State Agrarian university[2]
All-Russian Research Institute of Technology repair and maintenance tractor fleet (GOSNITI)[3]

E-mail: Saidchist@mail.ru

Abstract. At present, the state technical policy aimed at the modernization of worn equipment, including agriculture, based on the use of high-performance technology called nanotechnology. By upgrading worn-out equipment meant restoring it with the achievement of the above parameters passport. The existing traditional technologies are not suitable for the repair of worn-out equipment modernization. This is especially true of imported equipment. Out here alone - is the use of high-performance technologies. In this paper, we consider the use of vibro arc plasma for surface hardening of cutting elements of agricultural machinery.

1. Introduction.

The most suitable high-performance technologies when recovering of agricultural equipment are plasma technology [1]. For plasma technologies typically use low-temperature plasma gas discharge [2-11]. A wide range of energy, thermal and gas-dynamic characteristics of low temperature plasma gas discharge enables its use in a variety of technologies. This is - the technology of heating, cutting, welding, welding, spraying a variety of metal, non-metallic and dielectric coatings, thin film deposition, various plasma-chemical reactions to produce products with unusual properties, the impact on the surface for the purpose of decontamination, polishing, hardening, changes in the structure . The heat capacity of the plasma jet can be varied from tens of watts to tens of megawatts, the temperature - from a few hundred degrees to 50 000°S, effective efficiency Heating may be up to 90%, the continuous operation period - up to 1000 hours 12, 13]. In this plasma apparatus, called plasma torches are relatively simple in structure and easy to operate process.

Analysis of the plasma properties and possibilities of its application has shown that for hardening of cutting elements agricultural units the most appropriate technology is vibro arc plasma technology hardening [14-16]. There is applied to the surface of the reinforcing composite material comprising aluminum oxide Al^2O^3 and SiO^2 silicon, and boron carbide B4C. When vibro arc cladding formed on the surface coating containing ceramic superhard inclusion of boron carbide, corundum and carbo corundum.. Simultaneously, welding takes place hardenable surface doping boron and nitrogen due to dissociation boronitro containing composite components, and carbon through its diffusion from the sublimation of the graphite electrode arc during combustion. The hardened layer is composed of three zones: a reinforced base transition and the main upper. Hardened base is characterized by a pronounced phase changes caused by the diffusion of the elements that make up the composite material. The transition zone is an alloy melted the top layer of the composite parts and mother. The main upper band is the most solid and consists of a steel matrix containing FeC formed iron carbides. Fe^2 C^3, borides FeB and Fe^2B and ceramic phase - iron spinel FeO Al^2O^3, boron carbide B^4C, corundum and carbo corundum. With this treatment, the surface of the working bodies of machines becomes a hardness of about 70 HRC and their endurance increases by 2.0 - 2.5 times.

2. Materials and methods of research.

For research were used 45 samples of steel 4 mm thick. As the composite material was chosen paste of the following composition: Powder PG-10H-01 - 55%, boron carbide - 20% borax - 10%, cryolite - 8%, SiO^2 - 5%. Al - 2%. Binder - 20% water glass solution. Research Methodology: The sample is pressed into a disc, grinding the resulting ground and polished. Microhardness measurements performed on one instrument KMT-1 Vickers method. The load F = 1 H, the holding time t = 10 seconds. The increase of 370.

3. Results and discussion.

Sample without treatment. Figure 1 shows a photograph of the grinding surface of the starting material. To study the microhardness was stamped 16 prints. On the surface of the section it can be seen that the structure of the starting material is uniform. Differences in the indentation of the device is not noticeable. The microhardness values of the sample are shown in table. 1.

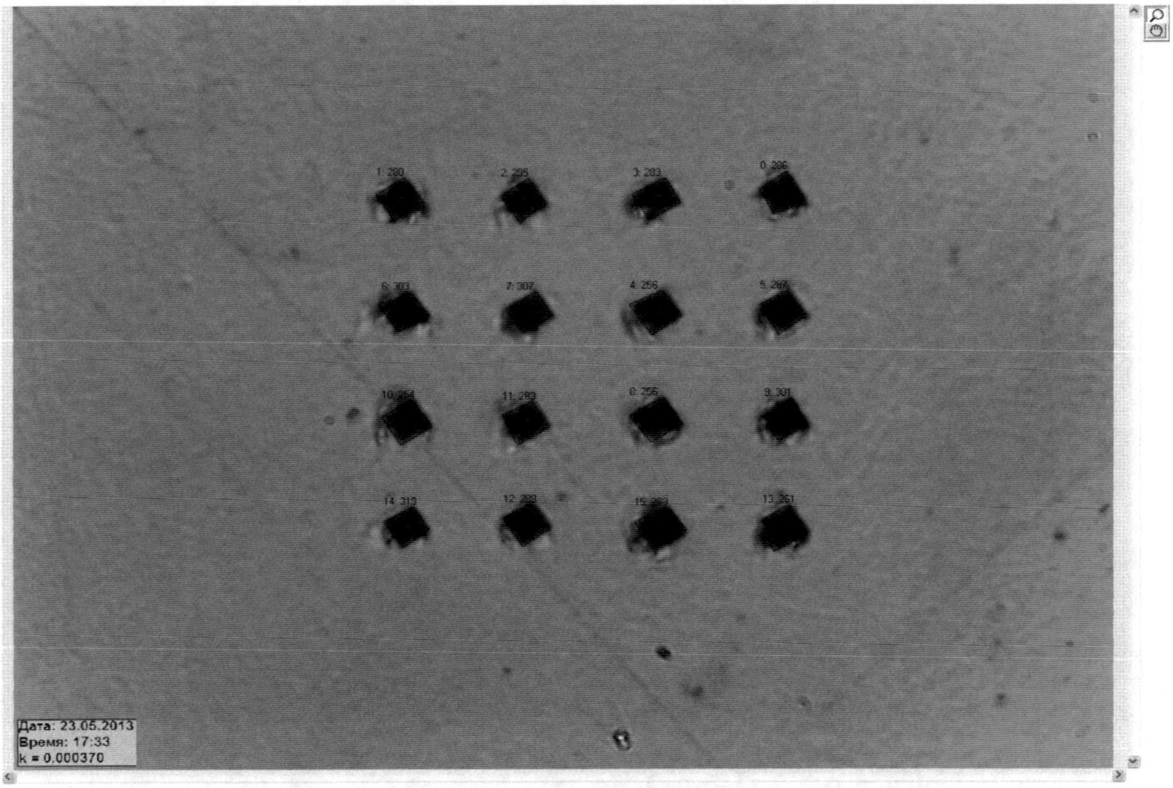

Figure 1 - Sample without treatment

Table 1 - The values of the microhardness of the initial sample

The study area	Microhardness, kgf/mm^2				Average value of microhardness, kgf/mm^2
The zone of the base material	280	295	283	286	286
	303	307	256	287	288
	254	283	256	301	274
	319	298	263	261	285
The mean value of microhardness of the base material, kgf/mm^2					283

Conclusion: The mean value of microhardness of the base material was 283 kgf/mm^2.

The sample after processing. Figure 2 shows a photograph of the grinding surface of the material after processing. To study the microhardness was stamped 36 prints: on the deposited layer 20 prints on the base material 16 prints. On the surface of the section it can be seen that the structure of alloy layer is homogeneous. The boundary of a clear transition. Microhardness was investigated on the weld layer, as well as on the base material. The microhardness values of the sample obtained in the border area of the two zones are shown in table. 2.

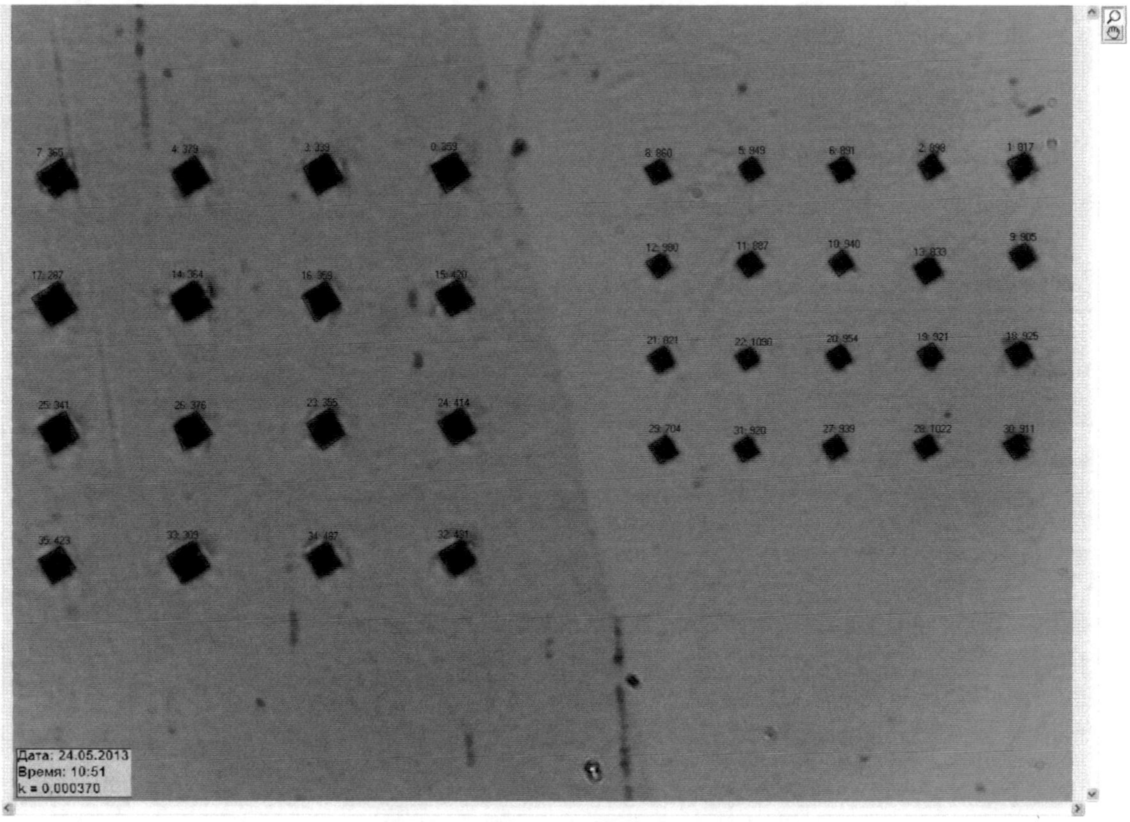

Figure 2 - The sample after processing

Table 2 - The microhardness values of the sample obtained in the border area of the two zones are shown

The study area	Microhardness, kgf/mm^2				Average value of microhardness, kgf/mm^2
The zone of the base material (hardened layer)	365	287	339	359	338
	379	364	359	420	381
	339	359	355	414	367
	359	420	487	431	424
The mean value of microhardness of the base material, kgf/mm^2					378
Coverage	860	980	821	704	841
	949	887	1090	920	962
	891	940	954	939	931
	898	833	921	1022	919
	817	905	925	911	890
The average microhardness of the coating material, kgf/mm^2					909

Conclusion: The mean value of microhardness of the base material near the surface of the sample was 378 kgf/mm^2. The mean value of microhardness of the coating material was 909 kgf/mm^2.

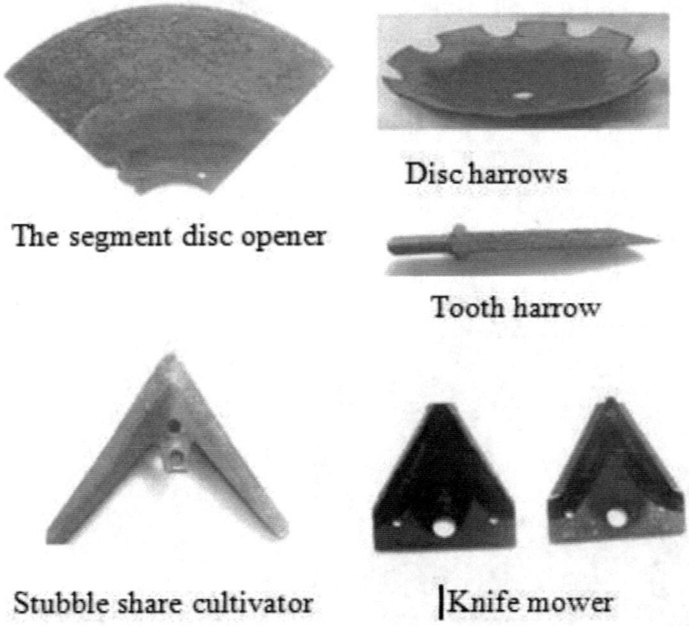

Figure 3 - Some hardened cutting elements agricultural units

Comparison of the results of processing the steel surface weld layer 45 of the composite material shows that the microhardness of the deposited layer is 3 times the microhardness of the substrate material, and the microhardness of the hardened backing layer - 1.5 times. Studies have shown that the hardening layer of the base material, depending on the injected power discharge may reach 1.5 - 2 mm thick, and in some cases up to 3 - 4 mm.

It was investigated the variation of microhardness steel surface 45 when processing vibro arc plasma without the use of a composite material. In this case, the depth of the foundations microhardness increased 1.5 times, and in the surface layer of the sintered to 4 times. Because of limited space complete these studies do not give.

Considered technology has been used for hardening elements such agricultural units, the disc coulter, claw cultivator, disc harrow, tooth harrow et al. (Figure 3). Processing is performed without the use of a composite material.

Below are photos of the surface of the disc opener to vibro arc after hardening (Figure 4). They show that after hardening sharply deteriorating the roughness of the original surface. However,

a

b

Figure 4 - the surface of the disc coulter images before (a) and after (b) vibro arc hardening. Increased 100 times. Photo size 1,6x1,2 mm

it must be assumed that such a deterioration of the surface will not affect the ability of the cutting tool. Figure 5 shows a photograph of the section opener after hardening. It can be seen that the upper layer of the molten alloy parts has a sharp boundary with the base.

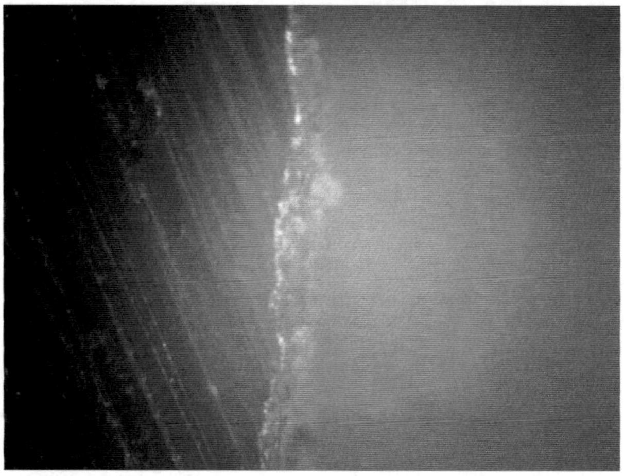

Figure 5 - Photo of the section opener. Increased 100 times. Photo size 1,6x1,2 mm

4. Conclusions.

1. A new enabling technology hardening surfaces cutting elements agricultural units, allowing 2.0 - 2.5 fold increase in their operational life;
2. The proposed technology has the versatility and can be used to upgrade technology.
3. Notwithstanding the applicability of the technology vibro arc hardening of surfaces in practice to this day remains to examine how the process of hardening and plasma-chemical processes occurring in the discharge.

Acknowledgments
The work is performed according to the Russian Government Program of Competitive Growth of Kazan Federal University.

References

[1] Adigamov NR. Modernization of agricultural equipment worn Tatarstan using nanotechnology /NR Adigamov, SN Sharifullin //Proceedings of the International scientific and practical conference "Engineering Science - Agriculture." - Kazan: Kazan State University of Agriculture, 2010. - Part 1. - Pp 44 - 45.
[2] SN Sharifullin. Increasing the operational reliability of the fuel pump you-high pressure automotive diesel engines. - Diss. Doctor. tehn. Sciences. -Moscow, 2009. - 368 p.
[3] Dautov G, Dautov I, Fayrushin I and Kashapov N 2013 *J. Phys.: Conf. Series* **479** 012001
[4] Galyautdinov R T, Kasparov N F and Luchkin G S 2002 *Inzhenerno-Fizicheskii Zhurnal* **75** 170–173
[5] Fadeev S A, Kashapov N F and Larionov V M 2014 *IOP Conf. Ser.: Mater. Sci. Eng.* **69** 012006
[6] Kashapov L N, Kashapov N F and Kashapov R N 2013 *J. Phys.: Conf. Series* **479** 012003
[7] Saifutdinov A I, Fadeev S A, Saifutdinova A A and Kashapov N F 2015 *JETP Lett.* **102** 637-642
[8] Gavrilova V A, Fazlyyyakhmatov M G and Kashapov N F 2013 *J. Phys.: Conf. Ser.* **479** 012010
[9] Kashapov L N, Kashapov N F and Kashapov R N 2013 *J. Phys.: Conf. Series* **479** 012005
[10] Israfilov Z K and Kashapov N F 1991 *Journal of Engineering Physics* **60** 364–368
[11] Dautov G, Kashapov N, Fayrushin I and Dautov I 2013 *J. Phys.: Conf. Series* **479** 012014
[12] Dautov GY. Plasma torches with stabilized electric arcs /GY Dautov, VL Dziuba, IN Karp. - Kiev, Naukova Dumka, 1984. - 168 p.
[13] Yurevich FB. Electric heating gas /FB Yurevich, VS Kulikov. Ed. corr. BSSR AG Shashkova. - Minsk, Science and Technology, 1973. - 189 p.
[14] VF Aulov. The results of field tests of hardened working bodies tillers /VF Villages, PV Luzhny, AV Kireynov //Proceedings of the CRP-STRINGS, 2013. - T. 113. - Pp. 300 - 309.
[15] Titov NV. Hardening of the working bodies of machines operating in an abrasive medium /Collection of scientific works SWorld. Proceedings of the international scientific and practical con Conference "Modern Problems and solutions in science, transport, manufacturing and Education 2012". - Issue 4. Volume 2 - Odessa: Kuprienko, 2012. - S. 46 - 48.
[16] Litovchenko NN Vibro arc surfacing graphite electrode nanometallicheskih /NN composite materials. Litovchenko, VN Kulikov, NV Titov. - M .: Svaroch-ing production, 2013. - # 2. - Pp. 51 - 53.

Journal of Physics: Conference Series **669** (2016) 012050 doi:10.1088/1742-6596/669/1/012050

Plasma technology for increase of operating high pressure fuel pump diesel engines

R Y Solovev[1], S N Sharifullin[2], N R Adigamov[3]
All-Russian Research Institute of Technology repair and maintenance tractor fleet (GOSNITI)[1]
Kazan Federal university[2]
Kazan State Agrarian university[3]

E-mail: Saidchist@mail.ru

Abstract. This paper presents the results of a change in the service life of high pressure fuel pumps of diesel engines on the working surface of the plunger which a wear resistant dielectric plasma coatings based on silicon oxycarbonitride. Such coatings possess high wear resistance, chemical inertness and low friction.

1. Introduction.

In [1] it was proved that reliability and durability of the high pressure fuel pumps (pump) diesel engines, primarily depends on the resources of their plunger. Unfortunately, the quality of domestic manufacturing plunger such that their share is less than 3 thousand operating hours, while that of foreign analogues it reaches 12 thousand or more. As was shown in [2], the resource is determined to develop the plunger of the pressure and wear of mating parts. Developing a pair of plunger pressure is directly dependent on the clearance between the sleeve and the plunger. The smaller the gap, the higher the pressure developed by plunger pair, therefore the greater share of work. Backlash in domestic plunger pairs is about 1.5 - 2.0 microns, and a foreign - one micron or less. In addition, the wear resistance of domestic plunger is 1.5 - 2.0 times lower than the wear resistance of the foreign. Studies have shown [2 - 4] that the achievement of a resource foreign counterparts plunger able to produce sputter of titanium nitride on the working surface of the plunger.

2. Materials and methods of research.

This article examines the possibility of creating another type of coating on the working surface of the plunger via plasma technology [5-13]. The coating produced in the plasma unit UFPU-111, the development of Scientific and Production Company "Plasmatsentr". It provides a coating based on silicon oxycarbonitride [14]. These coatings are a class of diamond-like and have high abrasion resistance, chemical inertness and low friction. According to these parameters, they are slightly higher than the coating of titanium nitride [14, p. 376]. In addition, non-vacuum plasma thin-film coatings are dielectric. This eliminates the chance of setting precision of mobile connections at the point overheating the workpiece surface.

For the experiments were chosen 4 new models of pump plunger pairs 33-02 Engine KAMAZ 5320. As well pump elements can be restored. The parameter for measuring chosen gidro density plunger. Meters on the device CP-1640A. Increase the pressure on the plunger end depends on the weight of the cargo and its location on the lever device. In our case, the cargo and its location on the lever of the device have been chosen in such a way as to increase the pressure on the plunger end was 500 N.

3. Results and discussion.

Measurement results gidro density plunger before and after application of a plasma coating on the working surface of the plunger shown in Table. 1. Here the plunger with the numbers 1 and 2, the coating thickness was about 0.5 microns, and with the numbers 3 and 4 of the order of 1.0 microns. However, we must bear in mind that the total elimination of the gap in the mating surface of the plunger assembly of parts by covering fraught with consequences. In particular, it became clear from the plunger rooms 3 and

Table 1 – Gidro density plunger before and after plasma coating on the working surface of the plunger, second

The surface condition of the plunger	Non plungers			
	1	2	3	4
Original	15	18	20	16
Plasma coating	28	30	140	120

4. When an ambient temperature of 25 ° C for the movement of plungers slidably sleeves occurred, and at a temperature of 10 ° C occurred their jamming.

Therefore, when developing the application of plasma coating technology must take into account this factor. The table above shows that the non-vacuum plasma thin-film coatings can increase gidro density plunger both new and refurbished.

There is a link between the size of the gap in the pump element and the fuel leak through the gap in the active course of the plunger. It can be expressed by formula [15]:

$$Q = \frac{\pi D p l^3}{12 \mu L}, \tag{1}$$

where Q - momentary amount of fuel flowing into the annulus; D- the diameter of the plunger; p- pressure difference; l- Radial clearance in the pump element; μ- the absolute viscosity of the fuel; L- length of the sealing surface.

From the above formula it is seen that the presence on the working surface of the plunger plasma non-vacuum thin film coating allows the 1.5 - 2.0 times to reduce the radial clearance in the pump element. This corresponds to approximately 5-fold increase in gidro density plunger assembly by non-vacuum plasma thin-film coating. In [2] determined to develop an empirical relationship between a pair of plunger pressure p and its gidro density τ

$$\frac{\tau_0 - \tau}{\tau} = 0{,}1(p_0 - p) \tag{2}$$

where τ_0 - the original density; p_0 - initial pressure (developed by a pair of pressure at the plunger τ_0); p - developed by a pair of pressure at the plunger τ.

From (2) it follows that at 6-fold increase in the difference gidro density plunger $(p_0 - p)$ assembly is 50 MPa. In our case, τ_0 and p_0 gidro density meet and develop a pair of plunger pressure plasma coating on the plunger, and p and τ – gidro density and develop a pair of pressure plunger without coating on the plunger (factory plunger pair). Normally developing pressure plunger assembly factory is about 110 MPa. Then a pair of pressure developed by plunger Plasma coating on the plunger will have a pressure of 160 MPa.

In [2] and obtained a law change promoted by a pair of pressure on the plunger of use

$$A^{kt} = 0.1\ (p_0 - p) + 1 \tag{3}$$

where A - constant characterizing the effect of initial density; k - coefficient, which takes into account the wear plunger; t - duration of the operation.

Journal of Physics: Conference Series **669** (2016) 012050 doi:10.1088/1742-6596/669/1/012050

Going from a plunger pump and a pair given that his time between working life is determined plunger by the formula (3) can be found, and life of the pump tn

$$t_{н} = \frac{lq\left\{0.1\left(p_0 - p_{\lim}\right)+1\right\}}{\kappa * lqA} \tag{4}$$

where p_{lim} - the maximum allowable operating pressure in the plunger assembly.

For high pressure fuel pumps p_{lim} = 20 MPa. In [2] it has been shown that the pattern of change of the operational characteristics of the pump operating time is saved and his life depends on the plunger developed by a pair of pressure and intensity of its wear. Using developed in [2] for the construction of a theoretical model of the operational characteristics of the pump with non-vacuum plasma thin-film coating the working surface of the plunger and the plunger to develop a pair of pressure of 160 MPa. Consider the basic model fuel pump Model 33-02 Engine KAMAZ 5320 with factory plunger pairs. The experimental data resource of the high pressure pump 3000 Moto hours at the initial developing a pair of plunger pressure of 70 MPa. Let the base model pump coefficient taking into account the wear rate of plunger is equal to one, k = 1. From (4) we find the constant A: lq A = $2{,}6 \cdot 10^{-4}$ or A = $10^{0.00026}$. This constant A will not change its value, since initial value is independent of the pump plunger assembly state (it is set by the operator). Details considered plunger assembly made of a material ShKh15. [4, p. 373] that the ratio of the wear rate of the material ShKh15 non-vacuum plasma and thin-film coating is 2.0. Then, using the formula (4) we find the life of the pump with the new plasma coating. It is equal to 9000 Moto hours. Next, using a theoretical model, we construct a characteristic pump plunger pairs, which have a plunger non-vacuum plasma thin film coating (Figure 1).

As seen from the graphs of Figure 1, plasma non-vacuum thin film coating on the plunger allows a 3-fold increase in service life of the fuel injection pump diesel engines.

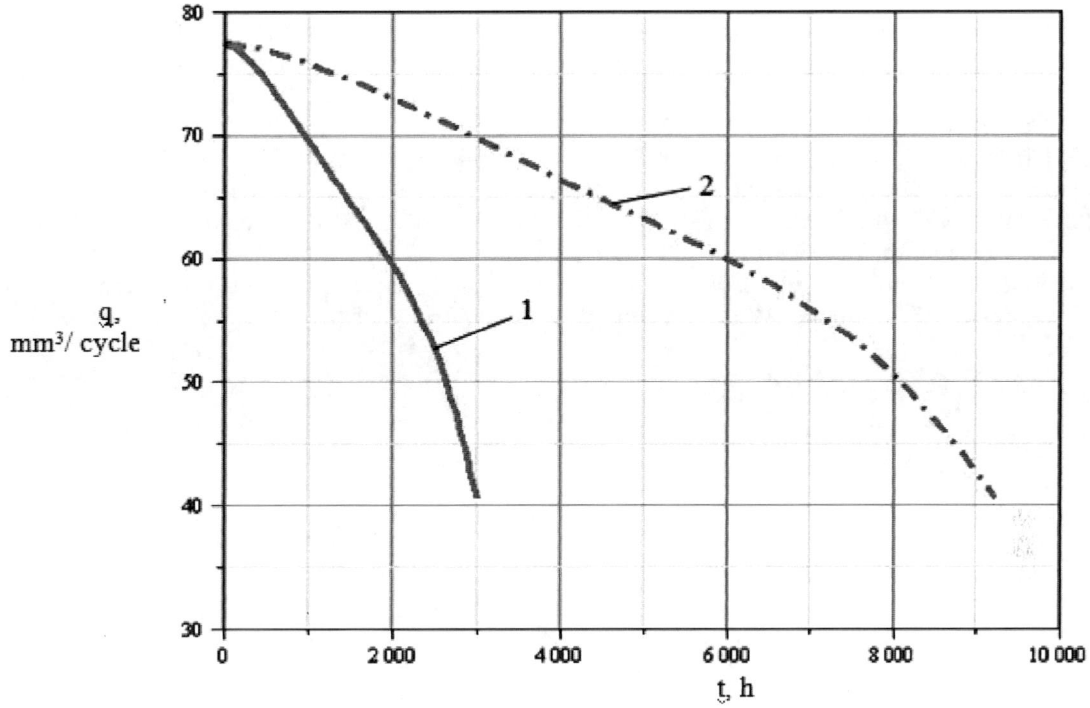

Figure 1 - Characteristics of the laws gov- erning working pump plunger from the factory with a pair of P_0 = 70 MPa (1) and a pair of plunger non-vacuum plasma thin-film coated plunger with P_0 = 160 Pa (2)

4. Conclusions.

1. Plasma non-vacuum thin-film coatings based on silicon oxycarbonitride allow increase gidro plunger both new and refurbished.
2. With the use of vacuum-free plasma thin-film coatings based on silicon oxycarbonitride possible in 2 - 3-fold increase in service life of the fuel pump of domestic production.

Acknowledgments
The work is performed according to the Russian Government Program of Competitive Growth of Kazan Federal University.

References
[1] Adigamov NR. Complex problem solving recovery of diesel fuel equipment /NR Adigamov, SN Sharifullin //Tractors and farm machinery, 2009. - # 3. – Pp. 38 - 40.
[2] SN Sharifullin. Increasing the operational reliability of high pressure fuel pumps automotive diesel engines. - Diss. Doctor. tehn. Sciences. - Moscow, 2009. - 368 p.
[3] Adigamov NR. Technology restore plunger high pressure fuel pumps without depersonalization /NR Adigamov, SN Sharifullin //Proceedings GOSNITI. - M., 2008. - Vol. 102. - Pp.106 - 108.
[4] Adigamov NR. High-technology to modernize the high pressure fuel pumps automotive diesel engines /NR Adigamov, SN Sharifullin, KS Podgaraya. Proceedings GOSNITI, 2013 - Vol. 113.- Pp 206 - 214.

[5] Galyautdinov R T, Kasparov N F and Luchkin G S 2002 Inzhenerno-Fizicheskii Zhurnal **75** 170–173
[6] Fadeev S A, Kashapov N F and Larionov V M 2014 IOP Conf. Ser.: Mater. Sci. Eng. **69** 012006
[7] Kashapov L N, Kashapov N F and Kashapov R N 2013 J. Phys.: Conf. Series **479** 012003
[8] Saifutdinov A I, Fadeev S A, Saifutdinova A A and Kashapov N F 2015 JETP Lett. **102** 637-642
[9] Gavrilova V A, Fazlyyyakhmatov M G and Kashapov N F 2013 J. Phys.: Conf. Ser. **479** 012010
[10] Israfilov Z K and Kashapov N F 1991 Journal of Engineering Physics **60** 364–368
[11] Dautov G, Kashapov N, Fayrushin I and Dautov I 2013 J. Phys.: Conf. Series **479** 012014
[12] Sosnin NA. Plasma technology /NA Sosnin, SA Ermakov, PA Topolyansky. - SPb.: ed. Polytechnic. University Press, 2008. - 406 p.
[13] Vladimir Antipov. Wear of precision components and characteristics of a violation diesel fuel equipment /V Antipov. - M .: Engineering, 1972. - 177 p.

Journal of Physics: Conference Series **669** (2016) 012051 doi:10.1088/1742-6596/669/1/012051

Opportunity some relationship tribo between plasma and service life of machines and mechanisms

V B Lomuhin[1], S N Sharifullin[2], I V Lapteva[3]
Novosibirsk Higher Military Command uchilische[1]
Kazan Federal University[2]
Novosibirsk State University of Civil Engineering (Sibstrin) [3]

E-mail: Saidchist@mail.ru

Abstract. The paper proposed the formula for calculating the residual life of the engines without disassembly using the method of spectral analysis of oil. The basis for calculation is the average change in the content of elements of deterioration in the operating oil for a certain period of time of the engine.

1. Introduction.

Service life of the machinery is directly dependent on the degree of wear of the friction units. Wear - that change the size, shape, weight or state of the surface of the product due to failure (deterioration) when its surface layer friction. Basically there are 3 types of wear. It - corrosive wear, mechanical wear and tear under the influence of an electric discharge. If the first 2 types of wear are fairly well known and sufficiently studied, the last kind few people paid attention.

In recent decades, there have been some studies that reveal the essence of the processes taking place in the friction units of machines and mechanisms in the presence of them tribo drugs [1 - 4]. It turned out that these processes play a major role in the resulting contact points mating surfaces friction unit electrical discharges. The lifetime of a single spot can be 10^{-7} – 10^{-8} s [5]. On a single spot having short-term individual temperature flashes. As a result, a very short period of time, a thin layer of molten material, and even in the high-energy state, similar plasma or magma. This “MAGMA-plasma” later called “Triboplasma”. The energy of the electrons emitted by the separation of solid surfaces reaches relativistic values [6].

We must assume that triboplasma occurring at break of the contact mating surfaces will be the main component of the process of wear friction unit. And wear a determining operating life of machinery. If you predict the service life of the units of cars, depending on the wear and tear, you can look for ways to improve the use triboplasma resource characteristics of machines and mechanisms.

Questions to predict the service life of the machinery is widely discussed in [7 - 9].

2. Methods of research.

In this paper we used the method to study emission spectral analysis to predict performance without collapsible parts.

3. Results and discussion.

The initial data for forecasting the residual resource can have different parameters and indicators of the technical state of the system. However, the most frequently used parameters such as wear.

The basis for the prediction of wear are probabilistic mathematical model of the process of wear, which wear can be presented either as a function of the random variable $x = f(t)$ or a random function $x(t)$. Prediction of wear of machine parts for wear, represented as a random variable, and produced by m_x and $Д_x$ or σ_x. This prediction does not take into account the individual characteristics of each machine wear group, resulting in an increase in the accuracy of predictions of its duration is significantly reduced, especially if the nature of the process of wear between the machines in the group varies considerably. Prediction of wear on the random function that is obtained by mixing specific implementations, carried by $m_x(t)$, $Д_x(t)$ and $K_x(t,t_1)$ or $r_x(t,t_1)$. The normalized correlation function $r_x(t,t_1)$ is characterized by stable dynamics of wear each car group, in view of which significantly increases the accuracy of forecasting. If the value $r_x(t,t_1)$ is close to one, indicating that the weak mixing implementations, the prognosis can be given for a longer period. At small (vigorous stirring) prediction with the required accuracy can only be done for a short period of time. Taking this into account, when predicting the optimal solution should be found, ie, the duration of the forecast should be established of the conditions required accuracy of the forecast.

Currently probabilistic mathematical model of the process of wear are the way of statistical processing of the results of measuring the wear of machine parts with normal exploitation. The main difficulty in finding such models associated with the current measurement of wear and tear. At the widespread linear and weight measurement wear, as the most authentic and relatively simple method for estimating the wear, wear full dynamics can be set only for those parts of machines and mechanisms, the measurement of which is not related to the dismantling of the machine. In this case, you can get a family of implementations, while stirring, and which is a random function of wear, representing the most complete probabilistic mathematical model of the process. However, in most machines such measurement items without dismantling the machine can not be performed, so that the wear of these parts is evaluated by two measurements carried out at the beginning and end of the test. With such frequency measurements can not find concrete implementation, and hence it is impossible to describe the process of wear of a random function. In this case, the dynamics of the wear on the whole group studied cars are available only through the statistical processing of a large number of random measurements of parts obtained by dismantling the machine with different duration of work between their showdowns. This model $u = f(t)$ is characterized by deterioration of the wear process, depending on the operating time as a random variable, not as a random process. Such a description of the actual pattern of wear displays it in one or another group of machines, but this map is incomplete, as part of the probability characteristics remain uncertain. Such information about the possible nature of the flow dynamics of the growth of depreciation $u = f(t)$ are given in [10].

The initial data for the forecasting performance of parts of the machine for a certain period of time t is the measurement of wear to the original time t_1. However, such data can be obtained at any point in time we are interested only in the machinery, which is required to measure the wear full or partial dismantling of the machine. In this case, the prediction performance become difficult to achieve since any disassembly of the machine, including the associated only with the measurement, may cause some changes in the conjugation of the measured items and hence to a significant error in the prediction.

A promising and reliable method for predicting performance without collapsible parts is the method of "wear products in oil" and, in particular, the emission spectral analysis of oil (Esam) [9, 11]. This method makes it possible to describe the process of wear parts and measure the random function wear at any given time. However, the use of this method for predicting has its own characteristics and, in particular, especially in the initial estimate of the wear. If one line or the weight measurement can judge the degree of deterioration of the items, according to the result of the spectral analysis of the process during normal wear can not do this for the following reasons. Esam method determined by the mass of metal removed from the friction surfaces for the period after the change of oil to measure.

Therefore, the entire mass of metal removed from the friction surfaces, can be determined only if the measurement of wear to perform regularly with the initial moment of the machine.

Given these features of the method of "wear products in oil," prediction using this method can be performed to determine the amount of metal removed from the friction surfaces and the speed of its receipt in the working oil.

The basis of this prediction information on the amount of metal removed from the friction surfaces at any predetermined time. With the highest precision, such information when evaluating wear with spectral analysis of the lubricating oil can be obtained from the balance equation, consisting in equal number to the number worn layer of wear in the crankcase. In this case the evaluation of working capacity of details of a particular machine depends on the error description of the process of wear balance equation.

For such a prediction is necessary to have:

- Random function of wear $G(l)$ or $G(t)$;
- Information about limiting the amount of metal $G_{\lim}$ removed from the surfaces of the parts in which failure can occur.

For information about limiting the amount of metal $G_{\lim}$ can be obtained by experiment or calculation [9, 11, 12]. The difficulty in obtaining such information is as follows. Even with the normal processes of wear, wear identical machine parts is uneven. Depending on the engine design and requirements on its operation, the loss of efficiency of the machine can occur either when the maximum size of a single part, while others wear can be significantly lower, or on reaching all parts of limit or close to the sizes. In the first case, an amount of metal to be taken in the shortest possible $G_{\min}$, and in the second case - for the maximum possible $G_{\max}$. In the experimental finding the limits of these values must use the statistics measuring the wear and tear from vehicles entering the repair of failures. Calculated by weight of the metal is determined by volume, calculated in accordance with Diagrams surface wear and taking into account the permitted norms rejection. The accuracy of the method depends on the complexity of the geometric shape wear. A more reliable method is experimental. However it is better to combine both of these methods.

Persistence load and engine speeds allows you to approach the balance equation with aggregate. The basis for this decision put the average content of the elements of deterioration in the operating oil for a certain period of time of the engine.

The balance equation, with aggregate is as follows

$$G = \frac{G_M \cdot \frac{\sum_1^n C_i}{n}}{100} + \frac{t \cdot g_y \cdot \varphi \cdot \frac{\sum_1^n C_i}{n}}{100} + \frac{G_M \cdot \frac{\sum_1^n C_i}{n}}{100} \cdot t \cdot q_{ц\phi} \text{ , g.} \tag{1}$$

where G_M - the mass of oil, g; C_i - Concentration of the element i in the oil-measurement%; t - time of the engine, watch; g_y - specific oil consumption, kg / h; $q_{ц\phi}$ - the proportion of the number of his element in the oil centrifugal filter is delayed one hour; φ - coefficient taking into account carryover of elements from the engine; n - number of measurements.

To calculate the need to have information on the value of the coefficient, which can only be obtained from the measurements. Reliability of determination according to equation (1) depends on the number of measurements over time. Moreover, it is desirable to make the measurement in the middle of the life of the oil.

Consider forecasting algorithms depending on the amount of available information and present the accuracy of the forecast.

The easiest time of residual life may be determined by the following formula

$$t_{\text{east}} = \frac{G_{\text{riding}} \cdot t}{G} - (t + t_1) \text{ ,} \tag{2}$$

where G_{riding} - permissible mass element wear, g; G - the mass of the element of wear accumulated during the oil-t, g; t - follow-up the work of the engine in order to predict the resource per hour; t_1- the time of engine start up observation hour.

The algorithm of the forecast represented by the formula (2), based on the assumption that the rate of wear items arrive before the start of the observation will be the same as in the period of observation.

Therefore, the accuracy of this prediction will depend on the stability of the dynamics of wear each motor group. The reliability of the forecast can be increased if we know the point of history, ie, the values of the mass of metal lost in the previous maintenance period.

In this case, the prediction should be performed according to the formula

$$t_{ocm.} = \frac{\sum \frac{G_{\partial on.} \cdot t_i}{G_i}}{N} - \left(\sum t_i + t_1\right), \tag{3}$$

where t_i - the length of time of i-th season of operation or harvesting (planting) on the accumulation of weight lost metal Gi; N- the number of seasonal operation.

A significant way the accuracy of the prediction can be improved by the submission of the accumulation of elements of deterioration function $G = f(t)$ for each engine or a random function $G(t)$ for the engine group. These functions are obtained by fitting the results of observations by the method of least squares. The most common type of such models are linear

$$G = a_1 \cdot t + a_0, \tag{4}$$

or quadratic

$$G = a_2 \cdot t^2 + a_1 \cdot t + a_0. \tag{5}$$

In the absence of information on the amount of metal accumulated at the beginning of the observation time for the residual life of the linear model defined by the formula

$$t_{ocm} = \frac{\left(G_{\partial on.} - \frac{t + t_1}{t} \cdot G\right) - a_0}{a_1} \tag{6}$$

and for the quadratic model by the formula

$$t_{ocm} = -\frac{a_1}{2 \cdot a_2} \pm \sqrt{\frac{a_1^2}{16 \cdot a_2^2} - \frac{1}{a_2} \cdot \left[a_0 - \left(G_{\partial on.} - \frac{t + t_1}{t} \cdot G\right)\right]}. \tag{7}$$

If you have information about the number of accumulated metal, forecasting for the linear model should be carried out according to the formula

$$t_{ocm.} = \frac{\left(G_{\partial on.} - \sum_1^N G_i\right) - a_0}{a_1}, \tag{8}$$

and quadratic

$$t_{ocm.} = -\frac{a_1}{2 \cdot a_2} \pm \sqrt{\frac{a_1^2}{16 \cdot a_2^2} - \frac{1}{a_2} \cdot \left[a_0 - \left(G_{\partial on.} - \sum_1^N G_i\right)\right]}. \tag{9}$$

4. Conclusions.

To determine the residual life of the above formula, you must have information about, ie, wear thresholds. Then the residual life of the machinery can be calculated by the expression [11, 12]:

$$G_{nop} = K_{nop} \cdot Q, \tag{10}$$

$$t_{ocm} = \frac{K_{nop}}{G_i} \cdot t_i - (t_i + t_1), \tag{11}$$

$$t_{ocm} = \frac{K_{nop} \cdot Q}{(K_j + K_0) \cdot Q} \cdot t_i - (t_i + t_1) = \frac{K_{nop} \cdot t_i}{K_j + \dfrac{\sum K_i + \dfrac{q}{Q} \cdot \left(\sum (1 - e^{-\alpha_i}) \dfrac{t_i}{\alpha_i} \right)}{\sum e^{-\alpha_i}}} - (t_i + t_1) \, . \tag{12}$$

Acknowledgments
The work is performed according to the Russian Government Program of Competitive Growth of Kazan Federal University.

References

[1] Perepelitsyn MG. Study the impact of repair compounds on the surface of the friction on the example of the cam mechanisms of automotive engines: Author. Dis. on soisk. scientists. step. cand. tehn. Sciences. /MG Perepelitsyn. - Novosibirsk, 2009. - 16 p.
[2] Dolgopolov KN. Improving the performance of household appliances through intensification processes lubrication of friction pairs: Author. Dis. on soisk. scientists. step. cand. tehn. /KN Dolgopolov. - Mines, 2009. - 16 p.
[3] A. Dunayev. The modernization of worn equipment with tribo drug. Monograph /AV Dunayev, SN Sharifullin. - Kazan: Kazan. University Press, 2013. - 272 p. ISBN 978-5-00019-141-5.
[4] Anatoly Dounaev. Friction Surfaces Modification Using Tribo-Compounds /Anatoly Dounaev and Said Sharifullin //World Applied Sciences Journal. ISSN 1818-4952, 2014. - Vol. 31 (2). - Pp. 272 - 276.
[5] Garkunov DN. Tribotechnology /DN Garkunov. - M .: Engineering, 1985. - 424 p.
[6] Bocharov GS. The energy of the electrons emitted in the separation surfaces /GS Bocharov, AV Eletskii, VA Nikerov //Plasma Physics, 2011. - Vol. 37. - # 4. - Pp. 396 - 400.
[7] Mikhlin VM. Prediction of the technical condition of vehicles /VM Mikhlin. - M .: Kolos, 1976. -
1. 287 p.
[8] Belsky VI. Pre-repair diagnostics of agricultural machinery. /VI Belsky. - M .: Agrotransport, 1987. - 172 p.
[9] Kyureghyan SK. Assessment of the wear of internal combustion engines by spectral analysis /SK Kyureghyan. - M .: Engineering, 1976. - 152 p.
[10] Sokolov AI. Application of emission spectral analysis to assess oil wear properties of the working oil /A.I. Sokolov, NT Tishchenko. - Tomsk: Publishing House of Tomsk University. - 1979. - 208 p.
[11] Lomukhin VB.Diagnosing Analysis crankcase oil as component without collapsible parts technology repair machinery /V.B. Lomukhin. //Diesel power plants river vessels. - 2001. - S. 21 - 25.
[12] Cher LA. Standard enterprise. System diagnostics for diesel engines the method of comprehensive analysis of the lubricating oil STP 314. 0-01-88 /L.A. Sher, NN Sorokin, AF Kononov, VB Lomukhin. - Science, Novosibirsk, 1988. - 79 p.

Mathematical modeling of plasma deposition and hardening of coatings-switched electrical parameters

A M Kadyrmetov[1], S N Sharifullin[2], AS Pustovalov[1]
Voronezh State Forestry akademiya[1]
Kazan Federal University[2]

E-mail: Saidchist@mail.ru

Abstract. This paper presents the results of simulation of plasma deposition and hardening of coatings in modulating the electrical parameters. Mathematical models are based on physical models of gas-dynamic mechanisms more dynamic and thermal processes of the plasma jet. As an example the modeling of dynamic processes of heterogeneous plasma jet, modulated current pulses indirect arc plasma torch.

1. Introduction.

One of the most progressive and effective lines of creating parts with desired properties on their working surfaces is a protective coating of gas-thermal methods. In this group, one of the most effective and universal method is plasma spraying. Existing ways of improving the plasma spraying, aimed at improving the efficiency of stationary coating, almost exhausted, and limited to an insufficient level of physical-mechanical and tribological properties of coatings subjected to high dynamic alternating and shock loads performance. Some work is devoted to improving the plasma deposition methods dynamizing processes using acoustic impact on the plasma jet (SSI "Institute of Heat and Mass Transfer" NASB), and pulsed arc power bypass (SSTU, St. Petersburg, MSTU. NE Bauman). However, questions remained unstudied theoretical description of processes at the modulation parameters of the plasma spraying and methodical study of optimization.

Based on the study of general questions of plasma spraying process can be concluded that improving the efficiency of plasma spray technology is advantageously carried out by the improvement, development and universalization of both the technological operation spray coating and subsequent processing steps of hardening or combining with them [1].

Improvement of the process step of spraying may be achieved by modulating the parameters of the plasma torch, by adding additives to the plasma gas (the plasma for the air - by the addition of carbohydrate-gen-containing gas), gas dynamic processes, and by combining these methods using [2, 3].

Modulation of the electrical parameters of the plasma spraying (current arc plasma torch) is technological method to simply and effectively regulate the power and energy of the arc by increasing or decreasing the pulse arc power at a certain frequency. It allows you to: generate a plasma jet at low intensity shock waves and intense acoustic waves; increase the drag coefficient of the spray particles and heat them to two or more times and thus enhance the energy parameters of the spray particles (increase of the speed and temperature of 30 ... 50%) [2-8]. This ensures increasing the density and strength of the coating.

The technological method of adding to the hydrocarbon-containing gas plasma air can increase the enthalpy and the thermal conductivity of the plasma jet to 2 times, create a neutral and a reducing atmosphere; increase the thermal efficiency of the plasma torch; increase the temperature of the plasma jet (from 5300 °K air to 6200 °K for propane in air) and the speed of the plasma jet and the sprayed particles by 10 ... 30% [3, 9].

For the development of technical advice in the process of plasma deposition and hardening of coatings requires knowledge of patterns and relationships between technological factors in the production of coatings in the mode of modulation parameters and gas-dynamic and thermal coatings quality criteria that determine the level of residual stresses in the coatings, their strength and tribological characteristics.

To accomplish this task have been developed mathematical models describing the technological system during application and hardening coatings mode modulation parameters.

2. Research methods.

Methodological basis of theoretical studies have made the scientific basis of mechanical engineering, the theory of gas dynamics in terms of jet streams, traveling wave velocity and entropy waves temperature, the theory of thermal welding processes, methods of computational mathematics (finite difference method and discrete components), advanced graphics and computer systems for the computer.

3. Results.

Technological use of dormant plasma deposition of hardening coatings modulates electrical parameters based on physical models of gas-dynamic mechanisms more dynamic and thermal properties, which are caused by the conversion of electrical energy pulse current indirect or remote arc modulation thermal and gas-dynamic energy of the arc and the space around it. Physical processes in the plasma jet in the modulation of indirect arc plasma generator with a frequency modulation v_m duration τ_m, amplitude ΔN and steep power dN/dt pulses is to generate in the plasma jet of entropy heat waves with the amplitude of the temperature ΔT_p and shock waves with Mach number M, appropriate strong or weak shock waves [10]. The result of the impact of these waves on the plasma jet is heterogeneous: increased turbulence in the plasma jet, leading to an increase in the uniformity of its cross-profile A range of a temperature L_T and enthalpy $L_{\Delta H}$; increasing the drag coefficient C_D and particle heat transfer them α by the plasma jet; increasing the speed V_p particles and their heating rate dT_p /dt. Such an increase in the energy state of the particles provides improvement of physical contact of the particles with the substrate and connected to it by increasing the pressure and the pulse pressure of the particles to the substrate upon impact, which results in a decrease of porosity P and increase strength coating performance (adhesion and cohesion σ_k σ_a). Physical processes in the coating in the modulation of the direct arc torch with an amplitude ΔN_d and pulse time τ_m consist of fusion of the coating to the transition zone to the substrate and the formation of physical contact for the time t_c and chemical interaction covering the substrate in point local zones bind direct arc to the base (coating). On the scale of the entire surface with the influence of the speed V and feed S torch formed certain density distribution of discrete zones of penetration and thermal effect on the substrate and the coating that contribute to the uniformity of the temperature field bases and improving mechanical and tribological characteristics of the coating.

Formation conditions and the quality of the coating are determined by the energy impact on the formed coating through three channels: from the sprayed particles (parameters speed V_p, the temperature of T_p and heat flow q_p), gas phase plasma jet (the parameters of the speed V_g, the temperature T_g, heat flows heat q_g and cooling q_c) and thermal and electromechanical effects on the sprayed surface and coating (direct arc power parameter N_p and its amplitude, pressure roller and its amplitude ΔP, heat flows from the thermal q_t and electromechanical q_e effects, thermal conductivity λ and heat capacity c_p materials, radii shaft R_{sh} and run-roller Rr, shaft speed n_{sh} and pitch of turns spraying H). The model shows that for the process to ensure the quality of decisive importance is the effect of two physical components of the plasma process: the energy state of the particles of the coating material and the thermodynamic state of the "coating-base." Therefore, the complexity of the physical processes described in the "indirect and direct arc - plasma jet - Floor - base" in the modulation mode, the electrical parameters of the plasma torch arcs led to the need to provide the system of partial models. Their study allows us to solve the problem of getting better quality coatings in a complex mathematical models that include a description of the processes with the modulation parameters: gas-dynamic, temperature field distribution of the "coating - substrate" with regard to the cooling surface deposition, thermal and electro-coating treatment.

Here we consider one of the private mathematical models developed for the dynamic processes of heterogeneous plasma jet, modulated current pulses indirect arc plasma torch. It includes a stationary-phase empirical description of the plasma jet, an analytical description of the stationary distribution in the jet traveling wave velocity and temperature and entropy waves the description of the behavior of spray particles in modulating plasma jet.

On the basis of the equations of motion and continuity of the gas phase distribution of the plasma jet traveling wave velocity jet axis x stands for a solution:

$$x=\left(\frac{\chi+1}{2}\cdot V_{\mathrm{var}}+V_{fix}+c_{fix}\right)\cdot t+f(V_g)\ , \tag{1}$$

where the gas velocity is $V_g = V_{var} + V_{fix}$; V_{var}, V_{fix}- respectively variable and fixed components of the velocity of the plasma torch to cut, m/s; c_{fix} - the speed of sound in the undisturbed environment at $V_g = V_{fix}$, m/s; t – time, s; χ - polytropic index. Function speed $f\ (V_g)$ determined from the boundary conditions change the speed at the nozzle exit of the plasma torch generated triangular wave power indirect arc of dependencies for the first and second half pulses, respectively:

$$V_{g0}=V_{fix}+\frac{2V_{\mathrm{a}}}{\tau_{mom}}\cdot\tau\quad\text{at } n\cdot\tau_{per}<\tau<\frac{\tau_{mom}}{2}+n\cdot\tau_{per}\ ,$$

$$V_{g0}=V_{fix}-\frac{2V_{\mathrm{a}}}{\tau_{mom}}\cdot\tau+2V_{\mathrm{a}}\ \text{at } n\cdot\tau_{per}+\frac{\tau_{mom}}{2}<\tau<\tau_{mom}+n\cdot\tau_{per}\ , \tag{2}$$

where $n = 0, 1, 2, 3,\ldots$; V_{g0}, V_{a} - respectively, the gas velocity and amplitude of its fluctuations at the nozzle exit of the plasma torch, m/s; $\tau_{uмn}$, τ_n - respectively, the pulse duration and the period of its pulsations, p.

The initial conditions corresponding to the distribution of the parameters of the plasma jet along its x-axis describes the dependence:

$$V_m(x)=V_{fix}\cdot\sqrt{\pi\cdot\left[\phi(x)+\left(1-\phi(x)\right)\cdot\frac{T_{norm}-300}{T_{fix}-300}\right]}\ ,\ T_m(x)=\phi(x)\cdot T_{fix}+\left(1-\phi(x)\right)\cdot T_{norm}\ ,$$

$$\rho_m(x)=\frac{\rho_{fix}}{\phi(x)+\left(1-\phi(x)\right)\cdot\frac{T_{norm}-300}{T_{fix}-300}}\ ,\ c_m(x)=\sqrt{\chi\cdot P/\rho_m(x)} \tag{3}$$

where for the initial portion of the plasma jet $\varphi = 1$, $\pi = 1$; for the transition - $\varphi \sim 0{,}5$, $\pi \sim 0{,}6$; for the main - $\varphi \sim 0{,}1$, $\pi \sim 0{,}1$; $V_m\ (x)$ - velocity at the axis of the plasma jet in m/s; T_{norm} - normal temperature, $T_{norm} = 300$ K; T_{norm} (x) - the temperature on the axis of the plasma jet, K; $\rho_m\ (x)$ - the gas density at the jet axis, kg/m3; $c_m\ (x)$ - the sound velocity on the axis of the plasma jet in m/s; P - pressure, Pa.

For the initial portion of the plasma jet at intervals of time corresponding to the first half of the pulse arc power, based on the expressions (1) and (2) the dependence of the variable speed of the gas phase in the form:

$$V_{\mathrm{var}}(x,t)=\frac{V_{\mathrm{a}}\cdot t}{\tau_{mom}}-\frac{c_{fix}+V_{fix}}{\gamma+1}+\sqrt{\left(\frac{V_{\mathrm{a}}\cdot t}{\tau_{mom}}-\frac{c_{fix}+V_{fix}}{\gamma+1}\right)^2-\frac{4\cdot V_a}{\tau_{mom}\cdot(\gamma+1)}\cdot\left[x-\left(c_{fix}+V_{fix}\right)\cdot t\right]}\ ,$$

and for times corresponding to the second half of the pulse, - in the form:

$$V_{\text{var}}(x,t) = -\frac{V_{\text{a}}}{\tau_{mom}} \cdot t - \frac{c_{fix} + V_{fix}}{\gamma+1} + V_{\text{a}} + \left\{ \left(-\frac{V_{\text{a}}}{\tau_{mom}} \cdot t - \frac{c_{0m} + V_{0m}}{\gamma+1} + \upsilon_{\text{a}} \right)^2 + \frac{4 \cdot V_{\text{a}}}{\tau_{mom} \cdot (\gamma+1)} \cdot \left[x - \left(c_{fix} + V_{fix} \right) \cdot \left(t - \tau_{mom} \right) \right] \right\}^{\frac{1}{2}}.$$

Entropy waves carried along with the flow, so the general solution of the temperature distribution in the longitudinal coordinate stream of x and time t is given by:

$$T(x,t) = T\left(0, t - \frac{x}{V_{mom}}\right).$$

Based on the boundary conditions corresponding to the change in temperature at the nozzle exit of the plasma torch in the form of a triangular wave generated by the pulse power of indirect arc:

$$T_{g0} = T_{mom} + \frac{2T_{\text{a}}}{\tau_{mom}}\tau \quad \text{at } n \cdot \tau_{per} < \tau < \frac{\tau_{mom}}{2} + n \cdot \tau_{per},$$

$$T_{g0} = T_{fix} - \frac{2T_{\text{a}}}{\tau_{mom}} \cdot \tau + 2T_{\text{a}} \quad \text{at } n \cdot \tau_{per} + \frac{\tau_{mom}}{2} < \tau < \tau_{mom} + n \cdot \tau_{per},$$

$$T_{go} = T_{fix} \quad \text{at } n \cdot \tau_{per} + \tau_{mom} < \tau < (n+1) \cdot \tau_{per}$$

and initial conditions (3) analytical temperature versus distance *x* and time *t*, where *Ta* - the amplitude of the temperature, K.

In the initial section of the plasma jet for the time corresponding to the first half of the pulse power of the arc, the expression for the temperature of the gas phase is of the form:

$$T_{norm}(x,\ t) = T_{fix} + \frac{2 \cdot T_{\text{a}}}{\tau_{mom}} \cdot \left(t - \frac{x}{V_{fix}} \right),$$

for the time corresponding to the second half of the pulse - form:

$$T_{norm}(x,\ t) = T_{fix} + 2 \cdot T_{\text{a}} - \frac{2 \cdot T_{\text{a}}}{\tau_{mom}} \cdot \left(t - \frac{x}{V_{fix}} \right).$$

Similar expressions variable speed and temperature of the gas phase are obtained for the transition and main portions of the plasma jet.

Changing the speed and temperature of the particles described by the equations of motion and heat balance in the quasi-stationary approximation, corresponds to what the drag coefficient and the heat transfer particles depend only on the instantaneous values of the hydrodynamic parameters. This is justified by the smallness of the time of establishment of the stationary boundary layer of particles equal to 0.1-10 microseconds, compared to the duration of the modulating pulses of the order of 100-1000 microseconds.

4. Conclusions.

1. It is shown that the description of the process of application and hardening coatings mode modulation parameters is possible on the basis of mathematical models developed on the basis of physical models of gas-dynamic mechanisms more dynamic and thermal parameters of the plasma jet.
2. Consider one particular mathematical models developed for heterogeneous gas-dynamic processes of the plasma jet, modulated current pulses indirect arc plasma torch. A mathematical model of heterogeneous plasma jets is implemented numerically by the Euler 2nd order. Description of pressure fluctuations in the plasma jet simulation model is implemented in a computer program.

Acknowledgments

The work is performed according to the Russian Government Program of Competitive Growth of Kazan Federal University.

Journal of Physics: Conference Series **669** (2016) 012052 doi:10.1088/1742-6596/669/1/012052

References

[1] Stanchev DI. recovery of machine parts thermal spray methods: the manual /DI Stanchev, AM Kadyrmetov, VI Kluchnikov, KA Yakovlev. - Voronezh: VGLTA, 2002. - 83 p.
[2] Kadyrmetov AM. Features of the air-plasma deposition and hardening of coatings /AM Kadyrmetov, GA Suhochev //Hardening and coating technology. - 2009. - № 4 (52). - S. 25 - 28.
[3] Suhochev GA Experimental studies of the parameters of control of air-plasma deposition and hardening of coatings /GA Suhochev, AM Kadyrmetov // Hardening and coating technology. - 2008. - # 11 (47). - Pp. 53 - 56.
[4] Kashapov L N, Kashapov N F and Kashapov R N 2013 *J. Phys.: Conf. Series* **479** 012003
[5] Saifutdinov A I, Fadeev S A, Saifutdinova A A and Kashapov N F 2015 JETP Lett. **102** 637-642
[6] Israfilov Z K and Kashapov N F 1991 *Journal of Engineering Physics* **60** 364–368
[7] Xia G, Chen Z, Saifutdinov A I, Eliseev S, Hu Y and Kudryavtsev A A 2014 *IEEE Transaction on plasma science* **42** 2768–769
[8] Saifutdinov A I, Timerkaev B A and Zalyaliev B R 2014 *J. Phys.: Conf. Series* **567** 012031
[9] Klubnikin BC. Plasma spraying in the active media /VS Klubnikin, MV Karasev, GK Petrov. - AL: Oh of "Knowledge" of the RSFSR, Leningrad, LDNTP, 1990. - 20 p.
[10] Kadyrmetov AM The theoretical basis and technological support for the quality of plasma deposition and hardening of coatings modulated electrical parameters: Author. Dis. on soisk. scientists. step. Doctor. tehn. /AM Kadyrmetov. - Voronezh, 2014. - 32 p.

Journal of Physics: Conference Series **669** (2016) 012053 doi:10.1088/1742-6596/669/1/012053

Highly non-linear resonance fluctuations of gas in a closed pipe

R G Zaripov[1,2], N F Kashapov[2], L A Tkachenko[1,2] and L R Shaydullin[2]

[1] Institute of Mechanics and Engineering, Kazan Science Center, Russian Academy of Sciences, 2/31 Lobachevsky str., Kazan 420111, Russia
[2] Kazan (Volga Region) Federal University, 18 Kremlyovskaya str., Kazan 420111, Russia

E-mail: liqn@mail.ru

Abstract. Highly non-linear oscillations of gas in a closed pipe were researched experimentally. Dependencies of the pressure oscillation amplitude were obtained for various frequencies and amplitudes of the gas excitation near the first eigenfrequencies when transiting to the shock-wave oscillation mode.

1. Introduction

The research of the effect of the medium oscillation to the plasma has a great applied significance. Mostly, peculiarities of processes occurring in cases of high-frequency and ultrasonic oscillations have been studied (see, for instance, digest [1]). Thus, the works [2-4] researched the effect of acoustic oscillations on features of a glow discharge and presented a self-consistent model of a positive glow considering the effect of the acoustic oscillations. High-frequency longitudinal oscillations of gas were created with an acoustic electrodynamic radiator. A necessity has occurred to research the effect of the medium oscillations at low frequencies, in particular, in systems of a resonator-pipe type, which a widely used in technics. When the medium is excited, highly non-linear pressure waves emerge in such systems near the resonance, which waves transform into periodic shock waves in resonance, with the pressure differential up to 0.8 atm [5]. Detailed digests of theoretic and experimental researches of non-linear resonance oscillations of homogeneous and inhomogeneous gas in closed and open pipes when excited with a flan piston in either end are presented in works [6-8]. One of the latest experimental works in this respect was focusing on movement of particles in case of non-linear oscillations of gas in an open pipe in the non-shock-wave mode [9, 10].

This work, being a part of the research cycle for features of plasma interacting with acoustic oscillations, presents the gas-dynamic part of the research of gas oscillations in a resonance mode when transiting to the shock waves.

2. Experimental setup

The research was performed in an installation based on a vibration generator 1 of TIRAvib S 5220/LS make with a power amplifier 20 of BAA 1000-ET model by TIRA firm (Figure 1). Into the table 3 of

[1] To whom any correspondence should be addressed.

the vibration generator, a rod 5 was screwed, on which rod a flat piston 6 is attached with the diameter d = 100 mm; the total weight of the installation is 250 g.

The piston enters the cylinder 7 and oscillates according to the harmonic law with the set frequency ν in Hz. In order to prevent the transition of the stand vibration to the floor, the installation was placed on pneumatic spring cushions 25. Four clamp bands 2 were fixed on metal angles 24 screwed to the base of the vibration installation. The clamp bands hold a support 13 with a special lower head 8 and cylinder 7. Ends of a glass pipe 10 with the length L=916 mm, the inner diameter of d = 10.05 mm and the wall thickness of 2 mm were glued with epoxy resin in an opening of the upper head 8 and the lower head 12. The pipe was held vertical with clamp bands 9 fixed on the lower head 8 and the upper head 12. A cap 11 with the height of 22 mm and the inner diameter equal to that of the pipe was tightly screwed to the upper head. A cooling fan 26 was used to prevent overheating of the vibration coil and the field excitation coil.

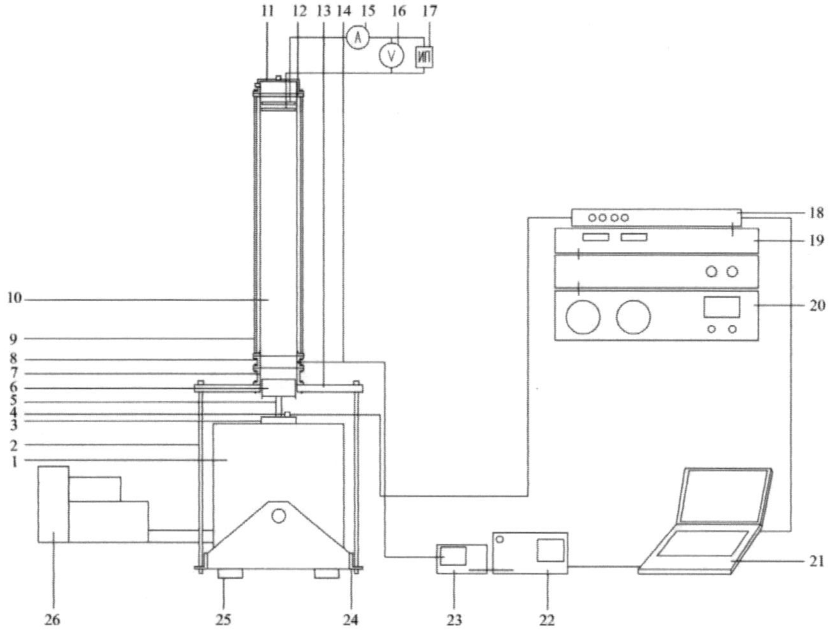

Figure 1. 1 – vibration generator, 2 – lower clamp band, 3 – vibration table, 4 – piezoelectric IEPE accelerometer, 5 – rod, 6 – flat piston, 7 – cylinder, 8 – lower head, 9 – upper clamp band, 10 – glass pipe, 11 – cap, 12 – upper head, 13 – support, 14 – pressure sensor, 15 – ammeter, 16 – voltmeter, 17 – power supply unit, 18 – controller, 19 – positioning controller TMS-2, 20 – power amplifier, 21 – control computer, 22 – oscilloscope, 23 – three-channel bridge voltage amplifier, 24 – metal angles, 25 – pneumatic spring cushions, 26 – cooling fan.

The vibration table was controlled via a computer 21 by means of a piezoelectric IEPE accelerometer 4 of 4513 model by Bruel & Kjaer firm and a controller 18 of VR9500 type by Vibration Research Corporation firm using a specialized VibrationVIEW software. The positioning controller TMS-2 19 was used to adjust the table of the vibration generator. A pressure sensor 14 of 8530C-15 model by Bruel & Kjaer firm was screwed into an opening of the head 8, and the signal from the sensor was fed through a three-channel bridge voltage amplifier 23 of ENDEVCO model 136 by Bruel & Kjaer firm to a digital oscilloscope 22 of DSO 3062A model by Agilent Technologies firm, and then via an RS-232 interface to the computer, where the data for the gas pressure p in bar can be observed and saved using a specialized DSO3000 software.

Sine oscillation of the piston near the first proper frequency ν_1 was set with the controller 18. Upon achieving the frequency and span of the piston oscillation $2l$ (where l is the span amplitude) with a certain value, a record of the pressure oscillogram was performed on the computer monitor. The voltage value ΔU in mV with the accuracy of 0.3 mV was converted into the value of the pressure oscillation span in bar using the conversion ratio ΔP (bar) = 0.00025 (bar/mV) × ΔU (mV).

In the course of further experiments, flat ring electrodes will be placed inside a closed pipe, as well as in the wave field near an open end of the pipe, which electrodes are planned to be supplied with a high direct voltage from the power supply unit 17. To research the features of the plasma emerging between the electrodes, current-voltage characteristics will be read from the electrodes using an ammeter 15 and a voltmeter 16 connected serial and parallel at different frequencies and amplitudes of the gas excitation.

3. Experimental results

The Figure 2 represents oscillograms with records of the gas pressure oscillations in time at different dimensionless excitation frequencies $\bar{\nu} = \nu/\nu_1$ and the piston span amplitude l = 0.375 mm. The transition through the resonance near the observed first eigenfrequencies ν_1 = 182 Hz was observed. It is obvious that away from the resonance ($\bar{\nu}$ = 0.97), the gas oscillates according to the harmonic law. Along with further increase of the frequency ($\bar{\nu}$ = 0.98, 0.99), a fracture is formed and the amplitude increases in the discharge area. An increase of the steepness of the rising edge of the pressure wave in observed. At the resonance ($\bar{\nu}$ = 1), the amplitude becomes maximum, and the front rising edge becomes close to the shock edge. After the resonance ($\bar{\nu}$ = 1.01), the amplitude decreases and a fracture is observed in the compression area. A further increase of the frequency $\bar{\nu}$ = 1.02) leads to a situation where the compression area in the pressure diagram is equal to the discharge area, but a fracture is observed in the crest of the wave. Away from the resonance ($\bar{\nu}$ = 1.03), the amplitude of the gas oscillation continues to decrease, and the gas oscillation once again assume a harmonic nature.

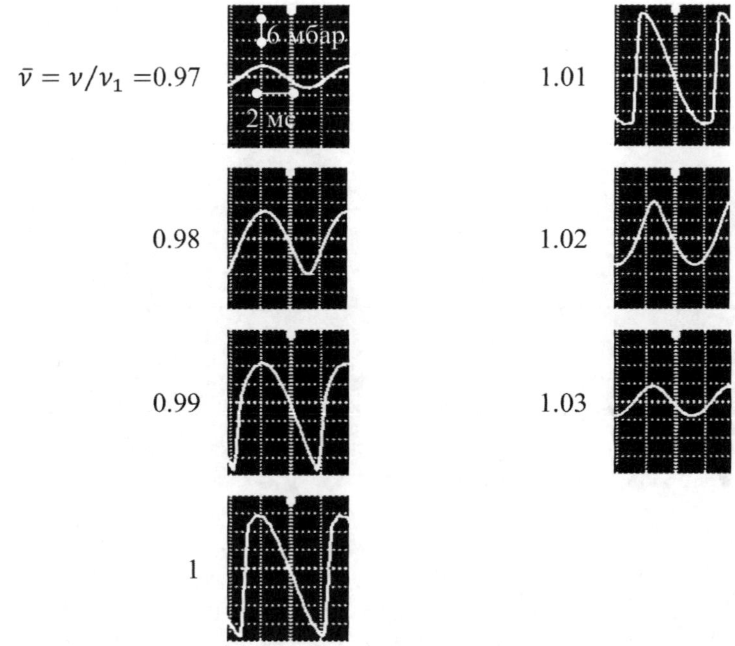

Figure 2. Oscillograms of the gas pressure oscillation at the amplitude of the piston shift l = 0.375 mm at different dimensionless excitation frequencies $\bar{\nu}$ near the first eigenfrequencies ν_1 = 182 Hz.

Quantitative results for the specified case are represented in Figure 3, where the amplitude-frequency curve is presented in the dimensionless form $\Delta\bar{p} = 10^2 \cdot \Delta p/p_0$ (p_0 – atmospheric pressure, $\Delta p = p_2 - p_1$ and p_1 are maximum and minimum values for the period of the piston oscillation). Notably, the value of the gas pressure amplitude at the resonance frequency is 4 times as high as away from the resonance. The theoretical resonance frequency is calculated according to the following formula [11]

$$\nu_1^* = \frac{c_0}{2L(1+\beta')}, \qquad \text{where } \beta' = \frac{1}{2}\left(1 + \frac{\varkappa-1}{\sqrt{\mathrm{Pr}}}\right)\frac{\delta}{R} \tag{1}$$

where c_0 is equilibrium sound velocity, and β' is absorption factor, $\varkappa$ is the adiabatic index, Pr is Prandtl number, δ is the acoustic interface thickness, R is the pipe radius, L is the pipe length.

There is a coincidence of values of the theoretic and the observed first and eigenfrequencies. The dotted line marks the ratio of the excitation frequency to the first eigenfrequencies calculated in consideration of the absorption according to the formula (1), the dashed line marks the ratio of the excitation frequency to that calculated without the consideration of the absorption. The observed resonance frequency has a lesser value than that calculated according to the linear acoustic theory (β'=0).

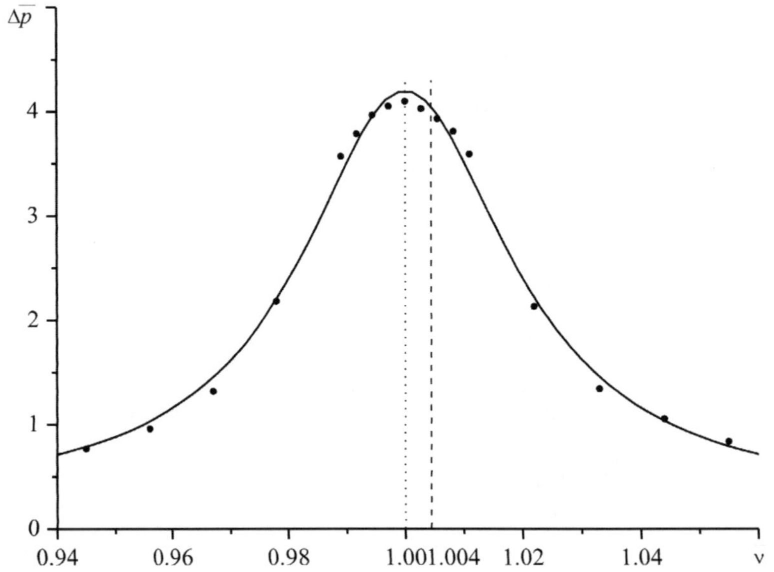

Figure 3. Dependency of the dimensionless span of the gas pressure oscillation on the dimensionless excitation frequency at the relative amplitude of the piston shift $\bar{l}$ = 3.99. Dots mark the experimental results, and the solid line is Lorentz approximation.

Figure 4 shows oscillograms of the gas pressure oscillation at the first eigenfrequencies ν_1 at different relative piston shift amplitudes. At $\bar{l}$ =1.06, the gas oscillates according to the sine law. Along with the increase of the amplitude ($\bar{l}$ = 2.66, 4.26, 5.86), a deformation of the gas pressure wave form is observed: the steepness of the front rising edge becomes greater than the back, and a fracture is observed. In this case, the shape of the wave remains continuous, and at $\bar{l}$ =7.46, a form close to that of a shock wave is observed.

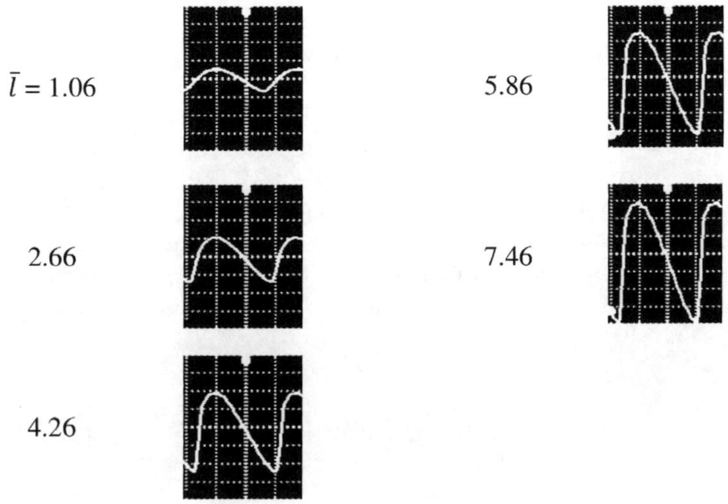

Figure 4. Oscillograms of gas pressure oscillation at the first eigenfrequencies at different relative amplitudes of the piston shift $\bar{l}$.

Figure 5 shows the dependence of the dimensionless span of the gas pressure oscillation in the resonance on the relative amplitude ($\bar{l}$ = $10^4 \cdot l/L$) of the piston oscillation, which characterizes the intensity of the oscillation. It is obvious that along with the increase of the piston oscillation amplitude, the span of the gas pressure oscillation increases. In these experiments, there is a power dependency $\Delta\bar{p} = al^n$ (where a = 0.82, n = 0.75). Notably, in the experiments with the emergence of shock waves, the power index in the resonance has the value of n = 0.5 [6].

Therefore, in these experiments, at small excitation amplitudes, where parietal losses are high compared to those for the gas compression in non-linear waves, there is a difference in the values of n.

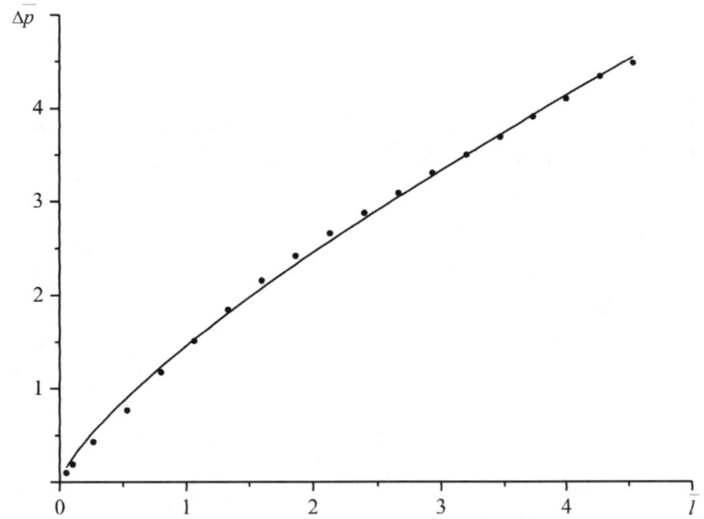

Figure 5. Dependency of the dimensionless span of the gas pressure oscillation on the relative amplitude of the piston shift at the first eigenfrequencies. Dots mark the experimental results, and the solid line is power approximation.

4. Conclusion

Peculiarities of non-linear gas pressure oscillation in a closed pipe in a wide range of frequencies and excitation amplitudes when transiting to the shock-wave oscillation mode in resonance were identified. Any increase of the excitation amplitude for all the frequencies in question leads to the increase of the span of the gas pressure oscillation, as well as to a deformation of the wave form: the steepness of the front rising edge becomes greater than that of the back one. The pressure oscillation has a continuous nature, but it becomes close to the interrupted one in the resonance. At the resonance frequency of the gas excitation, an increase of the piston stroke amplitude leads to the increase of the span of the gas pressure oscillation different from that observed in the mode of formation of shock waves. The experimental data obtained are necessary when researching the effect of acoustic waves with non-linear pressure forms on the plasma.

Acknowledgments

The work is performed according to the Russian Government Program of Competitive Growth of Kazan Federal University.

References

[1] Aramyan A R and Galechyan G A 2007 *Physics-Uspekhi* 50 1147-69
[2] Kashapov N F and Fadeev S A 2014 *Izvestiya Vysshikh Uchebnykh Zavedeni. Physics* **57**(3/2) 110-113
[3] Saifutdinov A I, Fadeev S A, Saifutdinova A A and Kashapov N F 2015 *JETP Lett.* **102**(10) 637-642
[4] Israfilov Z K and Kashapov N F 1991 *Journal of Engineering Physics* **60** 364–368
[5] Zaripov R G and Ilhamov M A 1976 *J. Sound and Vibration* **46**(2) 245-257
[6] Ilgamov M A, Zaripov R G, Galiullin R G, Repin V B 1996 *Applied Mechanics Reviews* **49**(3) 137-154
[7] Zaripov R G, Galiullin R G and Galiullina E R 2001 *Actual problems of continuum mechanics. On the 10th anniversary of IME KazSC RAS* 19-35 (Kazan: IME KazSC RAS)
[8] Tkachenko L A 2011 *Actual problems of continuum mechanics. On the 20th anniversary of IME KazSC RAS* 98-113 (Kazan: Foliant)
[9] Gubaidullin D A, Zaripov R G, Tkachenko L A and Shaidullin L R 2015 *J. of Engin. Physics and Thermophysics* **88**(4) 843-847
[10] Zaripov R G, Tkachenko L A and Shaidullin L R 2014 *J. Phys.: Conf. Series* **567** 012042
[11] Tkachenko L A and Zaripov R G 2013 *Russian Aeronautics* (1) 28-32

The effects of low-temperature plasma treatment on the capillary properties of inorganic fibers

A R Garifullin, I Sh Abdullin, E A Skidchenko, I V Krasina and M F Shaekhov
PNTMC Department, Kazan National Research Technological University, Kazan, 420015, Russia

uni0404@mail.ru

Abstract. Solving the problem of achieving high adhesion between the components in the polymeric composite material (PCM) based on carbon fibers (CF) and basalt fibers (BF) is proposed to use the radio-frequency (RF) plasma under lower pressure by virtue of efficiency, environmental friendliness and rationality of the method. The paper gives the results of studies of the properties of CF and BF after RF capacitive discharge plasma treatment. The plasma modification modes of carbon and basalt fiber were investigated. The efficiency of treatment tool in surface properties modification of carbon and basalt fibers was found, namely capillary properties of CF and BF were researched. The optimal treatment modes were selected. It was found that the method of plasma modification in the radio-frequency capacitive discharge under the lower pressure contributes enhancing the capillary properties of inorganic fibers, in particular carbon and basalt ones. It shows the tendency to increase of the adhesive properties in PCM, and, consequently, the increase of the physical and mechanical properties of the products.

1. Introduction

Technology of creating fibrous polymeric composite materials (FPCM), which are applied in the production of sports equipment and sports cars, includes activation of the fiber surfaces to enhance adhesion between fiber and matrix. Nowadays in the production of reinforced composites fiber is pretreated with chemical solutions which are toxic and aggressive substance, and thus require complex protection systems and disposing of as hazardous waste. The radio-frequency capacitive (RFC) plasma under lower pressure is the perspective treatment tool for materials with different nature [1]. This method makes it possible to treat organic and inorganic materials of various consist and structures. It is especially important to note the fact that the RFC plasma treatment technology is environmentally friendly [2].

Based on the results of previous researches [3, 4] in the study of the low-temperature plasma (LTP) effect on the properties of inorganic fibers and filaments to make materials hydrophilic treatment in RFC discharge under lower pressure in argon atmosphere is carried out.

2. Materials, methods and equipment

As study objects were considered three types of CF with different degree of graphitization and the structure of weaving: high modulus fiber Kulon 500/0.07, carbonized fibers LUP-0.2 and Ural T-2-22r; basalt twisted thread BS10-68Z40-KB-12.

Capillarity was measured by height of fluid lifting with the control and the treated by RFC plasma CF and BF samples.

To modify the fibers it was used the radio-frequency plasma unit of capacitive discharge with plane-parallel electrodes (figure 1).

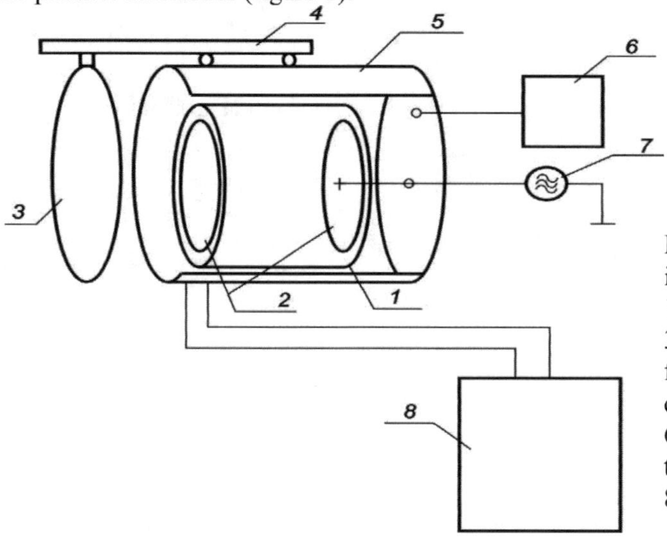

Figure 1. Scheme of experimental-industrial RF plasma unit: 1 - plexiglass drum; 2 - RF-electrodes; 3 - cap of vacuum chamber; 4 - console for opening the lid of the vacuum chamber; 5 - vacuum chamber; 6 - system of supply and regulation of the plasma gas; 7 – RF-generator; 8 - vacuum pumping post.

In this study the input parameters of the unit for CF treatment were set within the following: plasma gas flow G = 0.04 g/s, frequency f = 13.56 MHz, the pressure in the vacuum chamber P = 26.6 Pa, voltages U, amperage I and the treatment time t were varied from1.5 kV up to 5 kV, from 0.3 A up to 0.7 A and from 3 up to 7 min respectively. In the experiments technical argon was used as the plasma gas.

To research the effect of LTP on the capillary properties of BF samples were pretreated in the plasma unit with mode: the plasma gas argon, G = 0.04 g/s, f =13.56 MHz, P = 26.6 Pa, t = 3 min with variable voltage U from 2 kV up to 7 kV and amperage I from 0.3 A up to 1.0 A.

3. Results

After the plasma treatment the carbon fiber samples show the tendency to increase the amount of fluid lifting height. This indicates an improvement wettability of the carbon textile (figure 2).

It is found that at the equal time of plasma treatment t= 5 min in argon atmosphere increase of RF voltage on the electrodes leads to a decrease of the capillary rise of liquid by CF (figure 3).

On the one hand, increase the plasma treatment time at a fixed voltage U=1.5 kV leads to increase capillarity of CF. On the other hand, increase the plasma treatment time at a fixed voltage U=3 kV leads to its decrease (figure 4).

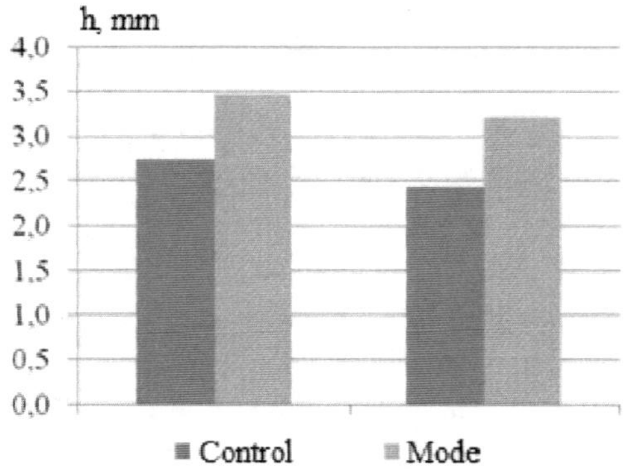

Figure 2. The effect of treatment on capillarity of CF: 1 – CF LUP-0.2; 2 – CF Kulon 500/0.07.

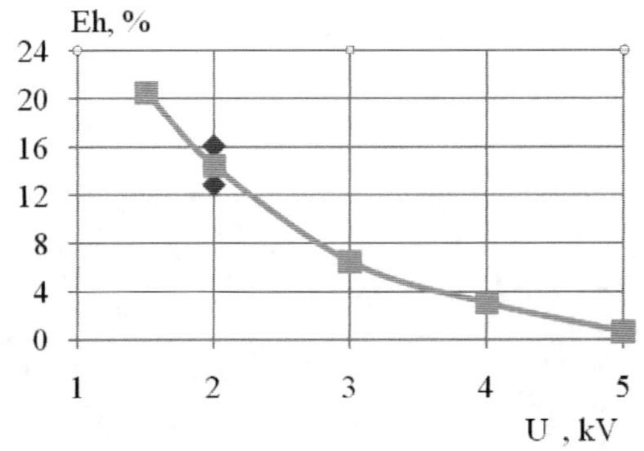

Figure 3. The effect of voltage between the plates on the capillary rise of liquid of CF Ural T-2-22r;

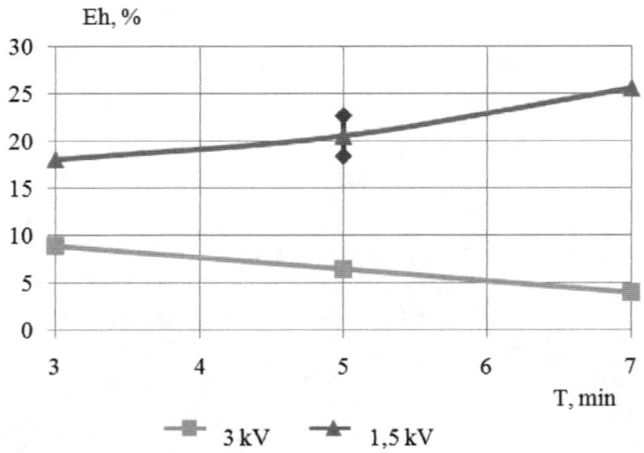

Figure 4. The effect of the treatment time on the capillarity of CF Ural T-2-22r.

The results of BV research are presented in table 1.

Table 1. The dependence of the BF capillarity from modes of plasma treatment

Voltage (U), kV	Amperage (I), A	Capillarity, mm
The control sample		34
2	0.3	60
3	0.4	64
4	0.6	74
5	0.7	67
6	0.85	62
7	1.0	62

Analysis showed that the maximum of capillarity of basalt thread is achieved with following LTP treatment mode: U=4 kV, I =0.6 A, t =3 min, G= 0.04 g/s, f=13.56 MHz, P=26.6 Pa. Capillarity due to this mode increases by 118%.

4. Conclusions

RFC plasma treatment under lower pressure in argon atmosphere enhances the capillary properties of CF and BF.

Accordingly the test results the tendency to increase adhesiveness of these fibers by LTP treatment was found. Projected, it will lead to increase the physical and mechanical properties of polymeric composite materials based on modified fiber.

Acknowledgments

The study was carried out with the financial support of the Russian Ministry of Education, the project number 2196 from 01.02.2014.

References

[1] Abdullin I Sh, Zheltukhin V S, Sagbiev I R and Shaekhov M F 2007 *Modifikatsiia nanosloev v vysokochastotnoi plazme ponizhennogo davleniia* [Modification of nanolayers in radio-frequency plasma of low pressure]. Kazan, (Izdatel'stvo

Journal of Physics: Conference Series **669** (2016) 012054 doi:10.1088/1742-6596/669/1/012054

Kazanskogo tekhnologicheskogo universiteta) p 356 (In Russian).

[2] Garifullin A R and Abdullin I Sh 2014 The present state of the problem of surface treatment of carbon fibers for further use in polymer composites as reinforcing elements (*Vestnik Kazanskogo tekhnologicheskogo universiteta* vol 7) pp. 80-85 (In Russian).

[3] Garifullin A R and Abdullin I Sh 2014 Plasma hydrophilize of carbon ribbon to create composite materials with high strength characteristics. (*Vestnik Kazanskogo tekhnologicheskogo universiteta* vol 17*)* pp. 101–102 (In Russian).

[4] Ershov I, Sergeeva E, Zenitova L and Abdullin I 2013 Effect of plasma treatment on the surface properties of glass fiber (Vestnik Kazanskogo tekhnologicheskogo universiteta vol 4) pp 97-99

Journal of Physics: Conference Series **669** (2016) 012055 doi:10.1088/1742-6596/669/1/012055

Fractal structure of low-temperature plasma of arc discharge as a consequence of the interaction of current sheets

N A Smolanov
Ogarev Mordovia State University

E-mail: smolanovna@yandex.ru

Abstract. The structure of the particles deposited from the plasma arc discharge were studied. The flow of plasma spreading from the cathode spot to the walls of the vacuum chamber. Electric and magnetic fields to influence the plasma flow. The fractal nature of the particles from the plasma identified by small-angle X-ray scattering. Possible cause of their formation is due to the instability of the growth front and nonequilibrium conditions for their production - a high speed transition of the vapor-liquid-solid or vapor - crystal. The hypothesis of a plasma arc containing dust particles current sheets was proposed.

Introduction. Vacuum arc technologies are the subject of many studies [1-3]. Interest in the processes in electric arcs grew after the discovery of carbon nanostructures [3]. The theoretical and experimental studies clearly observed two approaches to the processes in the plasma flow arc.

The first one takes into account only the ionic composition of the plasma and its behavior in external fields - electric and magnetic. It has been established that weak (10 mT) field leads to a focusing of the plasma flow, a further increase in magnetic field increases the ionic charges [4].

The second approach is more realistic. It is complicated by the consideration of the behavior of the plasma flow, which, in addition to the ion and electron components of the plasma, there is a drop fraction of arc discharge and solid pieces of the cathode. On these grounds it can be attributed to a dusty plasma.

It should be noted that from the cathode spot (CP), which is the basis of the plasma flow, until the condensed matter or cathode connections, is a whole series of processes of interaction of particles deposited - ions, electrons, dropping plasma fraction. It is known that metal droplets leaving the cathode rapidly evaporate and turn into plasma formation in the immediate vicinity of the CP [1]. These particles move in the imposed magnetic and electric fields and represent a superposition of interactions are usually gas-discharge plasma and plasma metal.

We have studied in detail the structure of condensed matter, deposited from the plasma stream near the cathode as a - (condensate dropping fractions), and on the walls of the vacuum chamber (soot) [5-9]. In our view, this approach allows to get closer to the problem of the relationship of processes in the plasma and the resulting structure. Methods of preparation and study of such structures are described in the works [5, 8, 9]. Of particular interest to the processes in the plasma arc appeared after the discovery of fractal structures in the products of deposition of plasma flow on the walls of the vacuum chamber [9].

A fractal is a shape, parts of which are in any way similar to the whole. Properties fractal defined by the expression: $M(\lambda,r)= \lambda^{D} M(r)$, where M - a "property" of the fractal, r - a unit of length, λ - coefficient of increasing spatial size of the fractal (r = λ r), D - Float (fractal dimension). If the fractal aggregate is formed from clusters, it has a value of fractal dimension D = 1,94. This is in line with the inter-cluster loose aggregates pores [10]. Fractal aggregates appear to volatile growth front when small perturbations of the front (interface) begin to grow much faster than the neighboring areas.

Work purpose. Thus, it is a process in a plasma arc stream as it moves from the cathode assembly of the vacuum plant type HHB-6 to the point of deposition and structures deposited from the plasma. In theory, magnetohydrodynamics cathode assembly is generally viewed as a coaxial plasma accelerator with a longitudinal magnetic field [11].

The presence of a longitudinal magnetic field in the plasma accelerator (cathode assembly) not only leads to a rotation of the plasma [11,12], but also to a number of interesting physical phenomena are not fully understood and provoke debate [13-18]. One of them is the retrograde motion of the cathode spot, and accordingly, the change of the direction of rotation of the plasma arc stream in general. Given the recent results of the study of the magnetic properties of the particles in the plasma and in the same retrograde character of their movement at certain critical values of the magnetic field [16-17], the question arises about the root cause of changes of direction - or cathode spot plasma flow?

Thus, in this paper is a continuation of studies [5-9], attempted to explain the formation and structure of condensed matter. They were precipitated from the plasma stream near the cathode (dropping condensate fraction) on the substrate in the center of the vacuum chamber (film) on the walls of the vacuum chamber (nanostructures, including fractal).

At the same time we are based on generalization of ideas and the results obtained in the study of current sheets in the plasma [19-21], including tokamak plasma, and their intermittency [22]. The analysis of the possible causes of retrograde motion of the cathode spot [14-15], rotation of the plasma [11-12]. Taken into account the results of the study of the spectral composition of the plasma arc, drip nature spots cathode spot [18] - (drip glowing spots).

Results. The surface morphology. The structure of the dust particles deposited from the plasma arc discharge to the metallic vacuum chamber walls cooled represents conglomerates interacting particles. The particles have a complex structure and in fact is the result of combining a large number of small particles of nanometer size (Figure 1).

Of particular note is that this dust ("metal snow" - Academician Kadomtsev) with a particle size of 10 to 1000 microns was formed as a result of hydrogen embrittlement of stainless steel at exploitation tokamak Globus-M [23]. The particles become magnetic stainless steel in the melting of its unipolar micro arcs. This adverse phenomenon (unipolar vacuum arc) leads to the destruction of one of the walls of the tokamak, facing the plasma.

We considered possible mechanisms of evolution of the dusty plasma cathode spot before its recombination on the substrate and the walls of the vacuum chamber in a simultaneous imposition of margins. Crystallization capillary plasma component and ion flux to the walls of the vacuum chamber leads to the formation of micro - and nanostructures, i.e. manifestation condensing agent in an excited state. At the same time there may be unknown to the structure modifications.

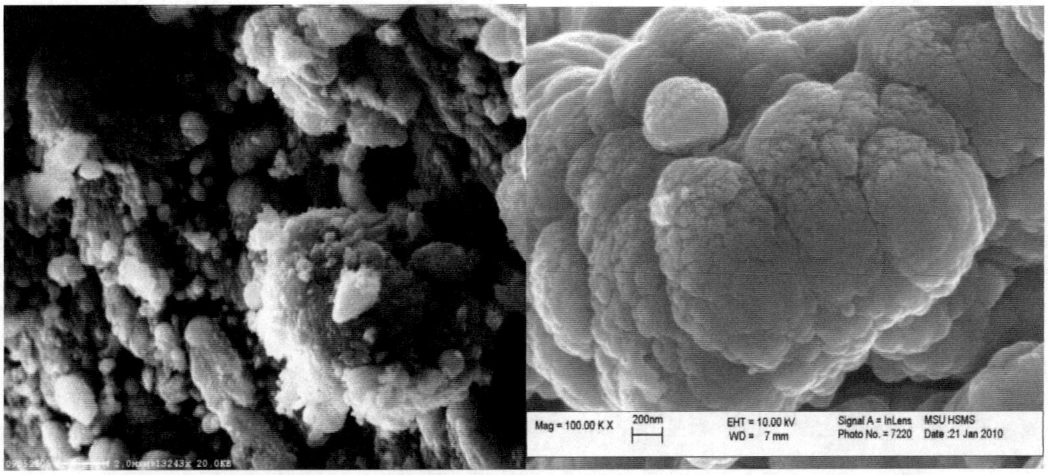

Fig. 1. The morphology of the powder magnetic and non-magnetic fraction

Small-angle x-ray scattering (SAXS). SAXS analysis results showed that the analyzed fine powder can be regarded as a structure of alternating metal-containing elemental and phase compositions from the fractal aggregates. It is found that the powder particles have structural studies in homogeneity nanometer scale, and their mean values lie within a fairly narrow range of about ~20-30 nm [8,9]. The fractal dimension of the particles of in homogeneities of different fractions (20, 63, 100, 140 and 180 microns) shown in Table 1 [9]. This corresponds to a loose aggregate of clusters with inter-cluster pores.

Table 1. The fractal dimension of in homogeneities [9]

Fractal, mcm	Linear scale of irregularities, nm		
	90-52	50-40	33-5
20	1,04	1,92	2,51
63	1,08	1,51	2,38
100	1,14	1,99	2,48
140	1,00	1,62	2,36
180	1,32	1,77	2,34

Fractal particles due to the conditions of formation of particulate matter, the composition of the plasma flow parameters of the electric and magnetic fields in the inter-electrode space. As already mentioned above, the formation of fractal aggregates similar occurs in conditions of instability of the growth front, when a small perturbation of the front (interface) begin to grow much faster than the neighboring areas [10]. We assume that we are studying fractal structures from the plasma arc are the result of the special properties of the current layers with intermittency [19-22], and the turbulence of the plasma flow [11,12].

Discussion. Possible causes of fractal particles. Flows of ions from the cathode spot moving in an arc discharge can be considered as interacting current layers. It is highly conducting layers in the plasma separating the magnetic fields of various kinds [19, 20]. Jump magnetic field leads to the appearance of a thin layer of an electric current. As a result of what is happening in the reconnection layer (intermittency) [22] the magnetic field lines of the magnetic field topology changes that accompanied the transition of its energy into heat radiation energy of magnetohydrodynamic flows and particle acceleration. Thus, the magnetic field generated by current layers, the source of energy dissipation which is the reconnection of magnetic field lines. The theory of current sheets requires a "special" magnetic field lines in a plasma, which arise under the influence (impact) disturbances at distances exceeding the dimensions of the neighborhood layer. Therefore, not every configuration of the magnetic field can create a special line, when the electric and magnetic field are parallel to each other [16]. It is possible that a similar situation may occur in the vicinity of the cathode spot of the arc discharge in a coaxial magnetic field, when a negative potential is applied to the substrate and an electric field in the electrode space. Then for dissipation of the magnetic field in a small region of the plasma, in particular in the area of the cathode spot, there are conditions for the growth of particles with a fractal structure with a highly developed surface.

Several studies have found that in laboratory conditions the plasma formed non-neutral - containing non-zero transverse $B_\perp$ and a longitudinal $B_{\parallel}$ magnetic field component (or the y- and z-components of the field). In this case, for the formation of the current layer need only one pair reconnect field components.

For an explanation on the basis of current layers of other effects, such as retrograde motion of the cathode spot and the change of plasma rotation at the critical value of the magnetic field, we note that the opening angle of the cathode spot products (ions, dropping fractions) near the cathode obeys the cosine. This can greatly affect the resulting magnetic field geometry.

Conclusion. Thus, the fractal structure of the dust (soot) particles deposited on the walls of the vacuum chamber, both in low-temperature plasma, and plasma fusion is due, in our view, a manifestation of (action) of the current alternating layers.

References

[1] Anders A. Cathodic Arcs: From Fractal Spots to Energetic Condensation. Berckey: Springer Science, 2008. 540 p.
[2] Lyubimov G A, Rakhovsky V I 1978 *Uspekhi Fizicheskikh Nauk*. **125** 665 (in Russian)
[3] Kratschmer W, Lamb L D, Fostiropoulos K 1990 *Nature* 347 354.
[4] Anders A, Oks Ye M, Yushkov G Yu 2006 *Journal of Technical Physics*. **76** 57. (in Russian)
[5] Smolanov N A , Pankin N A , Batin V V , Pavkin E P 2014 *Prikladnaya fizika* **1** 30 (in Russian)
[6] Smolanov N A , Pankin N A 2013 *Journal of Physics: Conference Series*. **479** 012012.
[7] Pankin N A , Smolanov N A 2013 *Journal of Physics: Conference Series* **479** 012006
[8] Smolanov N A, Pan'kin N A 2014 *Journal of Surface Investigation. X-ray, Synchrotron and Neutron Techniques*, **8**, 1089.
[9] Smolanov N A , Neverov V A 2015 *Pisma o materialakh* **5** 179 (in Russian)
[10] Cleman M , Lavrentovich O D , 2002 Soft matter physics: An introduction. NY-Berlin-Heidelberg: Springer, 637 p
[11] Morozov A I 2008 Vvedeniye v plazmodinamiku. M.: FIZMATLIT - 616.
[12] http://ufn.ru/ru/articles/2015/6/c/
[13] Zabello K K , Chaly A M , Shkolnik S M 2013 *Journal of Technical Physics* **39** 44 (in Russian)
[14] Bobrov Yu K , Bystrov V P , Rukhadze A A 2006 *Journal of Technical Physics*, **76** 35(in Russian)
[15] Andreev A A 2007 *Izv. Kharkov. University*. **32** 763. (in Ukraina)

[16] Diachkov L G. 2011 FNTP-2011. **2** 152(in Russian)
[17] Karasev V Yu, Dzliyeva E S 2011 FNTP-2011. **1** 147(in Russian)
[18] Shchurenkova S A. 2012 Avtoref. diss. k. t. n., Tomsk, 18 (in Russian)
[19] http://ufn.ru/ru/articles/2010/9/h/
[20] http://ufn.ru/ru/articles/2015/2/a/
[21] Ostrovsky G V 2013 Journal of Technical Physics **83** 59 (in Russian)
[22] Budaev V P , Khimchenko L N 2006 Fractal Grown of Deposited Films in Tokamak: *Preprint IAE-6404/7*.
[23] http://ufn.ru/ru/articles/2010/10/c/

Journal of Physics: Conference Series **669** (2016) 012056 doi:10.1088/1742-6596/669/1/012056

The features of high-current gas discharge in a narrow gap between the liquid electrolyte and solid electrode

A Kh Tazmeev, R N Tazmeeva and F S Sarvarov

Branch of Kazan Federal University in Naberezhnye Chelny, 10A Sujumbike av., Naberezhnye Chelny, 423812, Russia

E-mail: tazmeev@gmail.com

Abstract. Gas-dynamic phenomena on the boundary "plasma – liquid electrolyte" currents in the range (10 – 21) A were studied. The regularities of the influence of deformation of the surface of the electrolyte on the characteristics of the gas discharge were revealed.

1. Introduction

Gas discharges with liquid electrodes allows producing chemically active non-equilibrium plasma at atmospheric pressure directly to a relatively simple manner. Therefore, in recent years, a growing interest in the study of such discharges, and the possibilities of their practical application [1-7]. On the boundary "plasma-electrolyte" there are complex processes of interaction of the gas discharge with a liquid medium. Many aspects of these processes currently remain poorly understood.

In paper [8] it pointed out that under the influence of the high voltage applied to the discharge gap, the surface of the electrolyte rises. When the discharge is ignited, there is a reverse phenomenon - the surface of the electrolyte is pressed inward. Some regularities of this phenomenon at currents not exceeding 11 A, were studied in paper [9]. The aim of this work is an experimental study of gas-dynamic phenomena at the "plasma - the electrolyte" at higher discharge current.

2. Experiment

The gas-discharge unit of the experimental device is shown schematically in figure 1. The discharge burned between *1* flowing electrolyte *2* and a water-cooled metal anode *3*. The direction of flow of the electrolyte is indicated by an arrow. Electrolyte was served as a weakly concentrated solution of Glauber's salt in distilled water. The electrical conductivity σ of electrolyte at room temperature were in the range (1.0 – 2.5) mS/cm. Mass flow rate m of the electrolyte through the zone of the discharge was varied from 8 to 40 g/s.

The anode was located at the height H = 15 mm. The thickness of the layer of electrolyte over the current lead was 13.5±0.2 mm. Thus, prior to ignition of the discharge between the electrolyte and the anode there was remained an air gap equal to 1.5±0.2 mm. The discharge was ignited by high-voltage breakdown of the gap. Under the influence of the discharge in the electrolyte was formed a "pit". The depth of "pit" Δh was determined from measurements of the potential distribution inside the electrolyte φ by using a single electric probe. Same probe as in paper [9] was used in experiments. He moved in a vertical direction of the coordinate device. The mechanical movement was transformed into an electrical signal, and the signal was fed to the "*X*"-entrance tablet two-coordinate potentiometer brand PDP4-02. Signal from the probe via a high resistor «*Y*»-entrance was received.

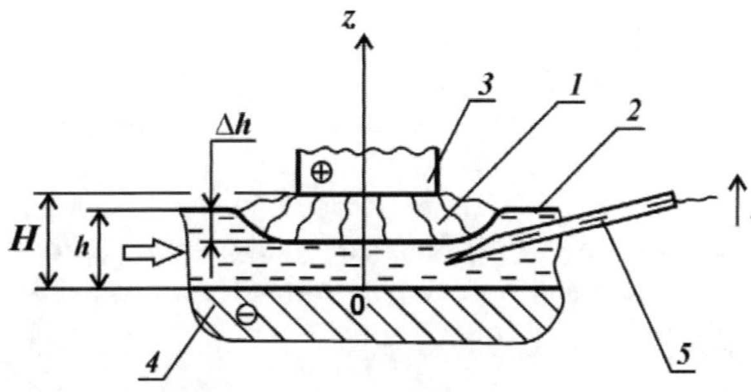

Figure 1. Gas discharge unit: *1* – discharge area; *2* – liquid cathode (electrolyte); *3* – anode; *4* – current lead; *5* – probe.

The research was carried out in the range of currents (10 – 21) A. The source of electrical power was served as a three-phase full-wave rectifier. The power circuit of the ballast resistor was missing. The power supply was grounded. To measure the discharge current and voltage on discharge unit were used pointer devices M1042 type of accuracy class 0.2. The temperature of the electrolyte was measured by a chromel-alumel thermocouple placed over the discharge area along the flow of the electrolyte.

3. The results of experiments, their processing and discussion

The graphs recorded on the tablet device PDP4-02 at different values of discharge current *I* and the mass flow rate *m* of the electrolyte are shown on figure 2. In this figure the vertical dotted line indicates the thickness *h* of the layer of electrolyte.

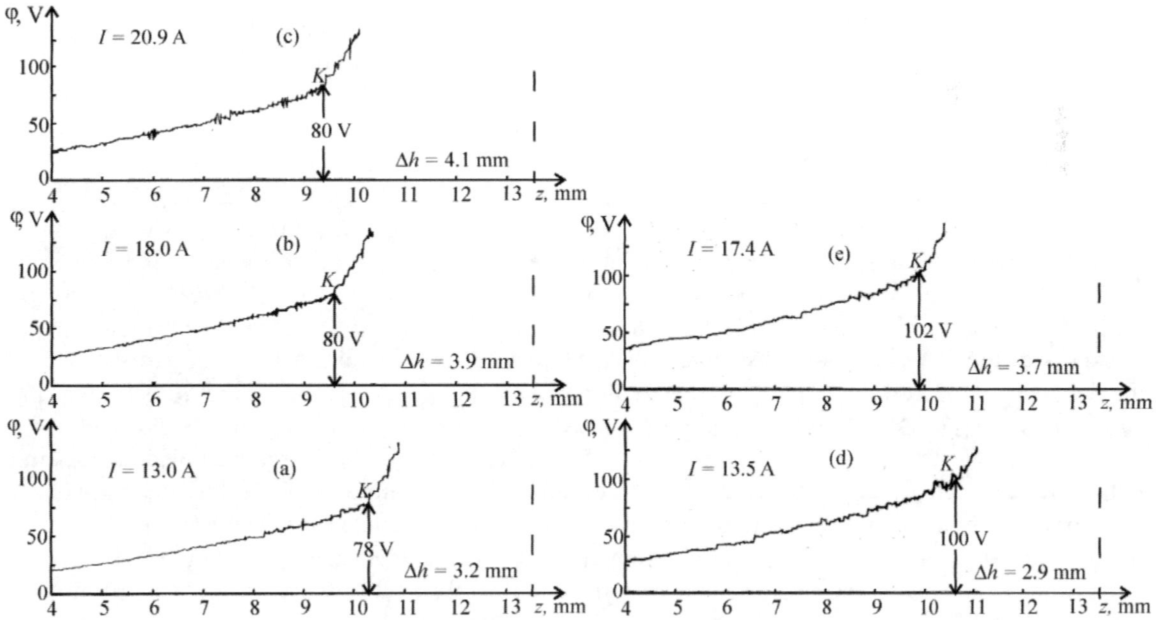

Figure 2. Potential distribution within the electrolyte of the cathode at low intensity of cooling. σ = (2.50±0.05) mS/cm. (a), (b), (c) – m = 10 g/s; (d), (e) – 40. Temperature of the electrolyte: (a) – *t* = 73 °C; (b) – 76; (c) – 79.5; (d) – 53; (e) – 61.

As can be seen, in all cases, inside the electrolyte the potential of φ grows linearly along a vertical coordinate *z* measured from the surface of the current lead (to point *K*). The values of φ, recorded by a probe at its exit from the electrolyte, are much higher (after a point *K*). Coordinate of a point *K* determines the position of the boundary "plasma – electrolyte". The depth of "pit" Δh is equal to the difference of the values of *h* and *z* coordinates of the point *K*. With increasing current, the depth of "pit" is growing. Such regularity is observed in all the investigated range of variation of the parameters *m* and σ.

Some of the results which obtained using an electrolyte with lower specific electrical conductivity equal to (1.35±0.05) mS/cm, are given in table 1. As follows from comparison of graphical and tabular data, the depth of "pit" is almost independent of the electrical conductivity of the electrolyte. The main parameter affecting the depth of "pit" Δh, is the discharge current *I*.

Table 1. The results of experiments

m, g/c	*I*, A	Δh, mm
10	13.4	2.7
	17.0	3.3
	19.7	4.1
40	12.9	2.9
	15.0	3.2
	20.2	3.9

The dynamic impact of the gas discharge with liquid electrolyte cathode with the formation on its surface "pits", results in the following remarkable phenomenon. It is as follows. At constant values of *m* and σ the value of potential φ corresponding to the point *K* practically remains constant. In other words this means that the voltage drop ΔU_k on the liquid electrolyte cathode is practically independent of current because the value of ΔU_k numerically equal to the potential corresponding point *K* on the graphs to the dependence of φ from *z* (figure 2). When current is increased, graphics is located above as noted earlier. However, the point *K* is shifted to the left, resulting in a value of φ, corresponding to this point does not change. From a physical point of view the constancy of ΔU_k is due to the following reasons. On the one hand, according to Ohm's law, with the increase in current increases the voltage drop in the electrolyte, which should lead to an increase in ΔU_k. On the other hand, reducing the thickness of the electrolyte layer by increasing the current is reduces electrolyte ohmic resistance of the cathode, and therefore the ΔU_k, is numerically equal to the product of the ohmic resistance at the current is decreases. Due to these mutually exclusive two factors ΔU_k remains almost unchanged.

As is known, the ohmic resistance of the electrolyte is significantly reduced with increasing temperature. In our experiments, with increase of current, temperature of electrolyte was increased, therefore the ohmic resistance decreased. And this, in turn, should lead to a decrease in ΔU_k. Thus, there is another, probably more important reason affecting to ΔU_k. It consists in heating the electrolyte. At high mass flow rates, the electrolyte heats little, and therefore with the growth of *m* is formed on the cathode a higher voltage drop ΔU_k, in particular it can be seen from comparison of the data presented in figure 2. If $m = 10$ g/s the voltage drop at the cathode is in the range of (78 – 80) V (figure 2(a), 2(b), 2(c)), and if $m = 40$ g/s it is (100 – 102) V (figure 2(d), 2(e)).

Deepening the "pit" with increasing current lengthens the discharge gap, which in turn leads to an increase in the discharge voltage U_d and with it the voltage at the terminals of a gas discharge device: $U = \Delta U_k + U_d$. For this reason, current voltage characteristics (CVC) are increased (figure 3).

The increasing character of the dependence of *U* and *I* is due to only one cause, namely, the increase of the discharge gap due to the formation of "pit" in the cathode electrolyte, as the voltage drop at the cathode is almost constant for all values of current. The same reason as the main considered in work [9] for the formation of increasing CVC in the current range (4 – 10) A. Thus, increasing CVC – is an inherent property of a gas discharge with a liquid electrolyte cathode with short distance between the electrodes.

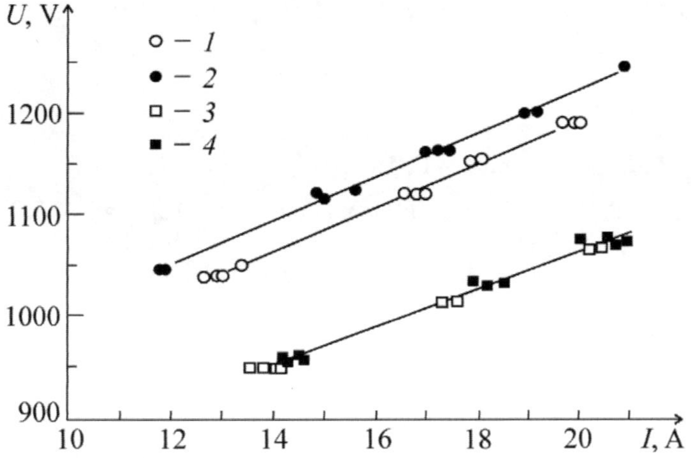

Figure 3. Current-voltage characteristics of the gas discharge device. *1, 2* – σ = (1.35±0.05) mS/cm; *3, 4* – (2.50±0.05) mS/cm. *1, 3* – *m* = 10 g/s; *2, 4* – 40.

4. Conclusions

The formation of "pits" in the electrolyte is a characteristic feature of a gas discharge with liquid electrolyte cathode at small distance between the electrodes, and the depth of "pit" grows with increasing current.

The voltage drop on the liquid electrolyte cathode is practically independent of current, due to the decrease of the ohmic resistance of the cathode at current increases due to the change in the thickness of the layer of the electrolyte and increasing its temperature.

References

[1] Maksimov A I and Khlyustova A V 2009 *High Energy Chemistry* **43** No. 3 149
[2] Rybkin V V, Smirnov S A, Titov V A and Arzhakov D A 2010 *High Temperature* **48** 476
[3] Levko D S, Shchedrin A I, Chernyak V Y, Olshvskii S V and Nedybalyuk O A 2010 *Technical Physics Letters* **36** 998
[4] Samukawa S, Hori M, Rauf S et el. 2012 *Phys. D: Appl. Phys.* **45** 253001
[5] Tazmeev Kh K, Arslanov I M and Tazmeev G Kh 2014 *J. Phys.: Conf. Ser.* **567** 012001
[6] Kashapov L N, Kashapov N F and Kashapov R N 2015 *IOP Conf. Ser.: Materials Science and Engineering* **86** 012019
[7] Valiev R A, Shakirov Yu I and Iliuchin A N 2014 *J. Phys.: Conf. Ser.* **567** 012040
[8] Lazarenko B R, Faktorovich A A, Duradzhi V N and Bryantsev I V 1970 *Eelektronnya Obrabotka Materialov* No. 4 18
[9] Miftachov M N, Tazmeev A Kh, Tazmeev Kh K and Fridland S V 2006 *Journal of Physics and Thermophysics* **79** 532

Journal of Physics: Conference Series **669** (2016) 012057 doi:10.1088/1742-6596/669/1/012057

The high-frequency current pulsations in the gas discharge with liquid electrode

G Kh Tazmeev[1], B A Timerkaev[1] and Kh K Tazmeev[2]

[1]Kazan National Research Technical University named after A.N. Tupolev, 10 K. Marx str., Kazan, 420111, Russia
[2]Branch of Kazan Federal University in Naberezhnye Chelny, 10A Sujumbike av., Naberezhnye Chelny, 423812, Russia

E-mail: tazmeevh@mail.ru

Abstract. The electrical discharges are experimentally investigated in the air between the liquid electrolyte and metal electrodes. Spectra of the current signals are registered and analyzed in the megahertz frequency range. It is established that one of the main causes of the high-frequency current pulsations is the flow of products and the erosion of the sputtering electrodes in the discharge region.

1. Introduction

Electrical discharges which contacted with the liquid are not uncommon. It is known the gas discharges which are burning between the solid electrodes in water or in a specially prepared electrolyte [1-3]. They are used in industry for the underwater stamping large-size thin-walled of sheet metal material, and also a plasma electrolytic modification of metal surfaces. Often there are some variations of a solid electrode placed above the liquid electrolyte [4-8]. They are perspective for initiating chemical processes in the electrolytes which are used in the textile industry [4-5]. These electrical discharges can be used in plasma-chemical reactors intended for the production of the syngas [7]. Low-current discharges can be applied in medicine [8]. However, in despite of the fairly widespread practice, electrical discharges contacted with the liquid poor investigated nowadays, especially in the low spatial and temporal intervals. One of the properties which are shown in small intervals of time is the presence of high-frequency current pulsations. They are fixed in various variants of electrical discharge with liquid electrolytes [9-12]. In the closed volume inside of the electrolyte can occur magnetosonic waves with frequencies in the megahertz range are assumed by authors of works [11]. This assumption is clearly inadequate to explain the reasons for the emergence of high frequency current pulsations in various embodiments, electrical discharges in contact with liquid electrolytes. The aim of this work was the experimental investigation of high frequency current pulsations in gas discharges which generated in the air gap between the liquid electrolyte and solid electrode.

2. Experiment

The power supply was a three-phase full-wave rectifier which was connected to the secondary windings of step-up transformer. The output voltage was smoothed by C-L-C filter. At idle, it is amounted to 1760 V. The current limit was made with by wire resistors.

Journal of Physics: Conference Series **669** (2016) 012057 doi:10.1088/1742-6596/669/1/012057

Oscillograms of current were recorded with an oscilloscope AKIP-15/1, 25 MHz bandwidth. The analysis of the current oscillations mode was performed with using fast Fourier transform of the same oscilloscope.

Solid electrodes are made of copper most often in the research practice. The solutions of sodium chloride in distilled water as a liquid electrolyte are used as the liquid electrolyte. Experimental results for such a combination of electrodes are provided on the figure 1*a*. The cathode is a liquid electrolyte, and an anode - a pointed cone at the copper rod with a diameter of 4.5 mm. On the figure outlines of the copper rod are marked with white lines. Specific electrical conductivity of the electrolyte was 10 mS/cm. It was poured into a porcelain vessel. Inside of the vessel was mounted graphite plate which was connected to the output terminal of the power source. A distance between the liquid electrolyte and a copper rod was set to 5 mm. The same interelectrode gap was maintained in all subsequent experiments with other electrodes.

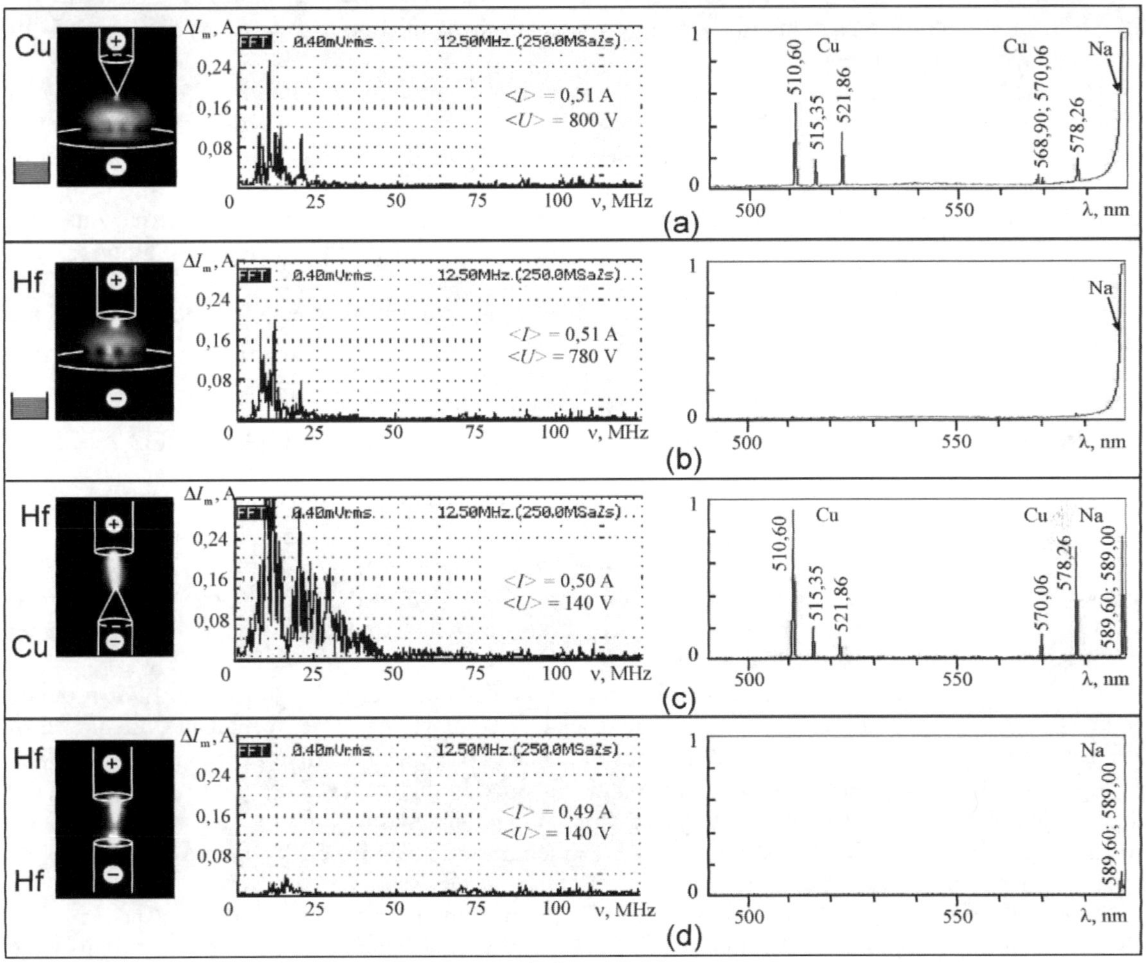

Figure 1. The variants of electrodes combinations. From left to right: instant photo of the electric discharge, the spectrum of pulsations and the emission spectrum.

Instant photos of discharge were obtained by using high-speed video camera Videoscan-401 which can record up to 25 frames per second with a minimum exposure of 1 μs. For all shown in figure 1 photos the diaphragm disclosed the same and selected the same exposure equal to 1 ms. As seen photos witch placed on the figure 1*a*, the discharge of liquid electrolyte cathode is a volumetric and zone of its binding to the electrolyte has a fairly large area. Plasma column discharge is colored in yellow, which indicates the flow of sodium from the electrolyte in the discharge area. The most likely mechanism for the portage of sodium is the spraying of the electrolyte. In fact, the spraying of the electrolyte was observed visually with the naked eye. The copper tip of the cone has a green color. This color indicates about the arrival of the copper atoms in the anode area of discharge.

Next to the pictures typical spectra of signals are located which they received by the oscilloscope. At the same place were got testimony switch devices M2016 and M2015, which was used as an ammeter and voltmeter for measure the discharge current $\langle I \rangle$ and voltage $\langle U \rangle$ between the cathode and the anode. The oscillograms were recorded at the same current in order to ensure a level playing field in the experiments. The signal for the registration of the discharge current was served as the voltage drop across the shunt with a resistance of 0.01 Ω. The spectral signals are submitted on the figure 1. As seen in the gas discharge which stimulated between the liquid electrolyte cathode and the copper anode are present the pulsations of high frequency current in the megahertz range. At that the effective amplitudes of pulsations ΔI_m are significant and their maximum values approaching 50% of the average discharge current $\langle I \rangle$.

The fragment of the emission spectrum of the electric discharge in the visible wavelength range is shown on the figure 1*a*. There contains the spectral lines of copper, which also confirms the receipt of the copper atoms in the discharge area. The curve line on the right edge is the result of saturation of the CDD-detector spectrometer. This picture confirms the intense flow of sodium in the plasma column with droplets of electrolyte. The recording of the emission spectrum was carried out with fiber-optic spectrometer AvaSpec-3648 with a resolution of 0.15 nm (diffraction grating 1200 lines/mm, the input optical gap of 10 microns). The spectra were recorded without averaging. The integration time of one record was 400 ms. The all subsequent entries (for all combinations of the electrodes and their relative position) was carried out in such a mode that allowed you to compare the intensity of spectral lines in spectra obtained.

Thus, the experiments showed that in the embodiment, "liquid electrolytic cathode - copper anode" high-frequency current pulsations with significant amplitude occur on the background of receipt of electrode materials (electrolyte, and copper) in the discharge area.

Repetition of the experiments in the embodiment of "liquid electrolytic cathode - anode hafnium" are showed that high frequency current pulsations remain, in spite of the elimination of one of the receipt of the electrodes, specifically, anode metal (copper). The results of these experiments are shown in figure 1*b*. In the emission spectrum (on the right edge of the second row) were recorded saturated radiation sodium atoms of CDD-signal detector, which indicates the intensity of receipt droplets of electrolyte in the discharge area.

The results which obtained in an embodiment "copper cathode - hafnium anode" are shown on the figure 1*c*. On the photo a greenish color in the discharge column, is visible. It indicates the arrival of copper atoms. The presence of the spectral lines in the emission spectrum of copper confirms this fact. Comparing the spectra of the current which signals obtained in this with the previous ones, we can see that the pulsations of current increased. This result can be explained by a more intensive introduction of the copper atoms in the discharge area. This assumption is confirmed by increasing the copper spectral lines of the emission spectrum.

The results which obtained in an embodiment "hafnium cathode - hafnium anode" are shown on the figure 1*d*. In this embodiment, the oscilloscope was fixed very weak pulsation, hardly distinguishable from the noise signal. In the spectrum of the radiation in the selected wavelength range was registered only the spectral lines of the "footprints" of sodium. In the experiments, a noticeable erosion of hafnium electrodes was not observed. Consequently, the flow of the material of the electrodes in the discharge area was practically absent. Thus, summarizing, we can conclude that the high-frequency current pulsations is occured when intensive inflow of materials of the electrodes in the discharge area.

In a gas discharge with contact to liquid electrolytes the cause of high-frequency current pulsations reason is the flow of droplets of the electrolyte in the interelectrode space.

The spraying of liquid electrolyte is increased with increasing current. However, in such combustion modes the electrolyte is heated and it requires forced cooling. The one method of heat removal from the electrolyte technically is implemented in this research [13]. In figure 2 is showed a photograph of the discharge between the flowing electrolytic cathode and cooled copper anode which produced by using the technical system [13]. The staffs are filmed with a Canon PowerShot S70. It was exhibited exposure equal to 1/2000 s. White lines are indicated the contours of the cathode and anode. Electrolyte was flowed (aqueous salt solution) from a vertical cylindrical passage with an external diameter of 75 mm. Its specific electric conductivity was 2.5 mS/cm. Discharge current $\langle I \rangle$ = 5.5 A. The interelectrode distance is 5 cm. The voltage between the cathode and anode $\langle U \rangle$ = 2320 V. In the photo clearly distinguished yellow "bubbles", which is likely to constitute a spherical formations formed as a result of falling droplets of electrolyte in the discharge gap. The green coloring on the photos is indicating on flow of products of erosion of the copper anode to the discharge gap interval.

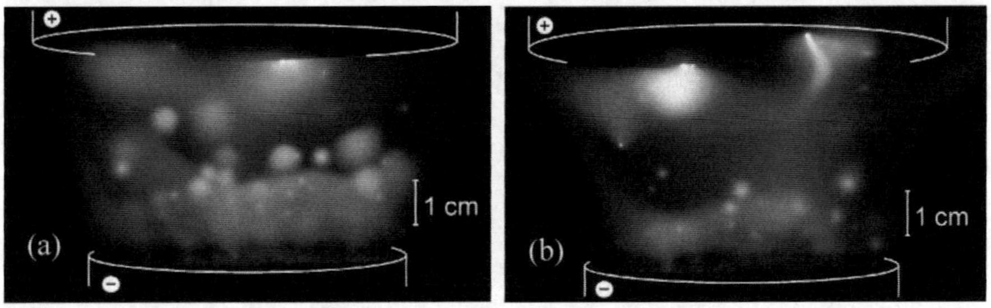

Figure 2. The droplets of electrolyte (yellow "bubbles") in the discharge area: *a* – more; *b* - less.

3. Conclusions

The main cause of high- frequency current pulsations in the electric discharge contacted with the liquid electrolytes is a spray of the electrolyte and flowing droplets to the discharge area. In the result of the rapid thermal decomposition of liquid droplets of electrolyte significant number of ions release which leads to local bursts of carrier concentration of current.

References

[1] Kosenkov V M 2011 *Technical Physics* **56** 1513
[2] Akhmetov N D, Gimadeev M M, Drulis V N et el. 2011 *Russian Aeronautics* **54** (1) 108
[3] Kashapov L N, Kashapov N F and Kashapov R N 2013 *J. Phys.: Conf. Ser.* **479** 012003
[4] Titova Y V, Stokozenko V G, Aleksakhina E L and Maksimov A I 2012 *Surfase Engineering and Applied Electrochemistry* **48** (4) 355
[5] Maksimov A I and Khlyustova A V 2009 *High Energiy Chemistry* **43** (3) 149
[6] Miftachov M N, Tazmeev A Kh, Tazmeev Kh K and Fridland S V 2006 *Journal of Physics and Thermophysics* **79** (3) 532
[7] Tsymbalyuk A N, Levko D S, Chernyak V Y et el. 2013 *Technical Physics* **58** 1138
[8] Samukawa S, Hori M, Rauf S et el. 2012 *Phys. D: Appl. Phys.* **45** 253001
[9] Park S, Yoon S and Kim G. 2009 *Proc. 19 Int. Symp. on Plasma Chemistry* (Bochum: IPCS) p 647
[10] Fujita H, Kanazawa S, Ohtani K, Komiya A and Sato T 2013 *Proc. 21 Int. Symp. on Plasma Chemistry* (Queensland: IPCS) p 422
[11] Kirko D L, Savelov A S and Vizgalov I V 2013 *Russian Physics Journal* **55** (11) 1243
[12] Tazmeev Kh, Arslanov I and Tazmeev G 2012 *Proc.VII Int. Conf. Plasma Physics and Plasma Technology* (Minsk: Kovcheg LTD) p 72
[13] Tazmeev Kh K and Tazmeev A Kh 2014 *J. Phys.: Conf. Ser.* **567** 012035

Journal of Physics: Conference Series **669** (2016) 012058 doi:10.1088/1742-6596/669/1/012058

The influence of the method of cooling liquid electrolyte cathode on the energy balance in the gas discharge

Kh K Tazmeev[1], I M Arslanov[1] and G Kh Tazmeev[2]

[1] Branch of Kazan Federal University in Naberezhnye Chelny, 68/19 Mira Str., Naberezhnye Chelny, 423823, Russia
[2] Kazan National Research Technical University named after A.N.Tupolev, 10 K.Marx Str., Kazan, 420111, Russia

E-mail: tazmeevh@mail.ru

Abstract. Experimentally investigated the energy balance in a gas discharge between a flowing electrolyte cathode and a metal anode at an power of tens of kilowatts. The discharge was burning in the air in the electrode gap with a height of 10 cm. The electrolyte was a solution of salt in distilled water. The concentration of the solution by weight was 5.5 g/l. The regularities of the influence of electrolyte mass flow through the flowing cathode on the energy characteristics of the discharge were studied. The modes of the discharge, whereby the energy balance of the portion of heat losses for heating of the electrolyte reaches a minimum were identified.

1. Introduction

Gas discharges with liquid electrolytic electrodes have a sufficient range for practical applications. They are promising for use in such processes as bacteriological water purification [1], the treatment of textiles in aqueous solutions [2], plasma-electrolytic processing of materials [3], the receipt of ferromagnetic powders [4]. In the power ranges that comprise tens of kilowatts, the gas discharge with liquid electrolyte cathode may serve as a source of energy in the plasma technologies for processing wastes [5].

With a large contribution to energy in the gas discharge liquid electrolyte it is heated and there is a need for cooling. Heating occurs due to the Joule heat within the electrolyte and the energy input from the gas-discharge plasma. The heat exchange between the electrolyte and the plasma takes place in difficult conditions. The electrolyte is evaporated and subjected to sputtering. In these conditions the cooling effect of the electrolyte significantly affects the energy characteristics of a gas discharge, which was experimentally found in [6] at currents of 7-22 A and the power of 8-33 kW. In particular, a reduction in thermal losses in flow of the cathode with decreasing mass flow rate of the electrolyte was shown. In this paper, we continued the study with a smaller flow of electrolyte in the zone of contact with the discharge.

2. Experiment

The experiments were performed on the installation, which was used in [6]. Part of constructive changes has been made in the gas-discharge unit. The cylindrical vessel 1, which implies the electrolyte was divided into horizontal partition 2 into two parts (figure 1). Such modernization is allowed to carry out experiments in the maximum permissible conditions under which the electrolyte flows from the neck of a cylindrical vessel with a temperature very close to 100 °C.

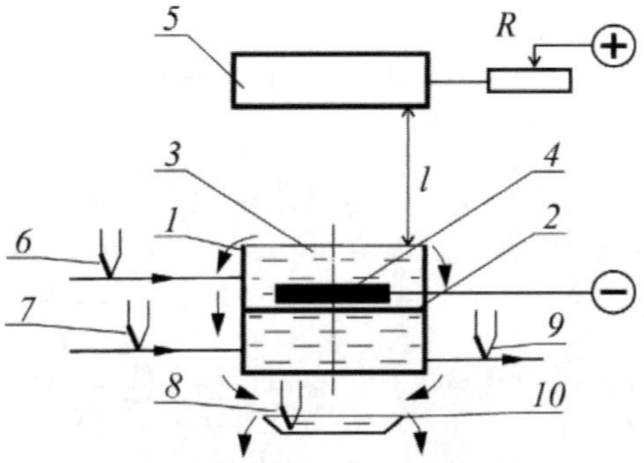

Figure 1. Gas-discharge unit. *1* - cylindrical vessel, *2* - partition, *3* - liquid electrolyte, *4* - current lead, *5* - anode, *6, 7, 8, 9* – thermocouple, *10* - collecting tank.

Joule heat generated inside the electrolyte, was calculated by the formula

$$Q_{\mathrm{j}} = I \cdot \Delta U_k, \tag{1}$$

where ΔU_{k} – the voltage drop within the electrolyte, measured with a probe technique.

The calculations of heat loss from the heat capacity of the electrolyte was taken as that of water. Mass flow through the upper and lower portions of the cylindrical vessel, respectively m_1 и m_2, were measured and monitored by a float rotameter. The numerical values of thermal losses in heating the electrolyte escaping through the neck of the cylindrical vessel, calculated by the formula

$$Q_{k1} = c \cdot (m_1 - G) \cdot \Delta t_1, \tag{2}$$

where G – is the mass loss rate of the electrolyte under the influence of a gas discharge, Δt_1 – the difference between the temperature which detected by a thermocouple, one of which is installed at the entrance of the cylindrical vessel and the other in the collecting tank *10* (figure 1), which receives the electrolyte.

3. The results of experiments, their processing and discussion

In figure 2 presented the diagram characterizing the energy balance in a gas discharge with different values of mass flow rate m_1.

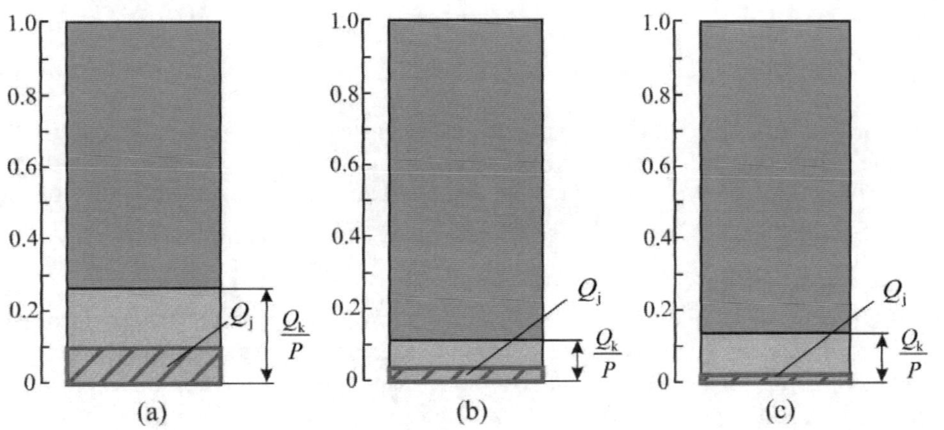

Figure 2. Heat losses at the electrolyte cathode in the energy balance of the gas discharge. The total flow of electrolyte m = 30 g/s. (a) – according to [6]; (b) – m_1 = 10 g/s; (c) – 5. Discharge current: (a) – <I> = 21.5 A; (b) – 18.1; (c) – 8.6.

The diagrams are built on experimental data obtained at zero value of the ballast resistor. In this case in addition to gas discharge devices the power source had no other external loads. Therefore, in all considered cases, the terminal voltage of discharge device was the same.

As can be seen from the diagrams, with decreasing mass flow of the flowing electrolyte, the portion of heat losses in the energy balance does not changing monotonically. At first it decreases and then increases slightly.

Part of the heat losses is Joule heat Q_j, which are generated within the electrolyte. In the presented diagrams there is a decrease in its share the energy balance. This situation is quite natural, since in these embodiments, a decrease in the mass flow of the electrolyte is accompanied by a decrease in the discharge current.

Changing the current by varying the mass flow rate of the electrolyte takes place at a constant voltage applied to the terminals of a gas discharge device. It can therefore be clearly confirm, that the discharge current is affected substantially by thermal phenomena on the electrolyte cathode.

At higher currents the electrolyte heats up and comes to a boil. The bubbles of steam which escaping to the surface is tearing discharge of the electrolyte. The area of contact of the discharge with the electrolyte in a liquid state is reduced. Apparently, due to such phenomena, there is a current limitation.

Boiling of the electrolyte leads to a complete break of current channels, and the discharge is extinguished when the current reaches some limit value. Thus, the smaller the mass flow rate of the electrolyte, the lower the value of current limit.

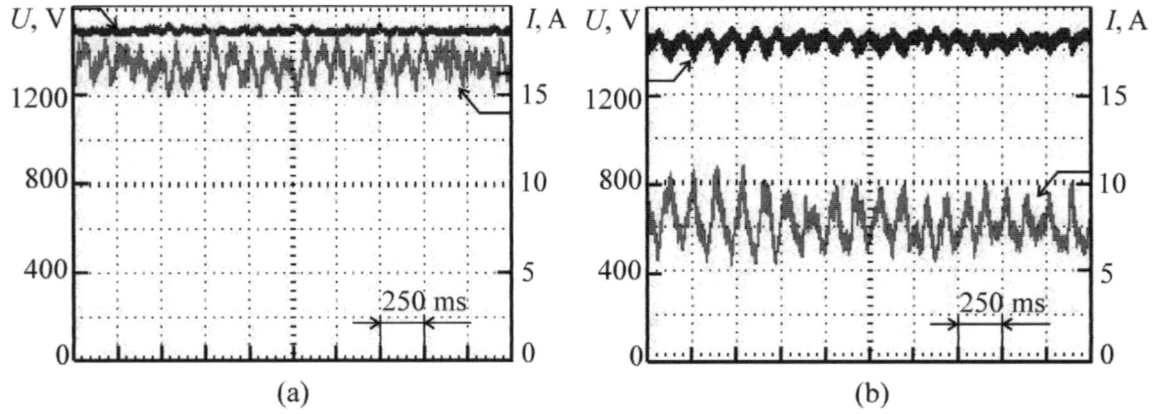

Figure 3. Waveforms of the currents and voltages. m = 30 g/s. (a) – m_1 = 10 g/s; (b) – 5.

The occurrence of instability of the burning discharge at low mass flow rate of the electrolyte was recorded on the waveforms of currents and voltages. In figure 3 shows waveforms obtained at two different mass flow rates of the electrolyte. As can be seen, with less consumption of electrolyte, vibration amplitude current and voltage is increasing.

The appearance of instability of combustion of a gas discharge at low flow rates of electrolyte was recorded on snapshots. In figure 4 is showed photos taken at two different mass flow rates of the electrolyte. As can be seen, at low flow rate of the electrolyte contact of the discharge with the electrolyte drastically changes its geometry and is moved across the surface of the electrolyte.

Thus, the mass flow rate of the electrolyte can be reduced only to a certain limit. At the same time measures should be taken to prevent boiling of the electrolyte in the binding zone of discharge.

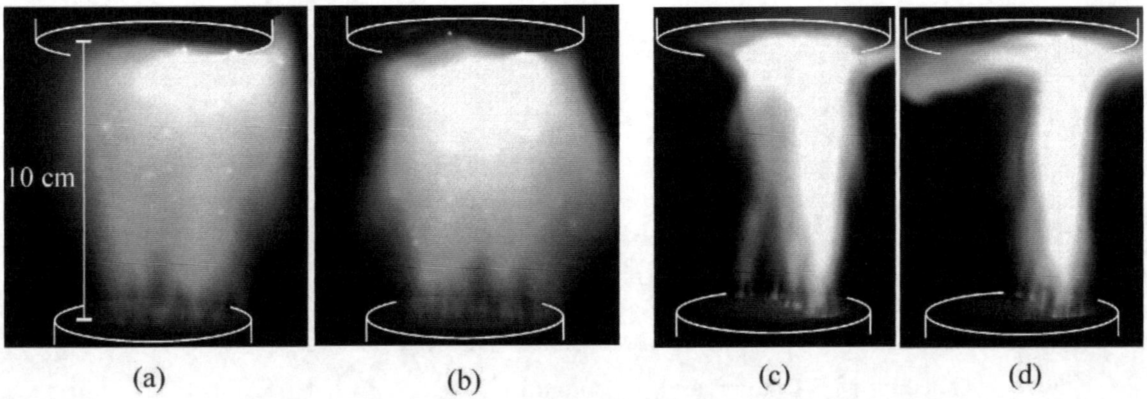

Figure 4. Snapshots of the gas discharge. The exposition is 200 μs. The interval between frames 125 ms. $m = 30$ g/s. (a), (b) – $m_1 = 10$ g/s and $\langle I \rangle = 18.1$ A; (c), (d) – 5 and 8.6.

4. Conclusions

The energy balance in the gas discharge with flowing electrolyte cathode essentially depends on the mass flow of the electrolyte. By varying the mass flow rate of the electrolyte can reduce heat loss at the cathode and reduce their share in the energy mix to 10 - 15% at the same time maintaining stable operation of the discharge.

References

[1] Barinov Y A, Kuzikova I L, Zinov`eva S V, Scol`nik S M and Medvedeva N G 2015 *Technical Physics Letters* **41** (9) 847

[2] Titova Y V, Stokozenko V G, Aleksakhina E L and Maksimov A I 2012 *Surfase Engineering and Applied Electrochemistry* **48** (4) 355

[3] Kashapov L N, Kashapov N F and Kashapov R N 2014 *J. Phys.: Conf. Ser.* **567** 012025

[4] Valiev R A, Shakirov Yu I and Iliuchin A N 2014 *J. Phys.: Conf. Ser.* **567** 012040

[5] Tazmeev A Kh, Fridland S V and Miftakhov M N 2006 *Vestnik Kazanskogo Technologicheskogo Universiteta* No. 6 43

[6] Tazmeev Kh K, Arslanov I M and Tazmeev G Kh 2014 *J. Phys.: Conf. Ser.* **567** 012001

The change in optical and electrical characteristics of tin oxide films under the action of argon ion beam

O Asainov[1], S Umnov[1] and V Temenkov[1]
[1]Physical Technical Institute,
Tomsk Polytechnic University,
634050, Lenina 2, bdg. 11, Tomsk, Russia

E-mail: usp@tpu.ru

Abstract. Thin films of tin oxide (TO) were deposited on the glass substrates at room temperature using reactive magnetron sputtering at various oxygen partial pressures. After the deposition the filmswere irradiated with argon ions beam. The change of the optical and electrical properties of the films depending on the irradiation time was studied. Films optical properties in the range of 300-1100 nm were investigated by photometry as well as their structural properties were studied using X-ray diffraction. Diffractometric research showed that the films, deposited on a substrate,have a crystal structure, and after argon ions irradiation they become quasi-crystalline (amorphous). It was found that the transmission increases proportionally with the irradiation time, but the surface resistance -disproportionally.

1. Introduction

The transparent conductive oxide (TCO) films are widely used in various industries. These optically transparent electrodes are for displays, solar panels, photoelectric devices, touch panels, etc. Currently, one of the most widely used TCO substances is indium tin oxide (ITO) because it has high electrical conductivity and optical transparency.

Recently, because of the high cost of indium, the research is conducted to find ITO alternative replacement. This is, in particular, the compounds based on tin oxide (dopant-Sb, F, As, Nb, Ta), zinc (dopant- Sn, Ge, Mo, F, Ti, Zr, Hf, Nb, W, Te), and others. Although, SnO2 and ZnO impurities increase their conductivity, but it does not reach the values obtained for the ITO films. However, it is not necessary in some practical applications (very large conductivity). So, in the paper [1] pure oxide films (SnO2) prepared by electron beam evaporation had transmittance of 90% and a surface resistance of 200 KOhm / □ (substrate temperature is 100oC). Originally, pure SnO2 films are the conductors if they are deposited in a shortage of oxygen, so they are non-stoichiometric.

For the deposition of thin TCO films various methods are applied, including thermal, electron- beam, laser vaporization, cathode, a high-frequency or magnetron sputtering and others. We would like to focus on the film deposition by magnetron sputtering with additional useof ion beam.The use of ion beams in the film coatings deposition leads to almost all films properties change [2-3].

[1] E-mail address: usp@tpu.ru .

The papers [4-8] present data showing that the use of ion assisting leads to the changes in the optical and electrical properties of the deposited ITO thin films. It means that ion beams allow achieving the optimum relationship of the metal and the oxygen components in the film. So, the transmission of the films treated in that wayis at 90%, and the surface resistance of the best samples is 20 Ohms/□. According to the authors, this is due to such factors as the changes in the mobility of the adatoms and crystal lattice damage.

In their works, the authors have used different types of ion sources that had been used at different deposition conditions (the ion beam parameters, the temperature of the substrate). In the papers mentioned above ,assisting ion beams were used in the deposition of the doped TCO oxides.

Whereas, the energy formation of oxygen vacancies and tin interstitials in SnO_2 is very low and, thus, these defects form readily, explaining oftenly observed high conductivity of pure, but non-stoichiometric, SnO_2 [9], so the ion beams could be a good tool to stimulate this process.

In this paper, we studied the influence of the irradiation of the pure tin oxide (without dopant) films with argon ions after its deposition on the optical and electrical properties.

2. Experimental details

Tin oxide films with thickness of about 30 nm were deposited by reactive magnetron sputtering of the target made of tin (99,5 wt.%) on the glass substrates at a room temperature. The glass substrates before the film deposition were purified by a chalk suspension in isopropyl alcohol.

The substrates were placed on a carrier that was able to carry out a reciprocating motion under the magnetron and the ion source in a vacuum chamber. The speed of carrier moving was 3.62 cm / second. The working gas was a mixture of oxygen and argon (partial pressure ratio is O_2 – 60%, Ar – 40%). The pressure in the vacuum chamber during sputtering was 0.3 Pa. The gas flow was controlled by an electronic meter and the gas pressure was measured by an ionization and thermocouple gauge. After the deposition, tin oxide films were irradiated with argon ions .

The ions were generated by the ion source with a closed electron drift with 2800 V voltage and a 0.28 a discharge current. The greatest number of argon ions in the beam has an energy of 400-600 eV. The chamber pressure in such case was 0.166 Pa. For measuring the surface electrical resistance Rsq. the samples were deposited with aluminum strips, which provided a reliable contact with the film. The thickness of the deposited films was measured using profilometer MicroMeasure 3D Station. To measure the transmission spectrum in the range of 300-1100 nm the spectrophotometer SF-256was used. The structure and morphology of the films was studied using the X-ray diffractometer XRD6000 and atomic-force microscope NT-MDT.

3. Results and discussion

All freshly prepared tin oxide films were transparent dielectrics. This is due to the deposition by reactive magnetron sputtering that was performed in an atmosphere with excess oxygen content. Under these conditions, there was enough oxygen to have all the metal oxidised to a state of complete oxide, so all the oxygen vacancies, creating conductivity, are busy and SnO_2 semiconductor film is deposited as a dielectric. After the tin oxide films deposition the samples were irradiated with argon ions. Figure 1 shows the results of measurement of the films transmittance both the non-irradiated and irradiated samples. As it seen from the figure the transmission is strongly dependent on exposure time. So, when the exposure time is less than ~ 4.5 sec. the transmission slightly improves. For longer times (≥ 8,8 sec.) the transmission is increased to 100% in almost the entire range of the wavelengths. It should be said, when the irradiation time is 13 - 22 seconds, there is the increase in transmittance at wavelengths less than 360 nm and a slight increase in the infrared region. The increase in transmission, especially in the short-wavelengths region, of the irradiated films is related to the change of the plasma frequency [10-12]. This change is a consequence of the electrons concentration increasing.

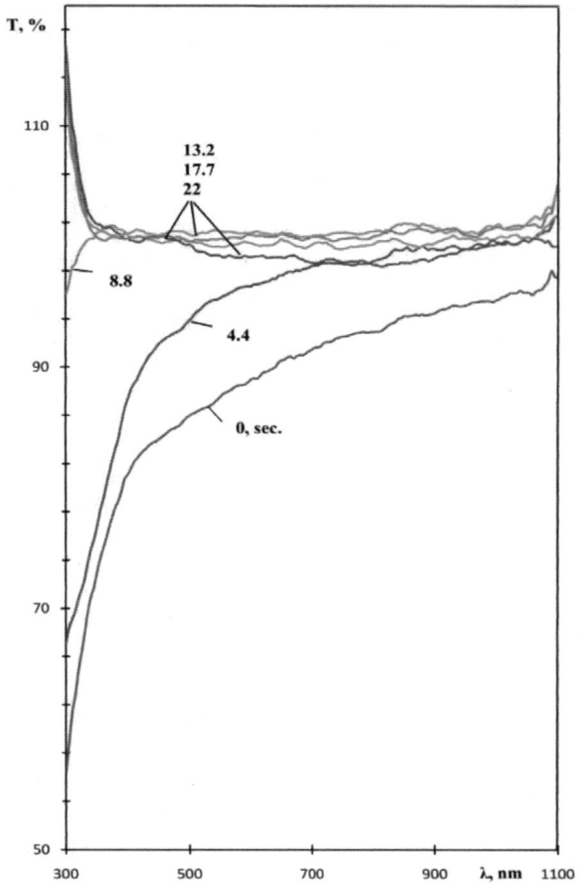

Figure 1. Dependence of the transmission on the irradiation time

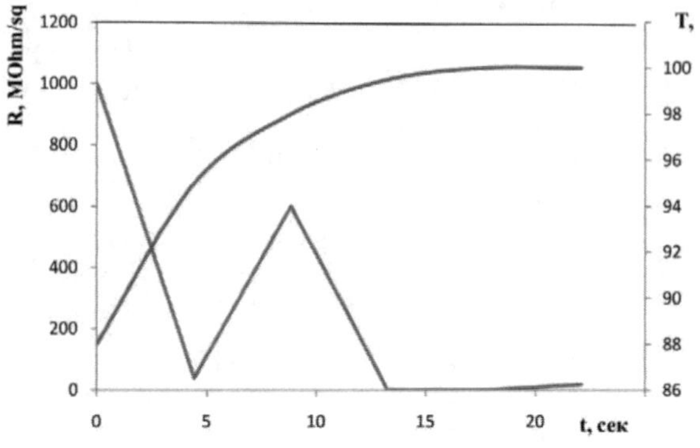

Figure 2. Dependence of the transmission on the irradiation time

However, the nature of the transmission is defined by the frequency of the incident light wave and the plasma frequency.

Figure 2 shows the results of the surface resistance measurements, an average transmittance value at various irradiation times. The table shows that the film surface resistance is disproportionately dependent on the exposure time. With exposure time increasing, it initially decreases and, then, increases. The film transmittance almost uniformly increases and quickly reaches 100%.

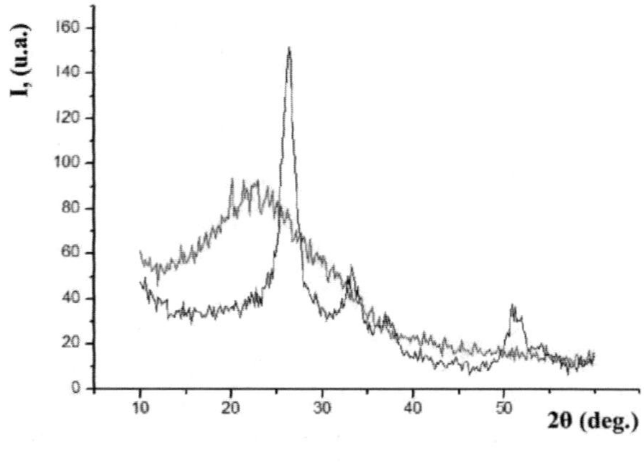

Figure 3. XRD diagram samples SnO_2

As the deposition was carried out at an excess oxygen content, the films were deposited in the SnO_2 stoichiometric form which is a good insulator.

During the Ar ions irradiation there is their collision with the film surface and the power transfer o in the local area to the neighbouring atoms of the crystal lattice, imbedded and adsorbed on the grains boundary. As a result, there are physical and chemical processes that lead to the break or reduction of chemical bonds, causing desorption or chemical reaction that alters the structure of the coating.

The concentration of the electrons increasing is connected both with the break of the Sn-O chemical bonds and desorption of oxygen, which leads to the formation of non-stoichiometric tin oxide. Thus, the film exposure leads to increase in the number of oxygen vacancies, which, in its turn, leads to reduce in the films electric resistance. However, the further irradiation time increase leads to the film resistance increase. This is, possibly, connected with the formation of Sn self-point defects [11-12]. Whereas, the point defects interact with all conductive electrons, so the electrons, located near the defects, are "bound" and do not participate in conductivity. As a result, the film resistance increases. Figure 3 shows the results of the study of the deposited films structure. The figures show that the deposited tin oxide films have a crystal structure (tetragonal) with the grain size bottom of ~ 9.8 nm. After argon ions irradiation the structure becomes quasi-crystalline (amorphous - the red curve). This is due to the opening of the crystal lattice chemical bonds and forming new ones, which lead to formation of quasi-crystalline structure.

4. Conclusions

Pure SnO_2 films were deposited at room temperature with reactive magnetron sputtering, followed by Ar ions beam irradiation with energy of 400-600 eV. It allows obtaining the films with 100% transmittance and satisfactory surface resistance of 100 KOhm/ □. The surface resistance value can be reduced by optimizing the deposition conditions and ion beam parameters. Thereby, transmission increases proportionally with irradiation time increasing, while the surface resistance - disproportionately. The crystal structure of SnO_2 film after ion irradiation is transformed into a quasi-crystalline.

References

[1] Shamala K S, Murthy L C and Narasimha Rao K 2004 *Bull. Mater. Sci.* **27** 295
[2] McNeil J R, Barron A C, Wilson A R, et al 1984 *Applied Optics* **23** 552
[3] McNeil J R, Al-Jumaily G A, Jungling K C et al 1985 *Applied Optics* **24** 486 [4] Meng L-J, Gao J, dos Santos M P, Wang X et al 2008 *Thin Solid Films* **516** 1365
[5] Yang Y, Huang Q, Metz A W et al 2004 *Advanced Materials* **16** 321
[6] Lee *S H*, Cho *H C*, Kim *H J et al* 2012 *Mol. Cryst. Liq. Cryst.* **564** 185
[7] Bae J W , Kim J S, Yeom G Y 2001 *Nuclear Instruments and Methods in Physics Research B* **178** 311
[8] Gilo M, Dahan R, Croitoru N 1999 *Opt. Eng.* **38(6)** 953
[9] Kılıç C and Zunger A 2002 *Phys. Rev. Lett.* 88, 095501
[10] Ginley D S and Bright C 2000 *MRS Bull.* **25** 15
[11] Umnov S and Asainov O 2015 *IOP Conf. Ser.: Mater. Sci. Engineering.* **81** 012001
[12] Asainov O and Umnov S 2012 *7th International Forum on Strategic Technology IFOST 2012* 6357554

Structure of the local environment of titanium atoms in multicomponent nitride coatings produced by plasma-ion techniques

O V Krysina[1,2], N A Timchenko[3], N N Koval[1,2], and Ya V Zubavichus[4]

[1]Institute of High Current Electronics SB RAS, Tomsk, 634055, Russia

[2]National Research Tomsk State University, Tomsk, 634050, Russia

[3]National Research Tomsk Polytechnic University, Tomsk, 634050, Russia

[4]Kurchatov centre of synchrotron radiation and nanotechnology, Moscow, 123182, Russia

E-mail: krysina_82@mail.ru

Abstract. An experiment was performed to examine the X-ray Absorption Near-Edge Structure (XANES) and the Extended X-ray Absorption Fine Structure (EXAFS) near the K-edge of titanium in nanocrystalline titanium nitride coatings containing additives of copper, silicon, and aluminum. Using the observation data, the structure parameters of the local environment of titanium atoms have been estimated for the coatings. According to crystallographic data, the Ti–N distance in the bulk phase of titanium nitride is 2.12 Å and the Ti–Ti distance is 3.0 Å. Nearly these values have been obtained for the respective parameters of the coatings. The presence of copper as an additive in a TiN coating increases the Ti–N distance inappreciably compared to that estimated for titanium nitride, whereas addition of silicon decreases the bond distance. It has been revealed that the copper and silicon atoms in Ti–Cu–N and Ti–Si–N coatings do not enter into the crystallographic phase of titanium nitride and do not form bonds with titanium and nitrogen, whereas the aluminum atoms in Ti–Al–N coatings form intermetallic phases with titanium and nitride phases.

1. Introduction

Modern protective coatings should match high requirements and have high physicomechanical and operational characteristics to provide reliable operation and long service life of tools, machine parts, mechanisms, and various products. The vacuum-arc deposition technology, owing to its irrefutable advantages [1], is most reasonable to use for the production of various functional coatings. In particular, it shows promise for synthesis of superhard nanocrystalline multiphase coatings which are formed on adding conventional TiN, ZrN, and AlN binary systems and various elemental additives to the base composition [2–4].

It is well known that the operational characteristics of a hardening coating strongly depend on the coating structure-phase state, structure imperfection, and crystallite size and on the level of internal stresses in the coating. It has been found that the decrease in crystallite size in titanium-nitride-base coatings caused by addition of Si, Cu, Al, etc. makes the coatings superhard (≥ 40 GPa) [4, 5] or even ultrahard (≥ 80 GPa), or imparts them with other useful properties [4–6].

An urgent problem in studying nanocrystalline multicomponent coatings is to reveal the mechanism of the transition of superhard coatings synthesized from multicomponent arc plasmas to the nanocrystalline state. The solution of this problem should partially or completely explain the hindered growth of crystallites on the nanometer scale in a coating during its deposition by the vacuum-arc technique.

The use of conventional X-ray methods to examine the features of the elemental composition and structure-phase state of multicomponent coatings during their synthesis is conjectural because of the specificity of the scattering of radiation beams by nanosized objects. The local crystalline structure of a nanomaterial can be different from the average structure, whereas just the local structure is responsible for the important physicochemical properties of materials, such as strength, hardness, thermal stability, etc. Therefore, to perform structure examinations of nanosized objects, XAFS (X-ray Absorption Fine Structure) spectroscopy [7] should be used alongside with high-resolution diffraction methods.

The work we present here was aimed at determining the structure parameters of the local environment of titanium atoms in titanium-nitride-based coatings containing additives of copper, silicon, and aluminum by using XAFS spectroscopy methods.

2. Equipment and procedures

The test objects were multicomponent titanium-nitride-based coatings containing no additive (TiN) and those containing additives of copper (Ti–Cu–N), silicon (Ti–Si–N), and aluminum (Ti–Al–N) [4, 8, 9]. The coatings were produced using plasma-assisted vacuum-arc deposition by evaporating sintered cathodes [10] in low-pressure arcs [2]. The plasma-assisted coating deposition was carried out on the QUINTA automated plasma-ion facility with the use of a DI-100 arc evaporator [11] and the PINK hot filament, hollow cathode plasma source [12]. Thus, the coatings were condensed from mixed metal- gas plasmas of density $\sim 10^{16}$ m^{-3}. The coating thickness was 3–5 μm. Beryllium foil of thickness 300 μm was used for the substrates.

To examine the fine structure of the coatings by XAFS methods (X-ray Absorption Near-Edge Structure (XANES) and Extended X-ray Absorption Fine Structure (EXAFS) spectroscopy), the equipment of the Structural Materials Science end-station of the Kurchatov Synchrotron Radiation Source (Kurchatov Institute, Moscow) [7] was used. The X-ray beam incident on a specimen was monochromated with (111)-oriented monocrystalline silicon shaped as a cutaway monoblock ("butterfly") which was mounted on a goniometric head. The energy resolution of the monochromator, $\Delta E/E$, was 10^{-4} and the working spectral range was 5–20 keV. Measurements were carried out in the transmission mode with the use of two ionization chambers filled with nitrogen. Current signals were recorded by Keithley digital picoammeters. Besides titanium-nitride-based coatings, pure titanium coatings were produced and examined to be used as reference objects.

The experimental spectra were processed using the IFEFFIT code [13]. The scattering phases and amplitudes of a photoelectron necessary for the EXAFS simulation were calculated by means of the FEFF code [14]. For the initial model of the local environment of titanium, crystallographic data for TiN were used (space group Fm3m, a = 4.239 Å; titanium atoms are in positions 4a (0, 0, 0) and nitrogen atoms in positions 4b (0.5, 0.5, 0.5) [15]). Nonlinear fitting was performed in the R space on the interval R = 1.0–3.0 Å with weight coefficient k^2, k = 2–12 $Å^{-1}$.

3. Results and discussion

3.1. Ti–Cu–N and Ti–Si–N coatings

The XANES spectra measured at the K-edge of titanium for test sets of Ti, TiN, Ti–Cu–N, and Ti–Si–N specimens are given in Fig. 1 and the Fourier transforms of the EXAFS spectra are given in Fig. 2.

The absorption spectra of the coatings are similar to each other and differ substantially from the spectra measured for the reference titanium specimen. Instead of the sharp pre-edge resonance peak at

4967 eV for the metal, the absorption spectra of the titanium-nitride coatings show diffuse structure in the energy range 4965–4973 eV. The spectral maximum is observed at 4985 eV, and the point of the fastest growth (the maximum in the spectrum of the absorption constant derivative with respect to energy) corresponds to 4980 eV (4966 eV for the metallic titanium).

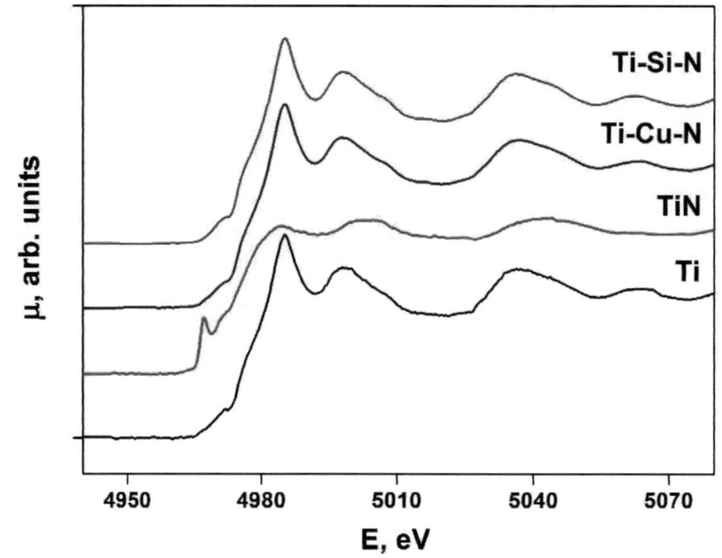

Figure 1. Experimental XANES spectra at the K-edge of titanium for test titanium-nitride coatings.

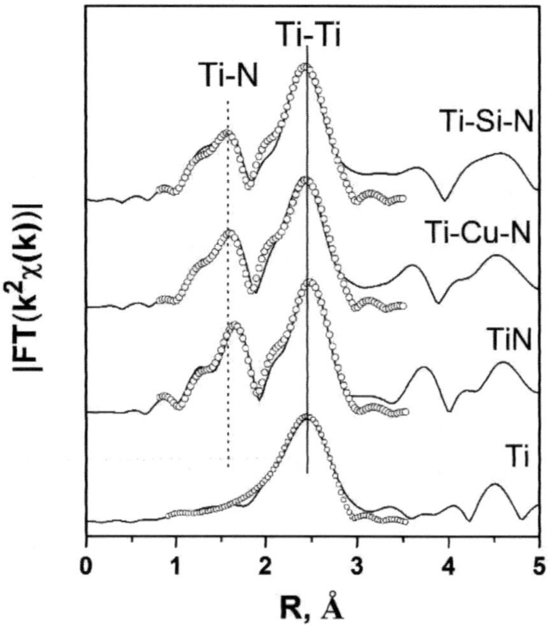

Figure 2. Fourier transforms of the EXAFS spectra for titanium-nitride coatings: the experimental and the optimized theoretical data are represented by solid lines and by circles, respectively.

For the test coatings, two dominant contributions corresponding to the first two coordination spheres of titanium atoms, Ti–N and Ti–Ti, are clearly identified. The parameters of the local environment of titanium atoms for the specimens, obtained by simulating the EXAFS spectra, are given in Table 1.

Table 1. EXAFS data on the parameters of the local environment of titanium atoms in copper- and silicon-containing titanium-nitride-based coatings produced by plasma-assisted vacuum-arc deposition with evaporation of composite cathodes.

Specimen	Ti–N		Ti–Ti	
	R, Å	σ^2, Å^2	R, Å	σ^2, Å^2
Ti	–	–	2.93	0.0089
TiN	2.09	0.0032	2.98	0.0070
Ti-Cu-N	2.10	0.0040	2.99	0.0061
Ti-Si-N	2.07	0.0054	2.98	0.0056

According to the crystallographic data for the bulk phase of titanium nitride, the Ti–N distance is 2.12 Å and the Ti–Ti distance is 3.0 Å. The respective parameters obtained for three specimens have nearly these values. Addition of copper to the composition of a titanium nitride coating slightly increases the Ti–N distance compared to the data obtained for pure titanium nitride, whereas addition of silicon decreases the bond distance.

The data obtained suggest that the titanium and nitrogen atoms in the synthesized Ti–Cu–N coatings are not related to the copper atoms, and this agrees with the previously obtained results [4].

3.2. Ti–Al–N coatings

As can be seen from Figs. 3 and 4, the XANES spectra at the K-edge of titanium for the Ti–Al–N coatings are substantially different from those for the coatings containing no additive and for the coatings containing copper. We see a smeared fine structure in the XANES spectra and no peaks associated with long-range coordination spheres in the EXAFS spectra. This points to amorphization of the coating material or to a great amount of aluminum in the coating (approximately 1:1 in proportion with titanium). In this case, the contributions of Ti–Al and Ti–Ti to the spectra could destructively interfere with each other.

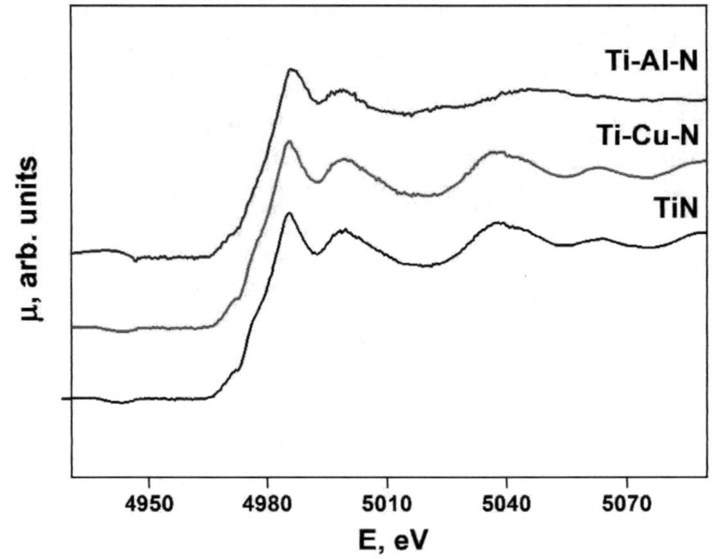

Figure 3. Experimental XANES spectra at the K-edge of titanium for TiN, Ti–Cu–N, and Ti–Al–N coatings.

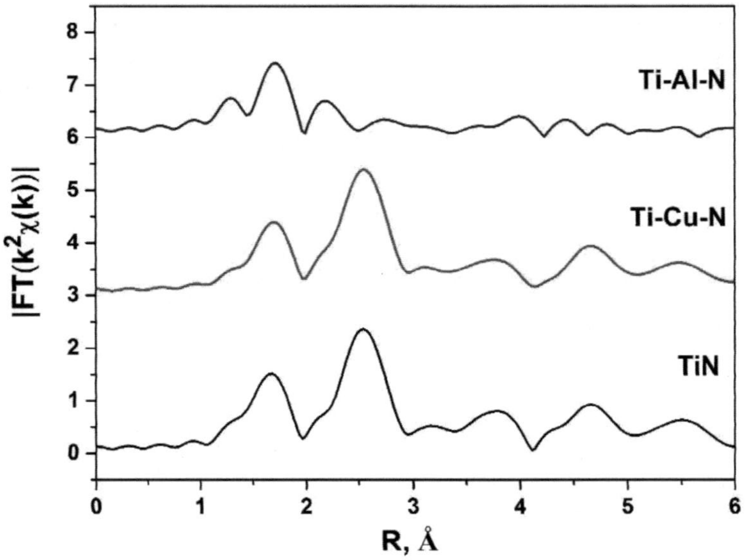

Figure 4. Fourier transforms of the EXAFS spectra for the test TiN, Ti–Cu–N, and Ti-Al–N coatings.

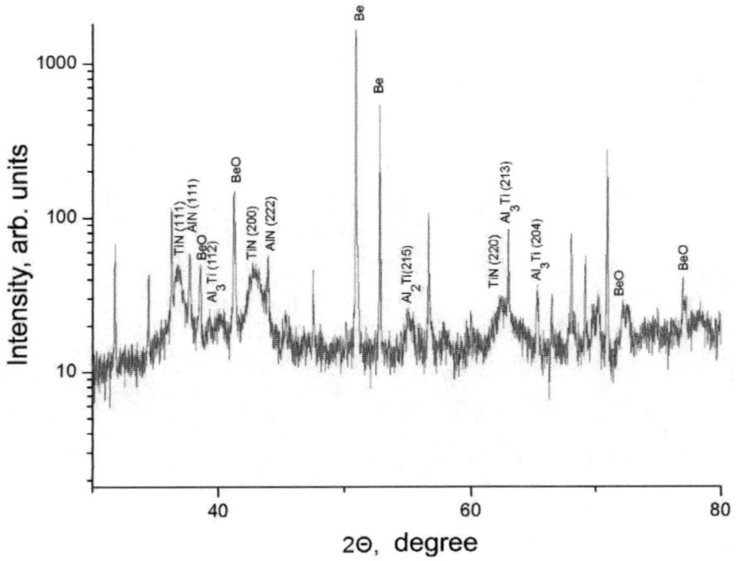

Figure 5. Diffraction pattern of a Ti–Al–N coating measured at a photon energy of 8 keV.

These data well correlate with diffraction investigations (Fig. 5), which have revealed the presence of TiN, AlN, Al_3Ti, and Al_2Ti phases, and point to a multiphase composition of the synthesized Ti–Al–N coatings.

4. Conclusion

X-ray absorption (XANES/EXAFS) spectra have been measured at the K-edge of titanium for titanium nitride coatings containing additives of copper, silicon, and aluminum. From the measured spectra, the structure parameters of the local environment of titanium atoms have been estimated for the test nanocrystalline coatings. It has been revealed that copper and silicon atoms in Ti–Cu–N and Ti–Si–N coatings do not enter into the crystallographic phase of titanium nitride and do not form compounds with titanium and nitrogen, whereas aluminum atoms in Ti–Al–N coatings form intermetallic phases with titanium and nitride phases. This is confirmed both by the XAFS spectroscopy data and by the structure-phase composition of the coatings revealed by X-ray diffraction analysis using synchrotron radiation, and these results agree with previous data of transmission electron microscopy and X-ray diffraction analysis [4, 9].

Acknowledgments

The work was supported in part by the Russian Foundation for Basic Research (Project No. 13-08-98108-r_sibir_a).

References

[1] Barvinok V A and Bogdanovich V I 1999 *Physical Grounds and Mathematical Simulation of Ion-plasma Evaporation* (Moscow: Mashinostroenie)
[2] Koval N N *et al* 2015 *Russian Journal of General Chemistry* **85** (5) 1326
[3] Ryabchikov A I *et al* 2005 *Vacuum* **78** 445
[4] Ivanov Yu F *et al* 2012 *Surf. Coat. Technol.* **207** 430
[5] Cavaleiro A and Hosson J T D 2006 *Nanostructure Coatings* (Springer Science+Business Media, LLC)

[6] Wei R 2008 *Handbook of plasma surface engineering research and its practical applications* (India: Research Singpost)
[7] Chernyshov A A *et al* 2009 *Nucl. Instr. Meth. Phys. Res.* A **603** 95
[8] Koval N N *et al* 2007 *Russ. Phys. J.* **50** 146
[9] Krysina O V *et al* 2013 *High Temperature Material Processes* **17** (2–3) 153
[10] Krysina O V *et al* 2013 *High Temperature Material Processes* **17** (2–3) 145
[11] Koval N N *et al* 2012 *Proceedings of the 25th International Symposium on Discharges and Electrical Insulation in Vacuum* (Tomsk, Russia) **2** 537
[12] Vintizenko L G *et al* 2001 *Russ. Phys. J.* **44** 927
[13] Ravel B and Newville M 2005 *J. Synchrotron Radiat.* **12** 537
[14] Ankudinov A L and Rehr J J 1997 *Phys. Rev.* B **56** R1712
[15] Christensen A N 1978 *Acta Chem. Scand.* A **32** 89

Decomposition of heavy hydrocarbons in argon arc with the sunken electrodes

B A Timerkaev[1], G R Ganieva

Kazan National Research Technical University, 420111, K. Marx St., 10, Kazan, Russia

btimerkaev@gmail.com

Abstract. The way of decomposition of heavy hydrocarbons in the argon arc with the sunken electrodes is proposed. The chromatographic analysis of the gases appeared in the chamber and the electronic and microscopic investigation of the carbonaceous deposits formed on electrodes during experiment is carried out.

In recent years much attention is paid to search for the effective plasmochemical ways of heavy hydrocarbons refining deepening [1-2]. In these works and works [3-4] it was shown that the plasmochemical influence on heavy hydrocarbons is efficient in case when the processing is made directly inside the processed materials. It is mentioned [1-5] that the discharges were organized between the electrodes sunken in hydrocarbons. Under these conditions the arch occurs in hydrocarbon vapors. In this case the hydrocarbon vapors directly interact with the electric arch. The plasma high-temperature influence leads to intensive decomposition of hydrocarbon molecules. During the thermic decomposition process the hydrocarbons molecules appeared in the arcing zone can be decomposed to the atomic state. Subsequently new molecules can gather from these atoms, such as methane, acetylene, hydrogen, carbon nanotubes, fullerenes etc. gather. However the formation of easy fractions requires more soft modes of plasma heavy hydrocarbons processing. The use of the discharge with the drowned electrodes in buffer gas can provide implementation of these terms.

The experimental equipment to provide the discharge with the drowned electrodes in buffer gas is presented in Fig. 1. The reactionary chamber 1 represents the ceramic capacity which is hermetically closed by a hermetically cap. Cover of the chamber has a branch pipe for extraction and the subsequent collection of the formed gases and light hydrocarbons. The electrodes 4, 5 and a pipe for buffer gas supply 2 are set in the chamber. Electrodes 4, 5 with a diameter of 5 mm are built in the cover of capacity 12. The cathode 4 is set on the support 11 and remains fixed. The anode is fixed on the holder 6. Construction of the holder allows to change the distance between electrodes and to provide arcing. The distance between the pipe 2 and the electrodes 4, 5 is 3-5 mm. In our experiments buffer gas was supplied from below to discharge area though gas supply sideways was also possible.

[1] Boris Timerkaev, Doctor of Physical and Mathematical Sciences, Professor of General Physics. E-mail: btimerkaev@gmail.com

This pipe was connected to a cylinder 3 with buffer gas. The reactionary chamber is filled with masut on two thirds. Thus the discharge area had to be at 3-5 cm below than a masut surface. In our experiments argon was used as buffer gas.

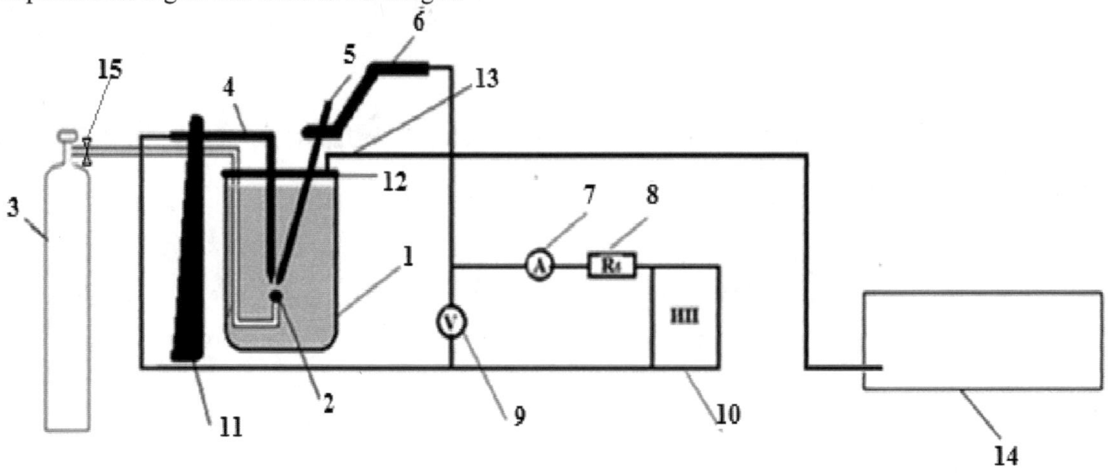

Figure 1. Experimental equipment for the decomposition of liquid hydrocarbon raw material in an argon arc with the sunken electrodes. 1 is the reactionary chamber; 2 is the pipe coupling for the buffer gas supply; 3 is the buffer gas cylinder; 4 is the cathode; 5 is the anode; 6 is the anode holder; 7 is an ammeter; 8 is ballast resistance; 9 is a voltmeter; 10 is a power supply; 11 is the cathode support; 12 is the cap; 13 is the gas extracting pipe; 14 is the capacity for gas collection; 15 is a valve.

The argon supply to discharge area has been provided before the arc ignition. The gas flow rate was 10 mg/s. Arc ignition was made by direct contact of electrodes. Arcing was followed by intensive extraction of gases from the chamber.

Burning of arc controlled by the testimonies of devices: ammeter 7, voltmeter 9 and by the intensive selection of gases from a reactionary chamber.

Let us give the analysis of the physical processes which happens when argon plasma interacts with heavy hydrocarbons. In electric arc a plasma cord argon heats up to 15000K, extends in volume and forms a gas bubble with walls of fuel oil. However on argon-fuel oil border argon temperature is not so much high. Separate atoms, ions of argon, and also electrons of plasma keep the energy sufficient for splitting the lengthiest molecules of hydrocarbons. At this time volatile and light distillates of hydrocarbons evaporate to the area of a gas bubble. The bubble is filled by gas-steam hydrocarbon fractions. The gas-steam bubble comes off area of the category and directs up. Energy of the electric field is spent for maintaining of electric arch including the argon heating, for argon ionization processes, for hydrocarbons molecules splitting, for heating of electrodes and of fuel oil.

Ethylene, Hydrogenium, methane and propane were the main gaseous products of plasmochemical decomposition of fuel oil in argon arch with the drowned electrodes. Their percentage ratios were similar to the case of decomposition without noble gas supply: ethylene (50%), Hydrogenium (30%), methane (9%) and propane (3%) in essence new at both. However, both essential increase in exit of gasoline fractions of hydrocarbons, and the structure of carbonaceous deposits on electrodes were essentially new exactly for the decomposition of hydrocarbonic raw materials in the environment of noble gas with the drowned electrodes.

Before the analysis carbonaceous deposits separately from the cathode and the anode were gathered in separate ceramic containers. Then deposites were carefully washed out in alcohol and dried in the furnace at temperature of 850 ºC ○ within 10 minutes. Further they were subjected to careful studying on the electronic scanning microscope LIBRA 120 PLUS. Fig. 2 – 4 are electronic and microscopic pictures of the received samples. Apparently from these pictures, carbonaceous deposits on the cathode consist of multilayer carbon nanotubes with the reference diameter of 9 - 15 nanometers. Their length reaches several microns.

Thus, it was shown that the electric arcing inside the heavy fuel oil allows to increase the depth of oil processing and to receive new valuable products.

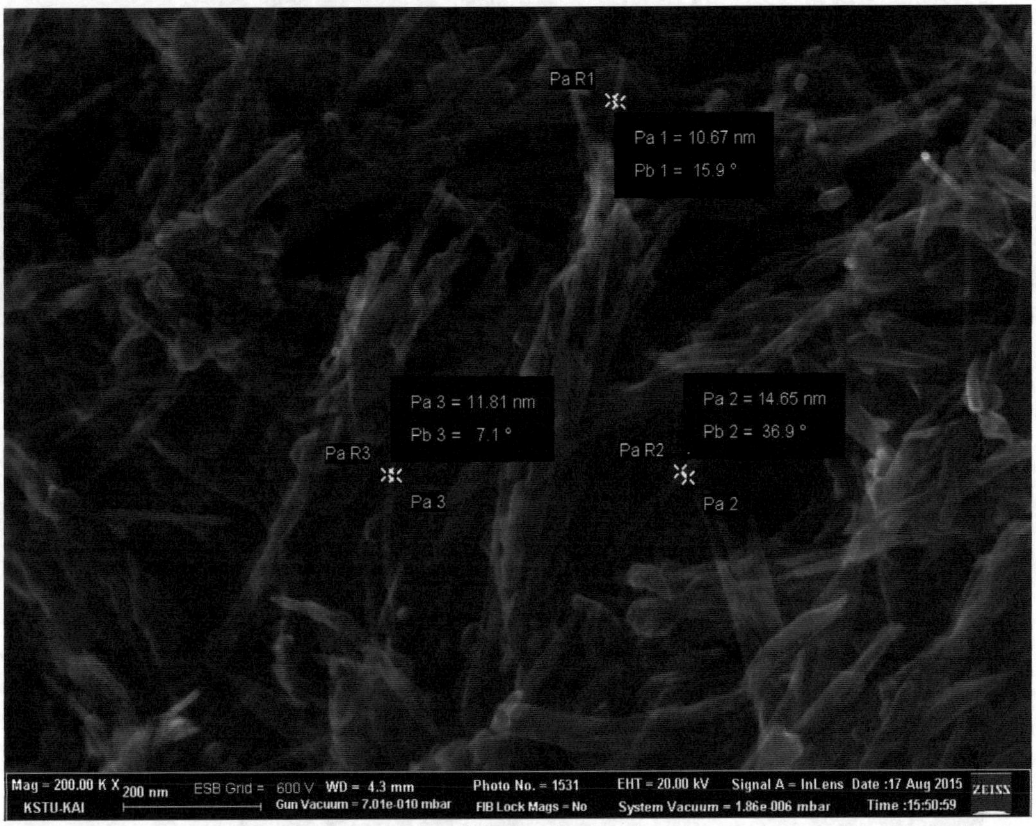

Figure 2.

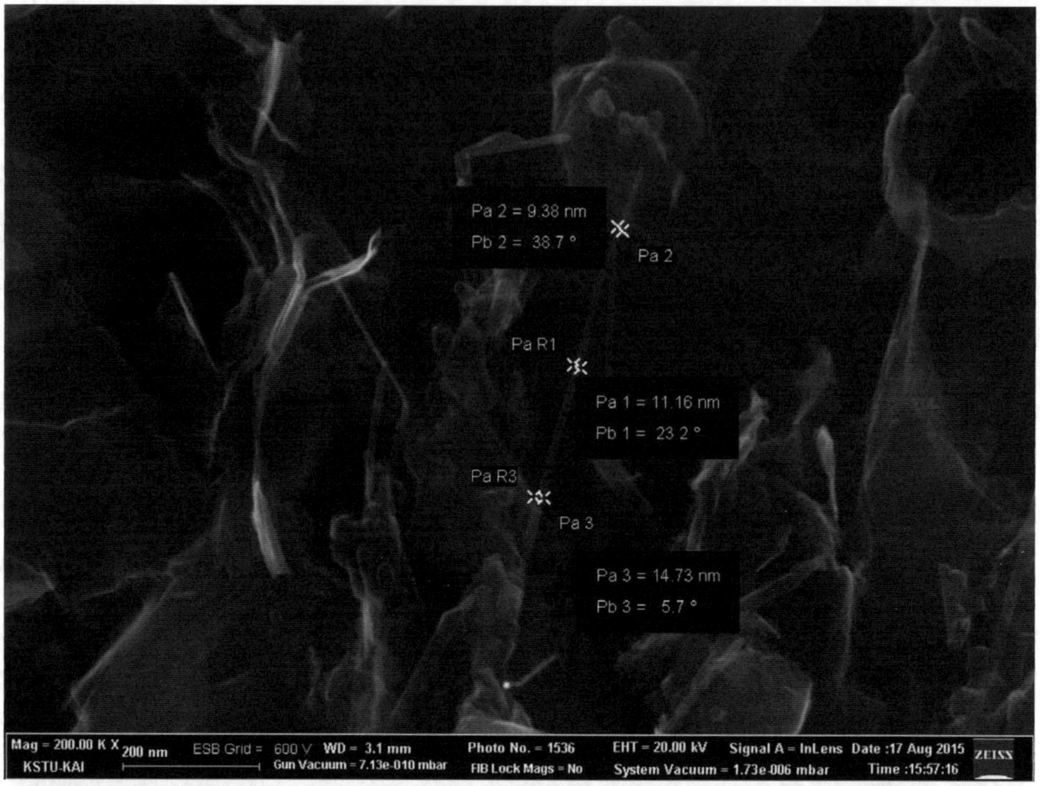

Figure 3

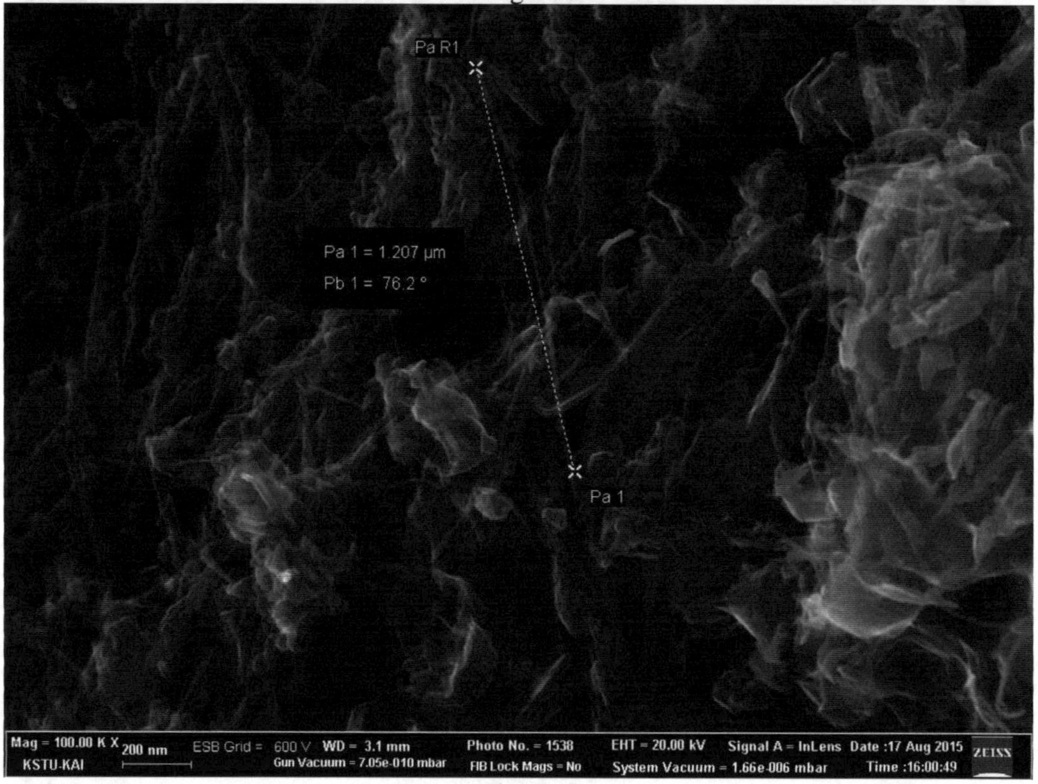

Figure 4.

Journal of Physics: Conference Series **669** (2016) 012061 doi:10.1088/1742-6596/669/1/012061

References

[1] Timerkaev B A, Ganieva G R, Ziganchin D I and AuhadeevM M May-June 2014 Electrical microdischarges in liquids and prospects of their application in plasma chemistry Journal of Engineering Physics and Thermophysics Vol 87, №3 (Minsk) pp 677-681

[2] Ganieva G R, Galeev I G, Gismatullin N K, Ziganshin D I, Tahautdinov R Sh and Timerkaev B A 2011 The decomposition of heavy hydrocarbons in the free electric arc Izv. Samara scientific.Center RAS Vol 13, №4 S pp 1156-1159

[3] Ganieva G R, Galeev I G, Gismatullin N K, Ziganshin D I and Timerkaev B A 2012 The decomposition of heavy hydrocarbons in a recessed arc. Vestn. KSTU Tupolev № 4 (Kazan) pp 184-188

[4] G.R.Ganieva, B.A.Timerkaev. Plasmachemical processing of raw liquid hydrocarbons by sunk microarc-discharge. IOP Conf. Series: Materials Science and Engineering 86 (2015) 012009. pp 1-3.

Carbon nanotubes formation in the decomposition of heavy hydrocarbons creeping along the surface of the glow discharge

B A Timerkaev[1], A O Sofronitskiy[1,2], A A Andreeva[1]

[1] Kazan National Research Technical University named by A.N.Tupolev, 420111, K.Marx str., 10, Kazan, Russian Federation
[2] Kazan Federal University, Institute of Physics, 420008, Kremlevskaya St, 18, Kazan, Russian Federation

E-mail: btimerkaev@gmail.com

Abstract. The possibility of heavy hydrocarbons decomposition in the plasma of the creeping glow discharge in a magnetic field is investigated. An electron microscopic analysis of carbon deposit on the electrodes and the walls of the discharge chamber, and gas chromatographic analysis formed in the course of the experiment is carried out. The grown carbon nanotubes have a length of about 6,17μm and a diameter of about 18nm.

1. Introduction

A high degree of disequilibrium is established in the glow discharge caused by a large detachment of the electron temperature from the gas one. Despite the fact that the temperature of the electrons in the glow discharge reaches many hundred thousands of degrees, the discharge itself is cold, and the gas temperature is only slightly different from ambient. In other words, in the glow discharge the electric field energy is transferred mainly to the electrons, which in elastic collisions with neutral particles do not practically lose their energy. Inelastic collisions, depending on the situation, can lead to the excitation of atoms or either ionization or to bonds rupture in the molecules. For the conversion of new hydrocarbon the bonds break between atoms is of particular attraction. The attractiveness of gas discharges for new compounds was confirmed in [1-4]. Modern hydrodynamic theory of such discharges is described in [5-6]. It is easy to provide selectivity of the reaction in the glow discharge by changing the reduced electric field. In this case there is no heating of the hydrocarbon mixture unlike thermal cracking. However, the organization of heavy hydrocarbons interaction with a glow discharge is not an easy task, as the glow discharge burns in a rarefied gas, but heavy hydrocarbons are in a liquid phase. The glow discharge arranged near the liquid surface practically does not interact with this fluid. In order to ensure the best possible interaction of the glow discharge with liquid hydrocarbons we decided to use a magnetic field.

2. A theoretical basis

The behavior of charged particles in the electric and magnetic fields is studied in details. As is known, an electric current in the gas is caused by the electrons and ions drift. Since opposite charged particles move in opposite directions in the electric field then the Lorentz force is routed in the same direction for all of them. By selecting the appropriate direction of the electric and magnetic fields, the discharge side of the deflection surface of the liquid hydrocarbons can be achieved.

Let us analyze the physical processes occurring in the glow discharge in a magnetic field near the surface of the liquid hydrocarbons. First, we estimate the mean free path of electrons by the formula $\lambda = 1/\sigma n$, where σ - the cross section area, $n = p/kT$ - the concentration of particles in the gas, p - pressure, T - the temperature of the gas, k - Boltzmann constant. For estimations let us assume that the pressure is 10 Torr, the temperature is 500 K, $\sigma = 7\cdot 10^{-20}$ m^2. Then the mean free path of electrons would be 0.074 mm and the potential drop in the positive column at this pressure corresponds to 10 V/mm. Thus, the mean free path of electron energy will be 0.74 eV, and the average velocity of the electron on the mean free path (i.e. the speed of the drift in the direction of the electric field) will be equal to $0.16\cdot 10^6$ m/s. At such speeds, the Larmor motion is of 1.5 microns. Therefore, if an electron in such magnetic fields begin its movement under the influence of an electric field, it will drift in the direction perpendicular to both the electric and magnetic fields. At the same time in such magnetic fields Larmor radius of the ions would be of 1 mm, and the ions in the mean free path will not have time to feel the effect of the magnetic field. This means that the electrons will carry out this difficult drift toward the surface of the liquid, and will also pull the plasma ions there. The discharge will be pressed against the liquid surface, i.e. will creep along the surface of the liquid. Therefore, this type of discharge can be called "creeping" discharge. The electrons of the discharge will effectively attack the surface of the liquid decomposing molecules of liquid hydrocarbons into lighter fractions. Due to the interaction with the plasma heavy hydrocarbon molecules can break off the liquid surface which will also be attacked by plasma electrons. Thus, creeping discharge is ideal for the interaction of the glow discharge with liquid hydrocarbons.

3. Experiment

For the experimental investigation of the glow discharge characteristics in the vicinity of the surface of the liquid hydrocarbon raw material in a magnetic field an experimental setup has been developed and constructed which consists of a vacuum system, electric power supply, the reaction chamber, as well as measuring and control instruments. Schematic diagram of the experimental setup is shown in Figure1.

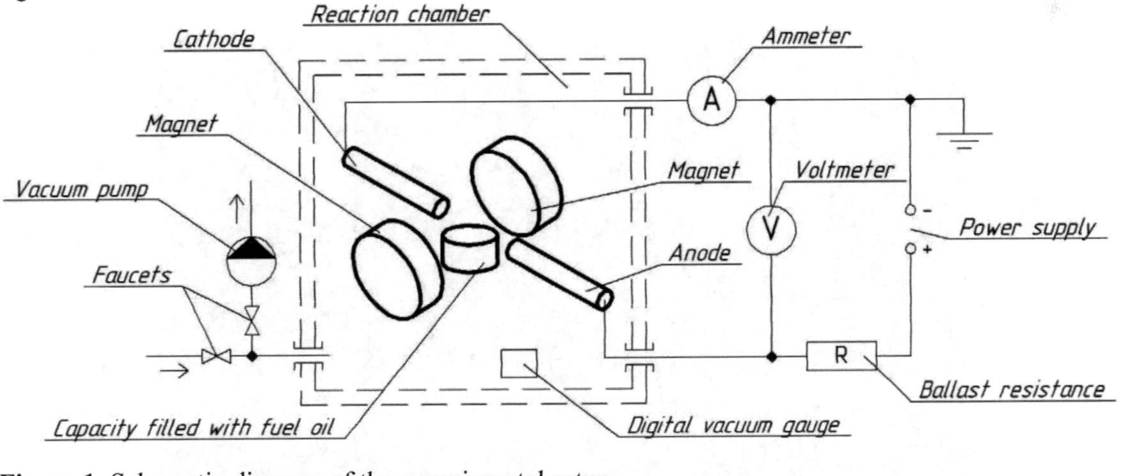

Figure 1. Schematic diagram of the experimental setup.

After loading all the necessary elements in the reaction chamber, the air is evacuated to a pressure of 1 Torr and is filled with argon to a pressure of 30 Torr. A glow discharge is energized and ignited on the electrodes. Gradually, adjusting the voltage, the current and the magnetic field the discharge with a creeping form is being achieved. During the experiment, the fuel oil in the tank turns into white gas-steam phase. This gas-vapor mixture is a product of decomposition of heavy hydrocarbons and a gaseous hydrocarbons and light oil. As chromatographic analysis shows, in the mixture of the gas-vapor phase of fuel oil decomposition products there are hydrogen, acetylene, methane, ethane, propane, butane, as well as a whole bunch of gasoline range. When the discharge burns, the carbon black is gradually formed on the electrodes. Whereas a portion of carbon atoms is directed to a cathode, and a part to the anode. The structure of the cathode carbon black differs from the carbon black structure on the anode.

4. Investigation of carbon black formed on the electrodes

Investigation of carbon black formed on the electrodes was carried out by the electron microscope AURIGA CARL ZEISS. In the presented electron micrographs of carbon black from the cathode (Figure 2-3) an ideal carbon nanotube length of about 6,17 μm and a diameter of about 18nm are seen.

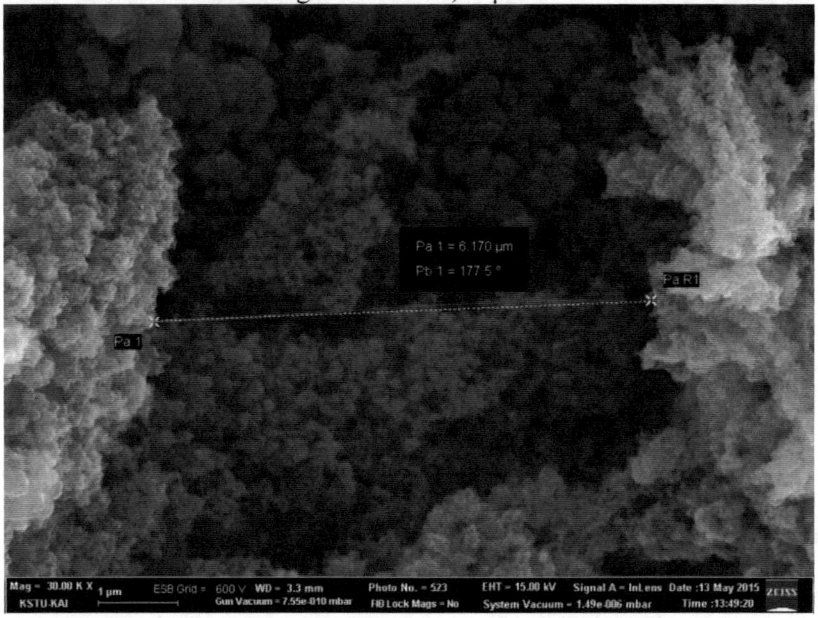

Figure 2. The length of the nanotubes 6,17μm.

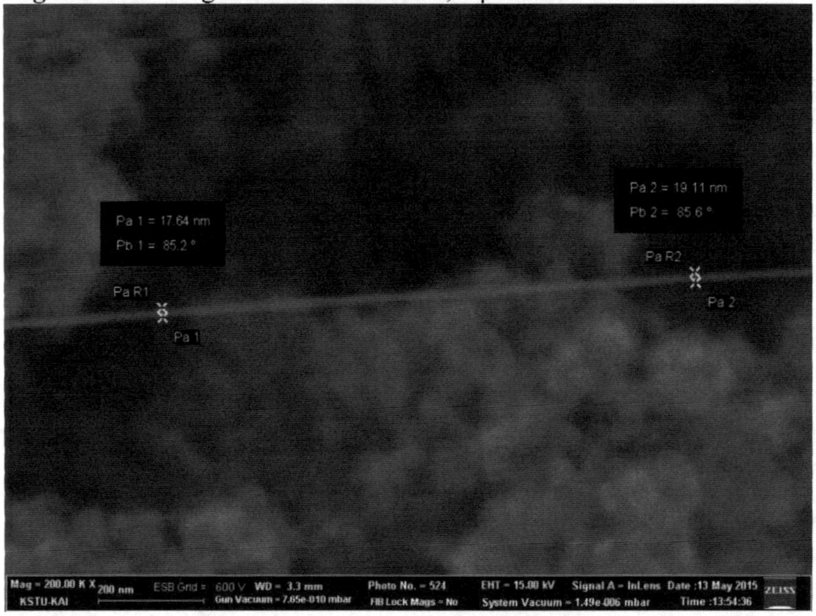

Figure 3. The diameter of the nanotubes about 18nm.

5. Conclusion

The From the research it follows that creeping glow discharge allows to make the deepening of fuel oil, the waste oil refining and oil sludge up to 100 percent. For the processing of 60 grams of fuel oil it requires about 0.2 kilowatt-hours of electricity (it is about 1 gram of fuel oil per minute when the power is 200 W). The target products of processing of heavy hydrocarbons are volatile and light oil fractions, as well as carbon nanotubes and a wide range of fullerene.

References

[1] Ganieva G.R., B.A. Timerkaev and etc .. The decomposition of heavy hydrocarbons in a free electric arc. Math. Samara scientific. Center RAS. 2011. T. 13, № 4. C. 1156-1159.

[2] Ganieva G.R., B.A. Timerkaev et al. Decomposition of the heavy hydrocarbons in a recessed arc. Vestnik. KSTU. Tupolev. 2012. № 4. S. 184-188

[3] Timerkaev B.A., Ganieva G.R., Ziganchin D.I., Auhadeev M.M. Electric microdischarges in liquids and the prospects for their use in plasma chemistry Journal of Engineering Physics and Thermophysics, Vol. 87, No.3, 2014.Inzhenerno Physics journal. V.87. No. 3. May June. S.677 -681. 2014

[4] G.R. Ganieva, B.A. Timerkaev and others. The arc discharge vapor hydrocarbons // Herald KSTU. A.N. Tupolev 2012 №4, Issue 1 Str.184-188

[5] Saifutdinov A.I., B.A. Timerkaev The drift model glow discharge with non-local values of the electric field in the ionization source .// Journal of Engineering Physics, Volume 85, №5, 2012 S.1104-1109. (/ Saifutdinov A.I., Timerkaev B.A. // Journal of Engineering Physics and Thermophysics, Vol. 85, No. 5, 2012, P. 1202-1207)

[6] Saifutdinov A.I., B.A. Timerkaev Non-local model of longitudinal structure of glow discharge and the role of diffusion of the charges along the current // Herald KSTU. A.N. Tupolev, 2012, №2, - S. 188-193

Heat Characteristics of Glow Discharge at Low Pressure with Supersonic Gas Flow

B A Timerkaev, D R Amirzyanov, D I Israfilov

Kazan National Research Technical University named by A.N.Tupolev, 420111, K.Marx str., 10, Kazan, Russian Federation

E-mail: btimerkaev@gmail.com

Abstract. Here, the results of experimental investigations on neutral gas temperature distribution over glow discharge plasma under transverse supersonic gas flow through a confined area of the discharge space. It is found that in the flow area the gas temperature decreases sharply. In addition, this leads to further increase of neutral particles concentration in flow area and, as a result, sustain existence of discharge at extremely low pressures.

1. Introduction

An impact of transverse supersonic gas flow on the glow discharge characteristics was observed in recent papers [1 – 4]. The effect opens up real opportunities to manage internal parameters of the discharge [5-8]. Namely, the glow discharge self-organization under proper supersonic gas flow will be determined by the flow area position in the discharge space. Supersonic gas pumping permits to obtain variable values of the neutral gas concentration over the discharge space. This effect is brighter with low pressures in the chamber. Certain advantages of such realization of the discharge are observed also with moderate and elevated vacuum pressures. The physics of processes with transverse supersonic gas flow pumping is determined by values of electric field strength. At very low pressures when the free path length of neutral particles is of the order of inter-electrode distance the self-contained glow discharge cannot operate: at these conditions the electrons on the path from cathode to anode cannot reproduce sufficient number of positive ions. Appearance of neutral particles in the confined flow area of the inter-electrode distance leads to generation of electron-ion pairs there due to collision ionization. Consequently, we have a peculiar distribution of the charged particles' concentration, electrons and ions will move off from the flow area with different velocities. As a result, in the area of higher concentration of neutral particles one has a higher concentration of positive ions. Elementary calculations shows that ions drift speeds are higher than a flow rate [3]. Therefore, ions are removed from this area by electric field, but not by gas flow. Both the internal field of space charges and the superposed external field provide a peculiar distribution of electric potential and field strengths. Electric field is slightly increasing till the boundary of supersonic stream area then it is sharply decreasing. Thus, all nearelectrode areas of a classical glow discharge are moved in a supersonic area [9-12].

2. Experement

Results of the experimental investigations of axial temperature distribution in discharge chamber are represented in this work. Electrical schematic diagram of the discharge chamber for neutral gas temperature measurements is shown in Fig. 1. The discharge chamber represents the cylindrical channel of molybdenum glass. The chamber has special branch pipes for installation of supersonic nozzle 5 to provide the supersonic stream and confusor 6 for gas evacuation. Cathode and anode during experiment are water-cooled. The supersonic nozzle and confusor made of epoxy, put on the corresponding branch pipes of a discharge chamber and hermetically sealed. Such nozzles are applied at not really larger velocities of supersonic flow. The chromel-copel thermocouple 4 was used for the temperature measurements. During experiment the thermocouple was moved along an axis of a discharge chamber. Movement of the thermocouple 4 was controlled by means of a vernier scale.

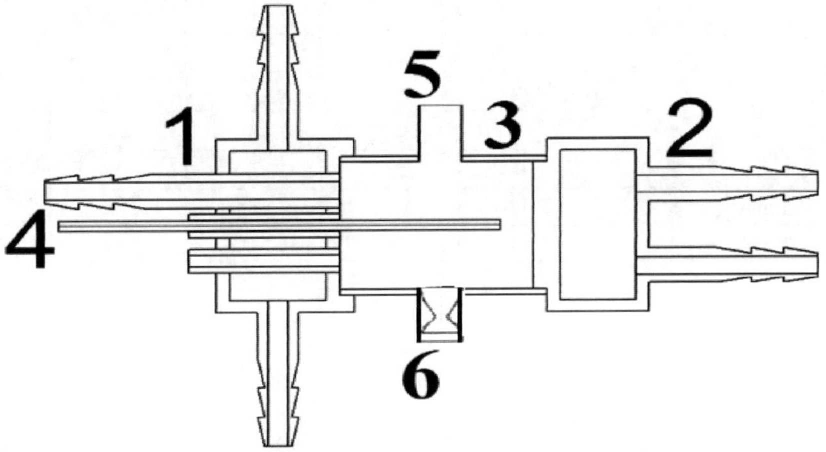

Fig. 1. Diagram of the probe researche discharge chamber axial section, 1 – cathode; 2 – anode; 3 – the interelectrode space isolated by molybdenic glass; 4 – thermocouple; 5 – the air evacuation pipe; 6 – the pipe for supply of gas in the chamber through the nozzle.

3. Discussing

There are many ways of gas heating in the glow discharge. As the ionic mobility is much less than electron mobility, the electric field energy is generally transferred to electronic gas. Afterwards part of this energy is transferred from electrons to neutral particles by elastic collisions, i.e. it is spent for direct gas heating [9]. The rest part is spent to excite the oscillating freedom degrees of molecules. Due to these processes, rather high vibrational temperatures are achieved in molecular gases. Just, the very small part of supplied energy is spent for electronic gas heating.

Experimental results of gas temperature measurement along the discharge chamber at various current values without supersonic gas flow are given in Fig. 2.

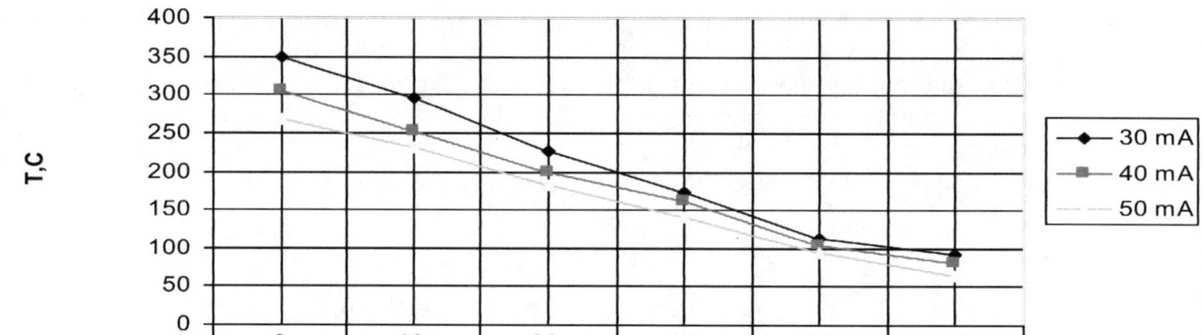

Z,mm	0	10	20	30	40	50
30 mA	350	295	226	173	114	92
40 mA	306	253	200	161	103	81
50 mA	269	231	182	140	95	65

Fig. 2. Temperature distribution in the positive column along the channel for various discharge current values.

As it is seen from the Fig.2, gas temperature smoothly decreases along a discharge chamber from cathode to anode.

The experimental results of gas temperature measurement along the discharge chamber at various current values with supersonic gas flow are given in Fig. 3.

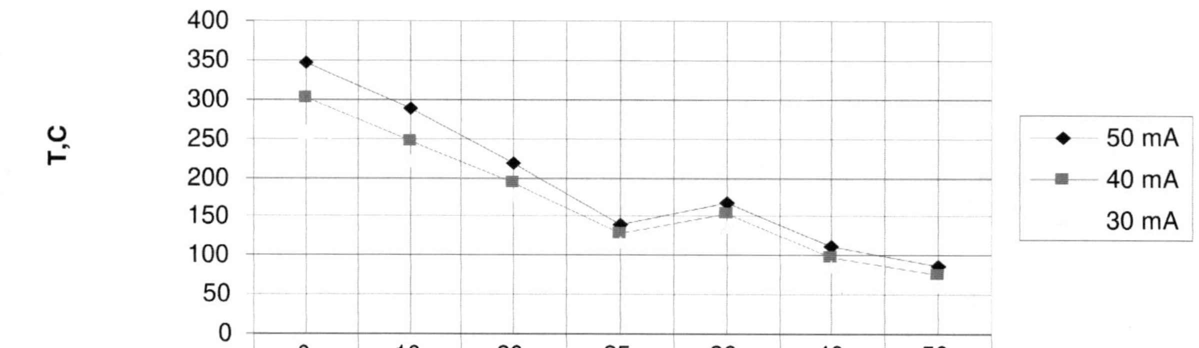

Z,mm	0	10	20	25	30	40	50
50 mA	347	288	218	141	167	112	88
40 mA	301	245	192	130	155	99	75
30 mA	258	220	176	117	136	87	62

Fig. 3. Axial temperature distribution in the discharge with supersonic gas flow.

4. Conclusion

Comparison of Fig. 2 and Fig. 3 shows that gas temperature in the field of a supersonic current (area from 23 mm to 28 mm) quite sharply decreases. Flow heat transfer and gas cooling due to the adiabatic expansion after supersonic nozzle can explain this.

References.

[1] Timerkaev B.A., Zalyaliev B.R., Karimov B.R., Israfilov D.I. Behavior of glow discharge in distributed supersonic gas flow in a plasma deposition devices. Vestnik KNRTU-KAI. 2013 г. №4.

[2] Timerkaev B.A., Zalyaliev B.R., Saifutdinov A.I. Control of glow discharge parameters using transverse supersonic gas flow - numerical experiment Journal of Physics: Conference Series 567 (2014) 012031.

[3] Timerkaev B.A., Zalyaliev B.R. Glow Discharge in a Transverse Supersonic Gas Flow at Low Pressures. High Temperature, 2014, Vol. 52, No. 4, pp. 471–474.

[4] Timerkaev B.A., Zalyaliev B.R., Saifutdinov A.I. Glow Discharge Characteristics in Transverse Supersonic Air Flow. Journal of Physics: Conference Series 567 (2014) 012032

[5] Israphilov I.H., Galiakbarov A.T., Israphilov D.I., Gabdrahmanov A.T., Samigullin A.D. Computer simulation of gas flow in the discharge chamber a pulsed plasma generator. Izvestia TulGU. Techical sciences. 2012. № 6. P. 90-97.

[6] Israphilov I.H., Israphilov D.I., Bashmakov D.A., Galiakbarov A.T., Samigullin A.D.Numerical analysis of temperature distribution in bottom electrode of dc arc furnace in process. Contemporary Engineering Sciences. 2014. T. 7. № 25-28. P. 1483-1491

[7] Apparatus for producing carbon nanostructures (variants). Israphilov Z.H., Israphilov D.I., Battalova A.R. Patent RU 112678 21.07.2011.

[8] I K Israfilov et al. The analysis of the bottom electrode thermal streams during work process in a melting furnace. 2014 IOP Conf. Ser.: Mater. Sci. Eng. 69 012015 doi:10.1088/1757-899X/69/1/012015

[9] Timerkaev B.A., Zalyaliev B.R., Ahmetov M.M., Petrova O.A., Israfilov D.I. self-organization of normal glow discharge. Vestnik KNRTU-KAI. 2014. №3 P. 158-163.

[10] Timerkaev B.A., Ahmetov M.M., Zalyaliev B.R., Petrova O.A., Israfilov D.I. longitudinal distribution of electrical parameters in normal glow discharge. Journal of Physics: Conference Series. 2014. T. 567. № 1. P. 012036.

[11] Timerkaev B.A., Saifutdinov A.I. Numerical simulation of regions of negative glow and furadeeva dark space in a glow discharge. Vestnik KNRTU-KAI. 2011. №2 P. 109-114.

[12] Saifutdinov A.I., Timerkaev B.A. Drift model of a glow discharge with account for the nonlocal value of the electric field strength in the ionization source. Journal of Engineering Physics and Thermophysics. 2012. T. 85. № 5. C. 1202-1207.

Journal of Physics: Conference Series **669** (2016) 012064 doi:10.1088/1742-6596/669/1/012064

Erosion of electrode metal in the electric discharge under the exposure of the electrolyte stream

Yu I Shakirov, R I Valiev[1], A A Khafizov, R A Valiev and R G Khakimov

Kazan Federal University Naberezhnye Chelny Institute, 68/19 Mira Street, Naberezhnye Chelny, 423810, Russian Federation

E-mail: rivaliev87@mail.ru

Abstract. The discharge between solid and liquid electrodes under the normal atmospheric conditions (SATP) was investigated. As electrodes, the metal materials and unsaturated salts solutions are used. Such discharge takes place in the range of voltage from 510 V to 525 V, a current from 0.2 A to 6 A, a electrolyte liquid flow rate from 2 cm^3/s to 14 cm^3/s, an electrolyte stream length from 5 mm to 40 mm. Increasing the flow rate of electrolyte liquid increases the discharge current. There are minimum and maximum critical values of the electrolyte flow rate of beyond which the existence of discharge is impossible. Specific values of the critical flow rate of the electrolyte depend on the nature, composition and stream length of the electrolyte. The electrical discharge between the electrolyte stream and a metal electrode is followed by erosion of the electrode material. The technique for local removal of metal by means of such a discharge is developed and the optimal values of erosion treatment parameters are determined. By using the discharge, technical processes of cleaning surface, deburring, drilling and cutting of metal materials were implemented.

1. Introduction

In the recent years, special attention is given to the study of the processes occurring in the electric discharge between solid and liquid electrodes [1 – 4]. The implementation of the plasma with various liquid and solid materials usage allows modifying and synthesizing products with a wide range of properties [5 – 7]. However, the absence of developed technologies delaying the implementation of these processes in production.

The usage of the electrolyte stream as a liquid electrode is attractive in terms of the local treatment of material by plasma. Such a discharge is accompanied by erosion of solid electrode material which can be successfully used for surface cleaning, deburring, drilling, cutting metal materials. Implementation of controlled erosion is limited by lack of studies of the processes occurring in the discharge between the solid electrode and the electrolyte stream.

The determination of electrical characteristics of the discharge, establishing their influence on the electrical and hydraulic properties of the electrolyte have a great importance for the development of erosion treatment technology.

2. Experiment

The discharge was done at atmospheric pressure between the liquid cathode and the solid anode with the values of voltage and current not exceeding 1 kV and 6 A, respectively. Experiments were

[1] To whom any correspondence should be addressed.

conducted with using various electrode materials. The aqueous solution of sodium chloride, copper sulphate, ammonium nitrate with different concentrations and industrial water were used as electrolytes. Copper, iron, titanium, and their alloys are used as the solid material of the electrodes. At the same time the solid electrode was used as an object of erosion machining.

For the erosion machining, electrolyte stream is directed to a certain area of the solid electrode. A stream of electrolyte is supplied through the ferrule with the dielectric nozzles. Current is fed to this ferrule. Stream length (L) is defined as the distance from the ferrule to the surface of the solid electrode. The diameter (D) of the stream depends on the size of the holes of the dielectric nozzles. To conduct the experiments a special tool to manage the supply (flow rate) of the electrolyte and to adjust the size (diameter and length) of its stream was made.

3. Results and their discussion

After getting out of the nozzle, electrolyte stream becomes a main element which supplies current to the metal electrode. It is known [8] that the electrical conductivity of the electrolyte is higher, as the higher the concentration of salts is in the solution, but it is seen only till a certain critical concentration. Typically, these values are slightly below the critical limits of solubility. Nevertheless, as shown by our experiments, the presence of undissolved crystal sediment in the electrolyte leads to instability of the discharge. These circumstances have been resulted into the choice of the unsaturated sulphate solutions as the electrolyte: from 5% to 15% for the NaCl, CuSO4, NaNO3. Another factor affecting the rate of current passage is the average rate of the electrolyte flow (or liquid flow rate Q).

The parameters such as power and electrolyte stream affect the discharge between the solid electrode and the electrolyte.

The critical current values (I_{cr}) and flow rate of the liquid electrode (Q_{cr}) for the occurrence of discharge were experimentally established. When currents and flow rate of the electrolyte are lower than the critical level, the discharge does not light, and there is a conventional electrolysis. The specific values of the critical current and critical flow rate of the electrolyte depend on the nature of the salts and its concentration in the solution, and on the temperature and length of the stream of electrolyte (Figure 1).

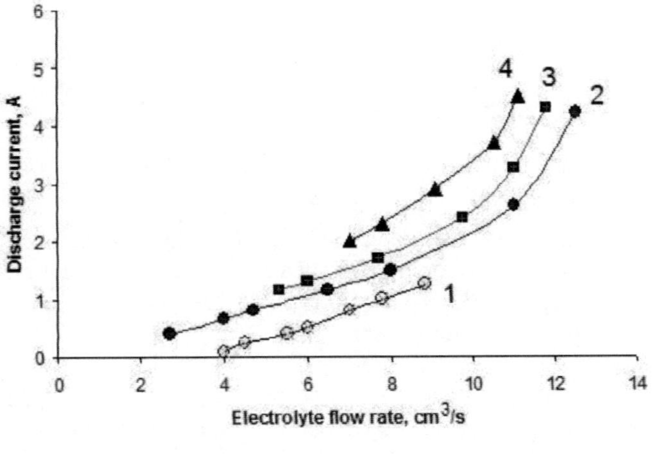

Figure 1. The dependence of the discharge current I on the flow rate of the electrolyte Q at different lengths of the stream L of electrolyte of process water (1 – L=18 mm) and 10% NaCl solution (2 – L=10 mm; 3 – L=14 mm; 4 – L=22 mm).

The increasing of the length of stream L increases not only the value of the critical current Icr, but also requires a larger flow rate of electrolyte Q_{cr} for the occurrence of discharge. For example, in the case of electrolyte of 10% sodium chloride solution, with increasing of stream length L from 10 mm to 22 mm, the critical current values I_{cr} and Q_{cr} electrolyte flow rate increase from 0.35 A to 1.75 A and from 2.7 cm^3/s to 6.6 cm^3/s, respectively. Such delay is supposed to be due to increased resistance of the stream and the voltage drop on it.

Increasing the flow rate of electrolyte increases the discharge current I, as this increases the number of current carriers per unit of time. However, after reaching a certain Q values, the discharge becomes unstable and the blanking is carried out. Thus, the discharge is characterized by another critical value of the electrolyte flow rate Q_{max}. This is the maximum flow rate of the electrolyte liquid at the given conditions, beyond the increasing of which, the discharge goes to electrolysis. This transition is accompanied by a spike increase of the current, flowing through the stream. Critical parameter Q_{max}, as the I_{cr} and the Q_{cr}, depends on the nature, composition and length of the electrolyte stream.

It should be noted that the electric discharge behaves similarly when using industrial water as an electrolyte. However, it is characterized by low values of discharge of current and blanked much earlier – at low flow rates of electrolyte.

Figure 2 shows the current-voltage characteristics of the discharge between the metal electrode and the electrolyte stream. Specifications are directly proportional and it can be considered almost horizontal. For electrolytes usage with normal atmospheric conditions (SATP), combustion of such discharge is possible in a range of voltage from 510 V to 525 V. It was determined that the location of the lines of current-voltage characteristics in the stress axis is influenced by both, the length of the stream and electrolyte conductivity.

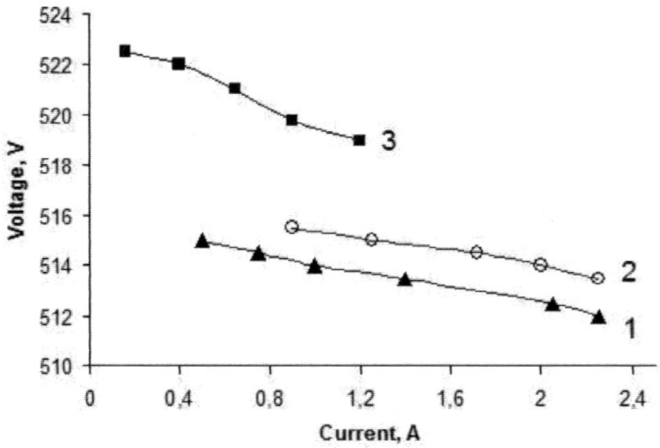

Figure 2. Current-voltage characteristics of the discharge between the metal electrode and the electrolyte stream with diameter of 3 mm for various lengths L of the stream of electrolyte from 10% NaCl solution (1 – L=10 mm; 2 – L=15 mm) and process water (3 – L=10 mm).

The practical interest is performed by the fact that the electric discharge between the electrolyte stream and a metal electrode is accompanied by erosion of the electrode material. This phenomenon can be successfully used to solve technological problems, such as surface cleaning, deburring, drilling, cutting of metal materials.

Based on these studies the technique of local metal removal by a discharge, combusting between the electrolyte stream and the solid electrode – workpiece. The technique has been tested on copper, titanium and steel parts. For example, a drilling hole of 2 mm diameter in a copper plate with a thickness of 5 mm at a discharge current of 2 A with a 10% solution of sodium chloride as electrolyte was performed in 8 minutes. The experiment showed that the higher the temperature of melting of the electrode material, the more time is spent on its drilling. Processing performance can be increased by increasing the discharge current. However, as shown above, when exceeding the critical values of the discharge is being blanked and the process of anodic dissolution is started. It is accompanied by a spike increase in electric power consumption and by a high heating of the electrolyte. In contrast to the processes of drilling and cutting of metal, the cleaning of metal surface requires softer treatment conditions. Apart from current density, the length of electrolyte stream influence on the effectiveness of the erosion treatment. Experimentally, the optimal values of L, which lie in the range from 5 mm to 8 mm, were determined. Outside this range, the discharge stability is being significantly reduced.

4. Conclusions

The parameters such as power and electrolyte flow rate affect the occurrence and combustion between the solid electrode and the electrolyte. Under the normal atmospheric conditions (SATP) and used as the electrolyte the unsaturated sodium chloride solution, copper sulfate discharge combusts in a range of interrelated basic parameters: voltage from 510 V to 525 V, a current from 0.2 A to 6 A, flow rate of electrolyte liquid from 2 cm^3/s to 14 cm^3/s, an electrolyte stream length from 5 mm to 40 mm.

Increasing of the flow rate of electrolyte liquid increases the discharge current. There are minimum and maximum values of the critical flow rate of electrolyte beyond which the existence of discharge is impossible. The specific values of the critical flow rate of the electrolyte depend on the nature, composition and length of the stream of the electrolyte.

The electrical discharge between the electrolyte stream and a metal electrode is followed by the erosion of the electrode material. Based on these studies, the technique of local metal removal with the help of such discharge is made and the optimal values of erosion treatment parameters are determined. It makes possible to solve the technological problems of surface cleaning, deburring, drilling and cutting metal materials.

References

[1] Kashapov L N, Kashapov N F and Kashapov R N 2015 The research of anodic microdischarges in plasma-electrolyte processing *IOP Conf. Ser.: Materials Science and Engineering* **86** 012019

[2] Dautov G Y, Nail K F, Viktor L, Renat G Z, Raphael G T, Fayrushin I and Kashapov R N 2013 Low Temperature Plasma in the Processes of Functional Coating Preparation *J. of Phys.: Conf. Ser.* **479** 79.

[3] Tazmeev Kh K and Tazmeev A Kh 2015 Gas discharge with liquid electrolyte cathode in the mode of occurrence of the constricted channels *J. of Phys.: Conf. Ser.* **567** 012035

[4] Tazmeev Kh K and Tazmeev A Kh 2005 Some specific features of heat and mass transfer of gas-discharge plasma with a liquid electrolytic cathode *Heat Transfer Research* **36** 623

[5] Valiev R A, Shakirov Yu I and Iliukhin A N 2014 Phase and structural conditions of low-temperature plasma interaction products with steel. *J. of Phys.: Conf. ser.* **567** 012040

[6] Valiev R A, Gaisin F M and Shakirov Yu I 1991 Special traits of powder obtained in discharge between steel electrode and electrolyte *Soviet Powder Metallurgy and Metal Ceramics* **30** 448

[7] Khafizov A A Valiev R I Shakirov Yu I and Valiev R A 2014 Steel surface modification with plasma spraying electrothermal installation using a liquid electrode. *J. of Phys.: Conf. ser.* **567** 012026

[8] Amitan G L, Baysunov I A and Baron Yu M 1988 *Spravochnik po elektrochimicheskim i elektrofizicheskim metodom obrabotki* (Leningrad: Mashinosroenie) p 719

Journal of Physics: Conference Series **669** (2016) 012065 doi:10.1088/1742-6596/669/1/012065

Plasma technologies application for building materials surface modification

G G Volokitin[1], N K Skripnikova[1], O G Volokitin[1], V V Shehovtzov[1], A G Luchkin[2] and N F Kashapov[2]

[1]Tomsk State University of Architecture and Building, 2 Solyanaya sq., Tomsk 634003 , Russian Federation

[2]Kazan (Volga Region) Federal University, 18 Kremljovskaja str., Kazan 420008, Russian Federation

E-mail: vgg-tomsk@mail.ru

Abstract. Low temperature arc plasma was used to process building surface materials, such as silicate brick, sand lime brick, concrete and wood. It was shown that building surface materials modification with low temperature plasma positively affects frost resistance, water permeability and chemical resistance with high adhesion strength. Short time plasma processing is rather economical than traditional processing thermic methods. Plasma processing makes wood surface uniquely waterproof and gives high operational properties, dimensional and geometrical stability. It also increases compression resistance and decreases inner tensions level in material.

1. Introduction

The problem of providing durability for buildings and constructions in difficult climatic conditions mostly depends on defending coating quality. Defending coatings prevent moisture penetration and protect buildings from UV, little air-bone dust particles, industrial steam, acid rains, and strengthen concrete or brick surface. Therefore structural elements of buildings and constructions are protected with special coatings.

In that reason it is necessary to find new methods of making defending coatings that meet requirements of durability and colour expression. The perspective solution of this problem lies in today's development and production of new building materials that have qualitatively new characteristics with the use of untraditional high technologies based on application of electrochemical, plasma, laser, radiation, electro-impulse, ultrasound and other highly effective material treatment methods for significant improvement of its properties and thus providing flow of non-equilibrium processes. Of great interest are studies [1] and practical realization of building materials and products surface thermic treatment methods. The most perspective method of building materials processing is plasma method.

There are several main building materials mostly used in construction industry. They are different types of concrete (cellular, heavy), kilns materials (ceramic brick), autoclaved material (sand-lime brick). The article studies defending-decorating coatings on building materials created with low temperature plasma.

2. Experiments

Materials

Substrate materials were cellular concrete B10-B15, heavy concrete, ceramic brick and silicate sand-lime brick, № 200-250. Decorative coatings were made from pastes of different compositions of binder-liquid sodium glass and filler - fine glass, granite or quartz sand. Coatings on wood were also investigated.

Processing.

Concrete products covered by paste were dried and processed with plasma (temperature 3000 - 5000°C, at a speed of 0,045 - 0,13 m/sec, power 40 - 55 kW) [2,4] as direct surface plasma fusion of product gives high porosity and decreases coating quality. Vitreous film has thickness of 0,15 – 0,4 mm and high adhesion strength.

For making protective-decorative coating on ceramic bricks by plasma deposition method powder particles were injected into plasma jet, melted, then achieved brick surface and got fixed on it. Figure 1 shows the scheme of the research facility for protective decorative coating of wood.

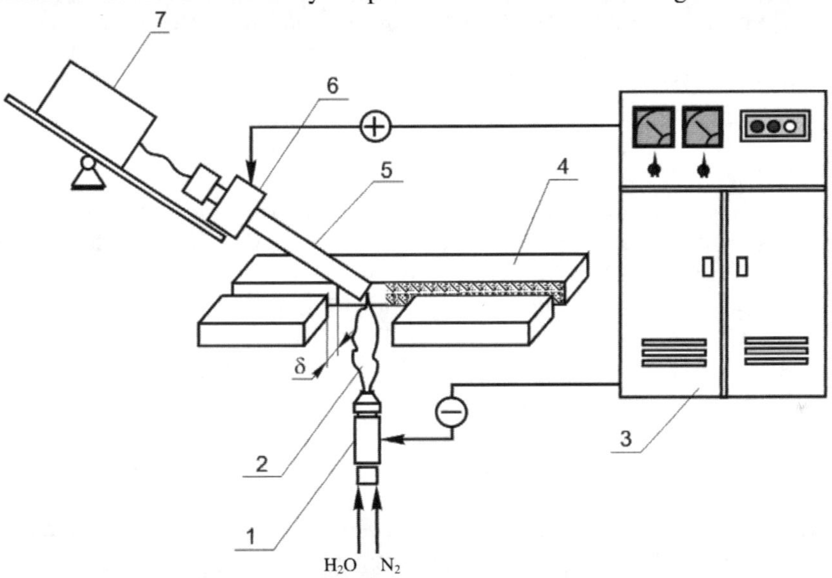

Figure 1. Facility for plasma processing of brick and wood surfaces: 1 – cathode; 2 – plasma jet; 3 – power source; 4 – sample; 5 – graphite anode; 6 – water-cooled tinsel-lead; 7 – electric drive of electrode.

The unit consists constructively of plasma generator 1, graphite anode 5 and DC power supply. Plasma generator with external anode is situated in the middle part of the conveyer, cathode is on the conveyer axis and located vertically near the table and with up oriented nozzle. External anode – graphite cylinder with diameter 70 mm, is fixed at an angle of 20° to the horizontal. Electric drive spins and moves graphite cylinder. For plasma cord generation graphite anode closes with plasmatron at a distance necessary for main plasma arc appearance. Conveyer slides provide constant gap (δ=15-20 mm) between oppositional located wood or brick samples during processing.

Properties study

Concrete quality is studied in accordance with state standard 25485-89. Coatings were tested for chemical stability, frost resistance and water permeability. Surface microstructure of fused building materials is studied with microscope JEOL JEM-2100 (Japan).

3. Results and discussions

Concretes

Test results [3] showed that plasma coating chemical resistance is significantly higher than that of base chemical resistance. Material frost resistance that mainly defines durability is satisfactory. Water permeability of plasma treated surface reduced by 1,5 – 2 times, open porosity also decreased several times. That allows apply walling from concrete with plasma coating based on pastes with different fillers for buildings that operate not only at normal temperature and humidity conditions, but also at mode of high humidity and aggressive actions.

At heavy concrete surface fusion there is no formation of uniform coating. Large fractions of gravel contained in heavy concrete and high temperature treatment causes thermal stress leading to cracking of the filler. In this regard, concrete surfaces finishing quality could be achieved by previous covering with concrete with porous aggregates. As porous aggregates in [2] expanded clay sand was used. At fusion of concrete with expanded clay sand the surface became relief and of dark green color. For decorating properties improvement of the finishing cover wastes of glass production, red and silicate brick wastes, porcelain chips were used. We found that the above mentioned additives infused into concrete surface layer defend it from calcium hydro silicates dehydration at plasma treatment and allows forming finishing layers of different textures and color on concrete. Simultaneously, there is a significant improvement of adhesion of the fused decorative layer with base. Vitreous film formed on concrete surface under plasma influence defends it from moisture, decreasing water absorption.

Bricks

At melting ceramic brick is forming finishing layer, its color being from gray to black. The surface texture depends on melting mode. Vitreous layer has high strength and high adhesion with substrate. For decorating film formation pastes deposition on brick surface is needed. Fusion begins after preliminary surface drying till residual moisture of about 10%.

The study [2] describes the plasma treatment influence on technological parameters of exploitative properties of ceramic brick with fused surface. The fused coating adhesion and frost resistance depend on fused layer thickness. Coating thickness on the surface layer of ceramic significantly depends on plasma jet temperature and processing time. The fused coating thickness increasing leads to adhesion decreasing because of stress between glass and ceramic resulting in cracking of surface layer and decreasing of products frost resistance.

After analysis of [1-3] we can note that coating adhesion as its most important parameter and frost resistance depend on plasma torch power and treatment speed.

Studies results [1,2] show that ceramic products with surface melted by plasma jet quite satisfy the requirements for building materials and at the same time have better outward appearance, lower water absorption and higher chemical resistance in accordance with state standard 530-2012.

Plasma interaction with silicate brick surface studies [2] allow applying high concentrated thermal jets to defend decorating coatings formation on building materials. Plasma treatment of silicate products surface leads to vitreous film appearance, its color depending on substrate composition. The product surface pretreatment with different chemical compound solutions results in vitreous color films formation. At plasma treatment on products surface high temperature processes occur. As pigments only heat-resistant compounds could be used. Figure 2 shows microstructure melted silicate brick.

At depth of 1 mm below the surface where temperature is about 600°C stable Calcium and Quartz silicates are forming. Results of electron microscopy showed identity of Calcium silicate structure in transition layer and Calcium hydro silicates. The [1] described a process of colored coatings formation by pretreatment with solutions of metal salts (i.e. cobalt, copper, iron, chromium). Pretreatment of silicate products with metal salts aqueous solutions provides not only coloring of vitreous films, but also adhesion increasing because of lower melt formation temperature and viscosity. Test results of silicate brick, ceramic brick, concretes and other products (chemical and frost resistance) showed that coating adhesion before and after test changed insignificantly, that indicates vitality and durability of products with defending decorating coating.

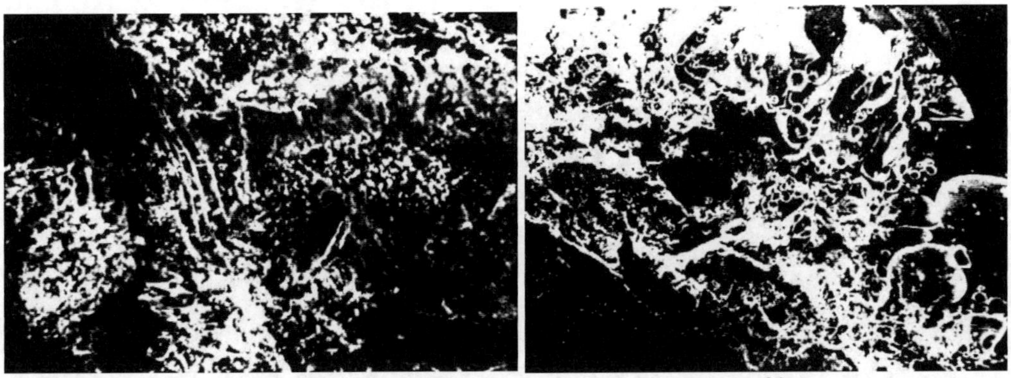

a) b)

Figure 2. Microstructure melted silicate brick: a) vitreous film; b) transition layer.

Wood

One of the priority areas of low temperature application is wood surface treatment for thick functional (hydrophobic) film formation that improves material physical properties and does not pass the liquid.

After wood plasma treatment experiments [5-7] samples surface changed colour into dark, golden brown. Wherein wood texture was clearly seen, growth rings became more visible. It is very useful at using samples in decorating. Moreover thermo treatment leads to significant changes in wood structure (Figures 3 and 4). While heating its structure evaporates resins, waxes, fats, phenols, glucose and hemicellulose components. As a result wood becomes stable to rot, mildew, defeat of various microorganisms and fungi. Material biological resistance increases. Usual drying can't achieve these effects [8].

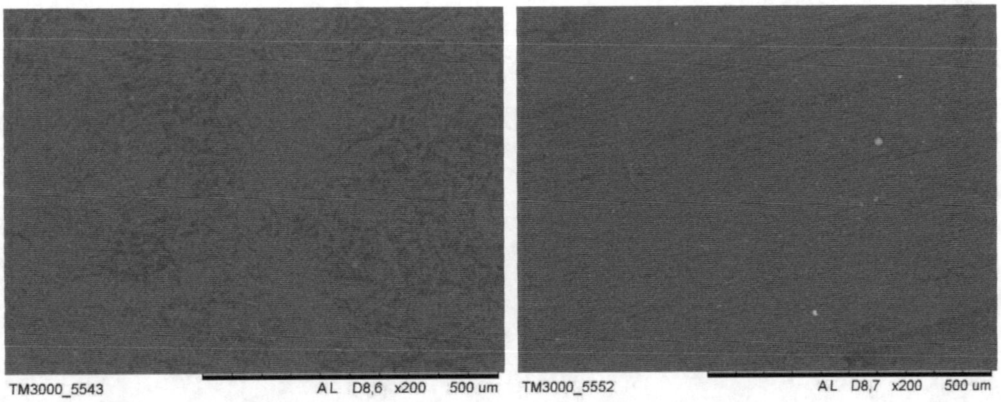

a) b)

Figure 3. Wood surface morphology: a) before and b) after plasma processing

Influence of processing regime on colour of wood surface presented in Table 1.

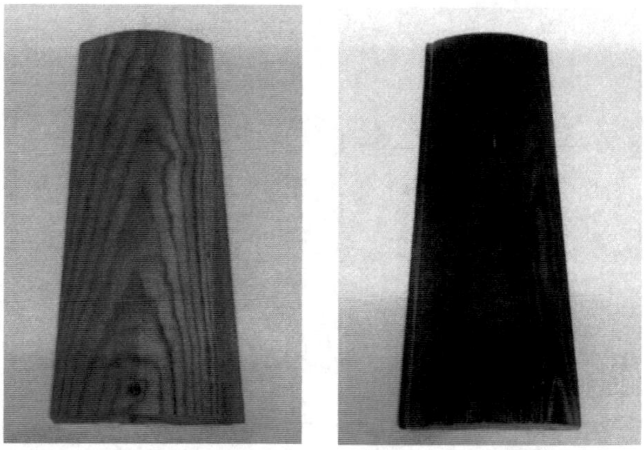

a) b)
Figure 4. Block-house appearance: a) before and b) after plasma processing

Table 1. Parameters of plasma processing and characteristics of wood surface

Plasmatron power, kW	I, A	U, V	Heat flux, W/m^2	Processing time, sec	Surface characteristic	Surface colour
20,8	130	160	$1{,}1\cdot10^6$	3,0	Incomplete combustion	Light brown
23,8	140	170	$1{,}5\cdot10^6$	2,0	Partly incomplete combustion	Light brown
35,2	200	176	$1{,}8\cdot10^6$	1,0	Uniform firing, depth 1 mm	Light brown
35,2	200	176	$1{,}8\cdot10^6$	2,0	Uniform firing, depth 1 mm	Golden brown
35,2	200	176	$1{,}8\cdot10^6$	3,0	Partly burnout	Dark brown
56	400	140	$2{,}6\cdot10^6$	1,0	Uniform firing, depth 2 mm	Golden brown
56	400	140	$2{,}6\cdot10^6$	2,0	Uniform firing, depth 2 mm	Dark brown
58,8	420	140	$2{,}8\cdot10^6$	1,0	Partly burnout	Dark brown
58,8	420	140	$2{,}8\cdot10^6$	2,0	Burnout	Black
59,4	440	135	$3{,}0\cdot10^6$	1,0	Burnout	Black

At the next stage of the experiments water permeability of samples before and after plasma treatment was studied. Water adsorption by wood was measured every 2 hours. Results are shown in Figure 5.

4. Conclusion

Analysis of the experiments allows us to make a conclusion that wood surface plasma treatment technology through little affecting time on product is rather economic as compared with traditional methods of thermal treatment. After processing the surface is more resistant to fungi, mildew and other wood defects [9]. The reason is in absence of their nutrient medium. Furthermore, wood samples after thermal treatment adsorb water several times slower than initial samples. Therefore wood threated by electro plasma technology could be used in climate with high humidity. Plasma treatment allows impart unique water resistant and exploitative properties, high dimensional stability, herewith increased wood compression stability and decreased level of internal stresses.

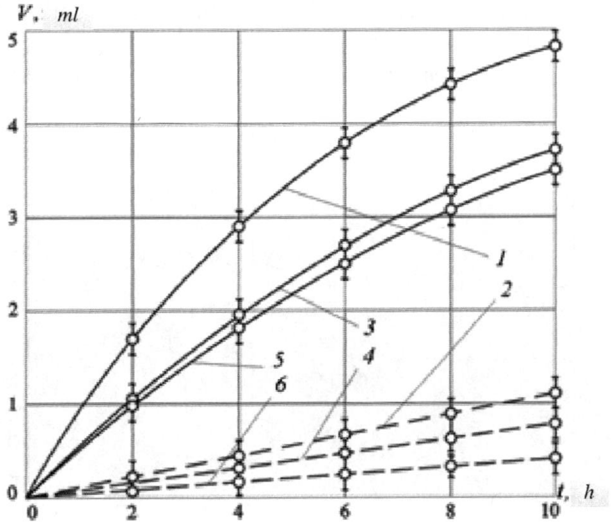

Figure 5. Water consumption (adsorption) in time. 1, 3, 5 – aspen, fir, pine before low temperature plasma treatment, 2, 4, 6 – aspen, fir, pine after low temperature plasma treatment.

References

[1] Volokitin G.G., Skripnikova N.K., Shilyayev A.M., Petrochenko V.V. Plazmennyye tekhnologii v stroitelstve [Plasma technologies in construction]. Tomsk : Vestnik of Tomsk State University of Architecture and Building. 2005. 291 p. (rus)

[2] Volokitin G.G., Lysak A.S., Anshakov A.S., Skripnikova N.K. Plazmennaya obrabotka materialov [Plasma treatment of materials]. Tomsk : Vestnik of Tomsk State University of Architecture and Building. 2009. 200 p. (rus)

[3] Anshakov A.S., Volokitin G.G., Volokitin O.G., Skripnikova N.K. Elektrotehnologicheskie ustanovki dlya plazmenno-termicheskoi obrabotki materialov [Electro-technological plants for materials plasma-thermo processing] Tomsk : Vestnik of Tomsk State University of Architecture and Building. 2014. 126 p. (rus)

[4] Bessmertnyi V.S. *Nauchnye osnovy formirovaniya potrebitelskih svoistv izdelii iz keramiki i stekla, obrabotannyh fakelom nizkotemperaturnoi plazmy [Scientific bases formation of consumer properties of ceramic and glass products processed by low temperature plasma torch]. Extended abstract of PhD dissertation*. Moscow University of Consumer Cooperatives, Moscow, 2004. (rus)

[5] Volokitin O.G., Tsvetkov N.A., Volokitin G.G. Perspektivy tekhnologii sozdaniya zashchitnodekorativnykh pokrytiy na poverkhnosti drevesiny s ispolzovaniyem plazmennoy tekhnologii [Perspectives of protective-decorative coating of wood surfaces using plasma technology]. Vestnik of Tomsk State University of Architecture and Building. 2012. No.1. Pp.112-116. (rus)

[6] Volokitin O.G. Plasma process modeling of wood surfaces / O.G. Volokitin, G.G. Volokitin, V.V. Shekhovcov, E.A. Maslov, N.V. Malyuga. // Vestnik TSUAB. 2013. p. 99-105. (rus)

[7] Volokitin O.G., Tsvetkov N.A., Volokitin G.G., Cherkashina N.A., RF Patent No. 2 484 951, Byull. Izobret., No. 10 (2002).

[8] Chukhchin D.G. *Resursosberegayushaya pererabotka drevesiny s ispolzovaniem plazmokhimicheskoi tehnologii [Resource saving wood processing using plasma chemical technology]*. PhD dissertation. Arkhangelsk State Technical University, Arkhangelsk, 1998. (rus)

[9] Onegin V.I., Udalov Y.P., Korsakov G.S., Lukin V.G. Plazmokhimicheskaya obrabotka poverkhnosti drevesiny dlya uluchsheniya ee rezonansnykh svoistv [Plasma chemical processing of wood surface for improvement its resonance properties] // Moscow: VINITI, 28.04.97, № 1414-B97, 1997, 10 p. (rus)

Journal of Physics: Conference Series **669** (2016) 012066 doi:10.1088/1742-6596/669/1/012066

Antifriction coatings based on a-C for biomedicine applications

Y N Yurjev[1], D V Kiseleva[1], D A Zaitcev[1], D V Sidelev[1] and O S Korneva[2]

[1] Department of Experimental Physics, National Research Tomsk Polytechnic University, Tomsk, 634050, Russian Federation

[2] Materials Properties Measurement Centre, National Research Tomsk Polytechnic University, Tomsk, 634050, Russian Federation

E-mail: sidelevdv@tpu.ru

Abstract. This article reports on the investigation of mechanical properties of carbon films deposited by dual magnetron sputtering system with closed and mirror magnetic field. There is shown that a-C films with predominantly sp^2-phase have relatively high hardness (up to 20 GPa) and low friction index (~0.01). The influence of magnetic field on friction index is determined. The analysis of experimental data shows the obtained a-C samples can be used for biomedicine applications.

1. Introduction

Interest in increase of life-time of medical implants by improving wear resistance has been enhanced. The promising approach in this way is surface modification by deposition thin films. The most medical applications require good adhesion and reduction of friction index of operation surface to 0.1 and lower.

Coatings based on a-C films have high mechanical and tribological properties, high wear resistance, biocompatibility and chemical inactivity. Such films can be used in biomedicine in case of modification surface of medical implant or prostheses [1-3]. Moreover, carbon films have properties of ideal lubricant and prevent the formation of wear products by friction. There are many ways to deposit such coatings with required mechanical properties: plasma assisted chemical vapor deposition [4], cathode vacuum arc deposition [2], magnetron sputtering [5] and etc.

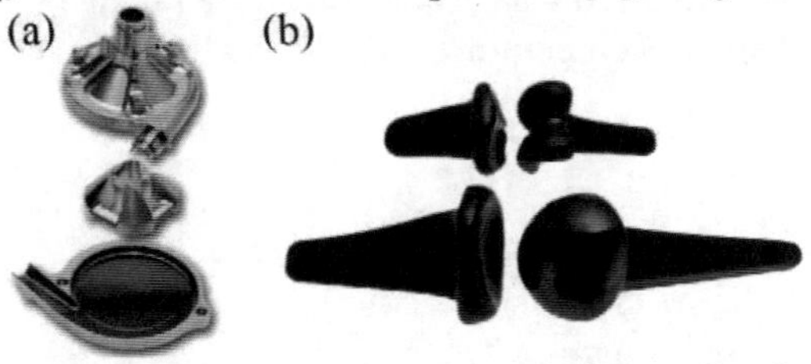

Figure 1. Carbon-coated products: (a) ventricular assist device heart pump, (b) artificial and hip joints.

High-quality antifriction coatings usually deposit by means of magnetron sputtering systems with advanced plasma density (10^{11}-10^{13} cm^3) [5]. For this purpose, magnetron systems with unbalanced magnetic field are often used. Such diode systems have relatively high density of ion current on substrate (more 5 mA/cm^2).

The dual magnetron sputtering system is an especial deposition tool. It is consisted of two planar magnetron systems operating in antiphase with bipolar mid-frequency power supply (~20-80 kHz). Dual magnetron sputtering systems have higher operation stability, provide increased energy of sputtering atoms (~10-50 eV) and improved adhesion of coatings to substrate. These aspects are important to biomedicine applications.

This article is devoted to investigation of mechanical properties of carbon films deposited by dual magnetron sputtering system in Ar atmosphere.

2. Experimental

2.1. Magnetron sputtering

The carbon films were deposited on steel substrate AISI 321(30x30x1 mm^3) by an ion-plasma installation «Yashma-5». The base pressure was 10^{-3} Pa. Steel substrates were treated in plasma of ion source with closed electron drift prior to deposition. The operation parameters of ion source were: voltage ~ 2.5 kV and current ~ 250 mA. For increase adhesion, sublayer Ti (100 nm) was deposited by planar magnetron sputtering system.

The deposition of a-C films (1 μm) were by sputtering carbon targets (MPG-6) in Ar atmosphere (99.95%) by dual magnetron sputtering system (Fig. 2). The distance between target and substrate was 100 mm. The a-C films were deposited at different values of pressure: from 0.1 to 1.6 Pa.

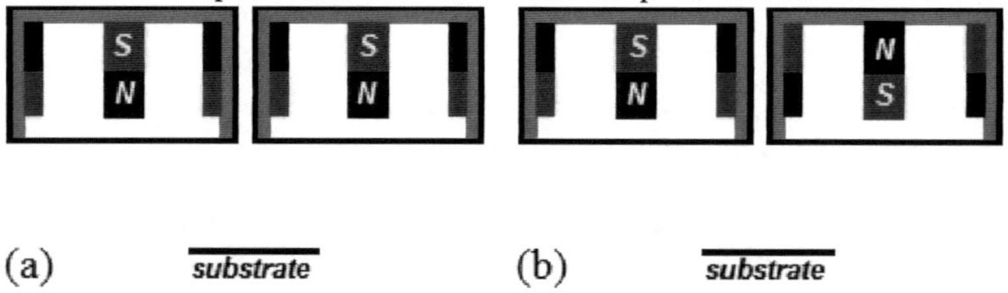

Figure 2. Dual magnetron sputtering systems with mirror (a) and closed (b) magnetic fields.

The dual magnetron sputtering system with mirror magnetic field has equivalent polarity of magnetic system [6]. For this reason, plasma flows bounce off relatively central axis. Conversely, the closed magnetic field dual magnetron has opposite polarity of magnetic system. Plasma flows in such case combine and create an additional magnetic trap for electrons. The change of magnetic system is resulted in variation of ion current density on substrate and impact of the growing film.

2.2. Films characterization

The coating thickness was controlled by quartz gauge «Micron-5». The microstructure of carbon samples was characterized by Raman spectroscopy (Centaur U-HR) in spectral range 100…2000 cm^{-1}. For resonance excitation, laser radiation (532 nm) was used. The film hardness was measured by Vickers method on Nano Hardness Tester (maximum load was 10 mN). The adhesion of coatings to substrate was characterized by scratching the surface by diamond indenter (Micro-Scratch Tester, CSEM). The friction index was tested by PC-Operated High Temperature Tribometer (CSEM) with friction pair ball-disk at atmospheric pressure.

3. Results and discussion

For microstructure identification, Raman spectra expand on two Gaussian components. The high intensities of D- (1310-1420 cm^{-1}) and G-lines (1500-1580 cm^{-1}) show content of graphite bonds (carbon in sp^2-hybridization). The data of samples structure is shown in Table 1.

According to research in the article of Chu P. K., Li L. [7]:

- cathode material is a commercial graphite (equivalent of MPG-6);
- experimental samples are a-C coatings.

Table 1. Structure characterization of carbon films.

magnetic field	P, Pa	G-line		D-line		I_D [c]/I_G [d]
		ν [a], cm^{-1}	$\Delta\nu_{1/2}$ [b], cm^{-1}	ν, cm^{-1}	$\Delta\nu_{1/2}$, cm^{-1}	
closed	0.1	1535	176	1347	390	0.88
	0.13	1551	232	1379	350	0.89
	0.18	1569	140	1372	369	0.87
	1.1	1578	101	1356	250	0.86
mirror	0.1	1511	179	1357	365	0.9
	0.13	1501	206	1328	418	0.89
	0.18	1506	194	1324	431	0.88
	1.1	1517	183	1338	405	0.88
cathode	-	1544	29	1315	47	0.85

[a] ν – frequency of Raman spectrum line
[b] $\Delta\nu_{1/2}$ – half-width of line
[c] I_D –intensity of D-line
[d] I_G – intensity of G-line

The shift of G-line maximum in 1581 cm^{-1} and higher shows disintegration of crystal grain (up to 2.5 nm). In contrast, the change of D-line maximum in lower frequency indicates growing size of graphite clusters [8]. According to experimental data, samples deposited by mirror magnetic field dual magnetron have higher crystal grain size. This effect is related to imbalance of the magnetic field of the dual magnetron.

Carbon films deposited by magnetron sputtering systems can have properties of hard or mild materials. It depends on operation pressure in vacuum chamber during deposition process. Results of hardness measurements are presented in Fig. 3. The indenter penetration distance in carbon films is about 180 nm. The hardness of experimental samples is higher at lower operation pressures. This dependence is a nonlinear for both configurations of magnetic field of magnetron systems. The hardest coatings (up to 20 GPa) are deposited at 0.1-0.13 Pa. The cause of coatings hardness drop at higher pressure is Ar atoms building in coatings structure. It leads to formation of loose and more porous structure.

Results of friction index measurements are presented in Fig. 4. The operation load of Tribometer was 5 N. The friction index of original steel substrates was 0.55.

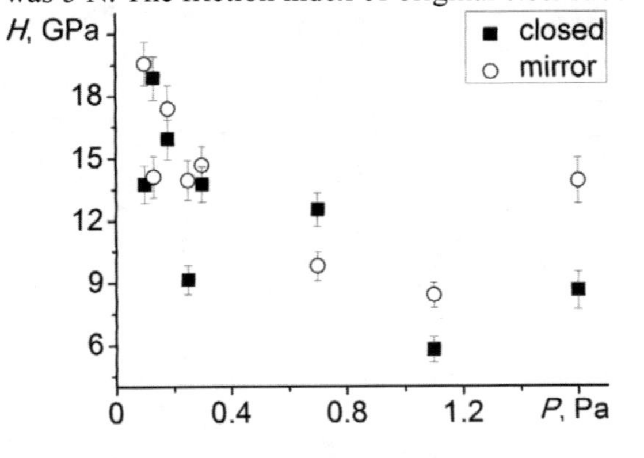

Figure 3. Hardness measurements of a-C films deposited by closed and mirror magnetic field magnetron.

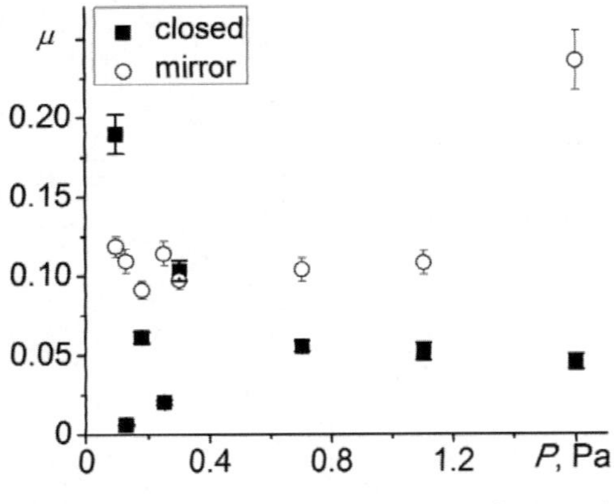

Figure 4. The friction index of experimental samples.

In accordance with Fig. 4, the friction index of a-C films doesn't change in the case of deposition by mirror magnetic field magnetron (0.08-0.12). For the second case, friction properties get worse at increase operation pressure. The average friction index for experimental samples is about 0.1.

Adhesion measurements demonstrate high results. The a-C coatings are stable at load 10 N. It explained by predominantly content of sp^2-phase in experimental samples. Otherwise, carbon films have high internal stress and it leads to decrease coatings stability.

According to results in [9,10], the friction index and hardness of human bone are 0.05 and 15-18 GPa, respectively. In this way, deposited a-C films satisfy requirements of the artificial impants.

4. Conclusions

Main results of our investigation can be summarized as follows:

- experimental samples have predominantly sp^2-phase and good adhesion properties;
- the hardness of a-C films is varied in 5…20 GPa and depends on operation pressure;
- the friction index of a-C films is lower for closed magnetic field magnetron sputtering system and the average friction index is about 0.1.

The deposited a-C films can be used to biomedicine applications.

References

[1] Thomson L A, Law F C, Rushton N and Franks J 1991 *Biomaterials* **12** 37
[2] Ren Y, Erdmann I, Khlopyanova V, Deuerler F and Buck V 2014 *Diamond and Related Materials* **44** 38
[3] Jones M I, McColl R, Grant D, Parker K G and Parker T L 1999 *Diamond and Related Materials* **8** 457
[4] Salvadori M C, Martins D R and Cattani M 2006 *Surface & Coatings Technology* **200** 5119
[5] Oskomov K V, Solov'ev A A and Rabotkin S V 2014 *Technical Physics* **59** 1811
[6] Musil J and Baroch P 2005 *IEEE Transactions on Plasma Science* **33** 338
[7] Chu P K and Li L 2006 *Materials Chemistry and Physics* **96** 253
[8] Ferrari A C, Meyer J C,Scardaci V, Casiraghi C, Lazzeri M, Mauri F, Piscanec S, Jiang D, Novoselov K S, Roth S and Geim A K 2006 *Physical Review Letters* **97** 187401
[9] Merkher Y, Sivan S, Etsion I, Maroudas A, Halperin G and Yosef A 2006 *Tribology Letters* **22** 29
[10] Pal S 2014 *Mechanical Properties of Biological Materials* **2** 23

Ion-plasma nitriding of machines and tools parts instrumental steels

V V Zvezdin[1], A A Spirin[2], Ruz R Saubanov [1], N M Zvezdina [1], A R Fayruzova[3]

[1]Naberezhnye Chelny institute (branch) of KFU, Naberezhnye Chelny, Russia
[2] Ford Sollers holding LLC, Naberezhnye Chelny, Russia
[3] LLC "Saturn", Naberezhnye Chelny, Russia

E-mail:irmaris@yandex.ru, rafisih88@mail.ru

Abstract. Here introduced features of formation diffusion bond during ion nitriding in glow discharge plasma in gaseous mediums (mixture of nitrogen and argon). It is shown, that argon existing in saturated medium changes the nitriding process kinetics and the phase composition of the outer zone. Here presented investigation results on ion-plasma nitriding of instrumental steels, focused on microstructure and tool areas phase composition change, operating in most difficult conditions.

1. Introduction

Physics of ion-plasma nitriding processes in gaseous medium is equal. However, quantitative and qualitative characteristics of separate steps of these processes have their own differences, that shows the influence on structure and the formation kinetics of nitriding case.

Ion nitriding process in hydrogenous mediums (ammonia, mixture of nitrogen and hydrogen) is well studied in Russian and foreign literature [1 - 3]. Hydrogen, being a good deoxidant, has a great impact on the nitriding case formation kinetics, intensifying the process. However, presence of hydrogen in the saturating medium in fair quantities causes surface embrittlement (especially sharp edges) and base weakening, which in certain conditions leads to strength and durability reduction of the construction elements. Adding argon into hydrogenous saturating medium acts nitriding case moldability to rise [4]. Replacement of hydrogen by argon in saturating medium (mixture of nitrogen and argon) excludes hydrogen embrittlement of nitriding case, but hereby the ion nitriding process energetics changes.

Tools strength and service characteristics to a great extend are defined by the surface condition and it's characteristics. One of the most effective ways for hard facing is ion-plasma nitriding.

2. Nitriding in the glow-discharge plasma

Plasma nitriding meets environmental requirements, a great number of installations is used for this purpose. The feature of plasma nitriding is that the process can be managed much easier in comparison with nitriding in gaseous medium. Processes of ion-plasma nitriding, that allow receiving hardened case in middle and low temperatures, see heavy use in different saturating atmospheres.

Journal of Physics: Conference Series **669** (2016) 012067 doi:10.1088/1742-6596/669/1/012067

During low temperature steel nitriding, in different saturating mediums, there happens preemptive nitrogen diffusion, and the structure and phase composition of diffusion layer defined by constitution diagram iron — nitrogen. During high temperature steel nitriding (>600°C) only nitrogen and ammonia are used as saturating mediums.

3. Method of formation anti-abrasion coating on the surface of products made from structural steel

Method of formation anti-abrasion coating on the surface of products made from structural steel includes ion-plasma nitriding in reagent gas medium — nitrogen, part surface cleaning and nitriding. Surface cleaning and nitriding are performed with the reagent gas pressure 5•10-3-2•10-2 mm of mercury, negative bias on parts 300-1000 V and ion current density 2 - 8 mA/cm2 during 30-90 min, cleaning is performed in inert gas plasma – argon with the pressure 3•10-4 - 7•10-4 mm of mercury and current density 3-5 mA/cm 2. Method allows to intensify the process and increase drawing dies service durability, which are under high unit loading in the triboprocess.

Modern devices for ion-plasma nitriding (Fig.1) have quite difficult heating system. Impulse voltage simultaneously and independently is fed to active screen and parts, which are under nitriding. This eliminated nearly all defects of traditional charts of ion nitriding, connected with the difficulty of necessary temperatures regulation, low flatness of heat field, arcing. Herewith, the surface cleaning time and heating to operating temperatures are sharply decreased.

While nitriding in coating surface there compression stresses are formed, which increases parts permanent strength.

4. Ion-plasma nitriding technology

Nitriding parts hardened case structure and characteristics depend on the following technology factors: strain between electrodes, structure of gas medium, degree of vacuum, operating temperature, process time, relative position of parts and electrodes. Nitriding parts are set in camera, connecting to negative electrode, camera is pressurized and the air is pumped out till the pressure 1 mm of mercury. The camera view is shown on figure 1.

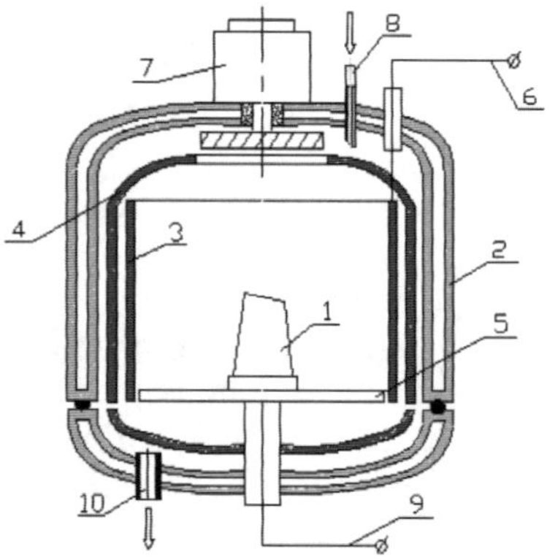

a)

б)

Figure 1. Installation diagram for ion-plasma nitriding (a) and it's view (б)
1 — hardening product; 2 — water-cooling camera; 3 — active screen; 4 — heating element; 5 — table; 6 — supply of high pressure to the screen; 7 — fan; 8 — supply of saturating gaseous medium; 9 — supply of high pressure to hardening product; 10 — to vacuum pump.

Selection of optimum pressure, which depends on parts configuration complexity and their reasonable positioning (Fig. 2), as with the change of pressure, the length of discharge cathode part changes. With increase of pressure from 1 to 10 mm of mercury the discharge cathode part area decreases from 10 to 1 mm. It must be taken into account to provide steadiness of diffusion coating on the product perimeter.

Figure 2. Parts position on the pallet for nitriding

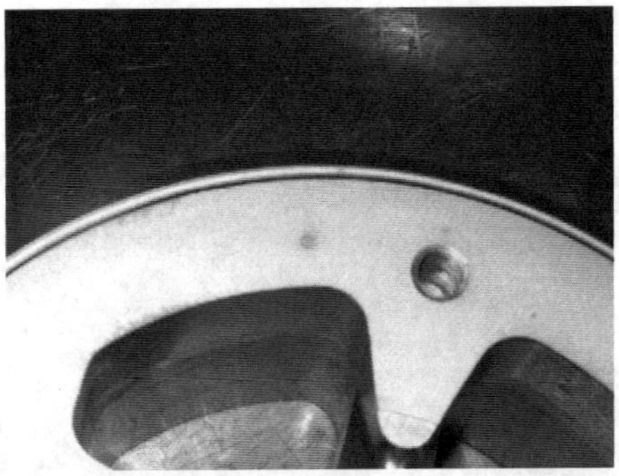

Figure 3. Fragment view of the parts after nitriding

Temperature of nitriding is usually of 470-580°C, voltage 400-1100 V, vacuum 1-10 mmHg. The operating pressure is limited by the glow discharge properties. With the pressure below 1 mmHg, the ions energy is not enough for heating of the processed part up to the operating temperature; with pressure above 10 mmHg, the discharge stability falls, the glow discharge turns into an arc one, and this process is accompanied by the melted microcraters appearance on the surface (Fig. 3).

Ammonia, nitrogen and a nitrogen plus hydrogen compound are used as nitrogen-bearing gases. Nitrogen plasma cannot contain oxygen as it decrease the operating atmosphere activity; hydrogen slightly influences the layer growth. The change of hydrogen concentration in the nitrogen-hydrogen compound within the limits from 1 : 9 to 9 : 1 does not influence the diffusion layer parameters. Ionic nitriding can be also performed in a hydrogen-free plasma. In order to regulate the composition of the nitride zone by carbon, carbon-bearing gases are added to the atmosphere.

Cathode spraying is conducted during 5-60 min with voltage of 1100-1400 V and pressure of 0,1-0,2 mmHg. During the cathode spraying, the part surface temperature does not go above 250°C. The operating parameters of the process with saturation: U = 400- 100 V, pressure 1-10 mmHg.

After the surface has been being processed during 5-60 min in the cathode spraying mode, the voltage will be increased up to the operational one, and the pressure will be increased up to 1-10 mmHg. With the pressure increase, the length of the cathode glow which is equally distributed on the part surface decreases repeating its contours. Cathode spraying can be conducted in a hydrogen atmosphere as well, the use of which is especially appropriate during austenite steel nitriding. The operating temperature of the process (470—580°C) is reached within 15-30 min. Heating speed is defined by the parts surface and their weight correlation.

After isothermal aging, the parts are cooled down to ambient temperature in vacuum. During this process the cooling speed is higher against the oven heating as during the ionic heating only the part will be heated, and the container walls (process booth) is not substantial due to the heat emission and convention in the vacuum conditions.

Steel 4X5MΦC is the main steel for various stamps, deformation of steel and non-ferrous metals and for molding types under pressure of aluminum and magnesium alloys with diameter up to 70-80 mm.

Consequently, the 4X5MFC (4X5B2FC) and 4X4VMFC steel forms durability will be higher by 1,5—2 times.

Thus, the nitride layer contains nitride compounds (Fe 2_3- e - phase,- Fe 4 N - /- phase) and internal nitriding zones. Figure 4 shows a photo pf microstructure and a graph showing changes in micro-hardness of the nitride layer of the steel 4X5MFC.

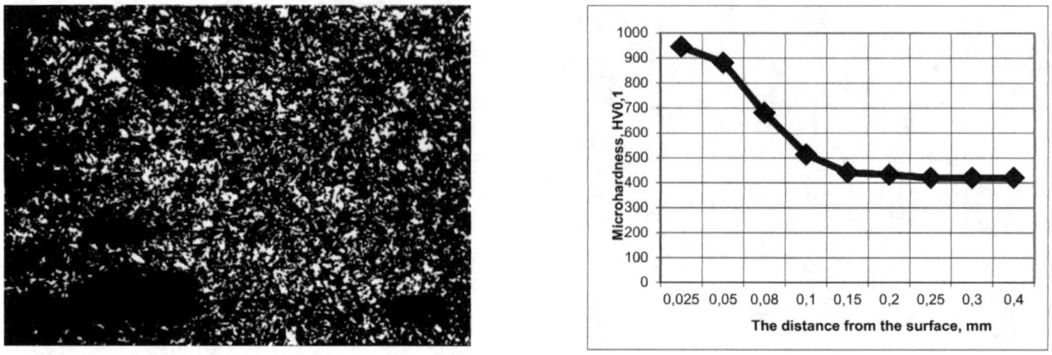

a) б)

Figure 4. Part microstructure, ×500; a) - distance is 5 mm from the bar end; b) - graph showing changes in micro-hardness of the nitride layer of the steel 4X5MFS

The austenite disintegration in the perlite and transition zones is preceded by the part of carbides emission, which takes place at the granules edges, which leads to viscosity decrease by 30-40 %.

5. Conclusion

Current density change within great limits (0,5—20 mA/sm^2) does not influence the nitriding process. Improvement of the ionic-and-plasma nitriding technical efficiency lies in optimization of the technological process parameters based on the weight-dimension and geometry features of a part. This is explained by the fact that these features influence the ionic gas density distribution evenness on the part surface as the plasma glow discharge intensity in the nitride atmosphere increases the sharp part edges.

References

[1] Arzamasov B.N., Bratukhin A.G., Eliseev Y.S., Panayoti T.A. Ion chemo-thermal treatment of alloys in gaseous medium. - M: Publishing house of Bauman MSTU, 1999. - 400 p.

[2] Korotchenko V., Bell T. Using if ion nitriding in manufacturing industry in Great Britain. Translation №Б-14287-M.: 1979.-29 p.Solodkin G. A. Ion nitriding of machine part and cutting tool. Synopsis of a thesis Candidate of technical Sciences- M.: 1979.-21 p.

[3] Investigations on ion-plasma nitriding of tool steels / Fayruzov R.S., Israfilov I.K.,Bashmakov D.A., Nugumanova A.I.// Interuniversity science book «Designing and research of engineering systems». - Naberezhnye Chelny : Publishing house of INECA, 2007. - №11.-C.100-106.

[4] Kaplun V.G., Pastukh I.M. Plasma-diffusion processes of materials surface modifications: technology and equipment//Works of academic and research symposium «Equipment and technologies of metals and alloys thermal treatment in machine building» - Kharkov (Ukraine),-2000.-C. 145-154.

Institute of Physics
Dirac House, Temple Back
Bristol BS1 6BE UK

ISSN: 1742-6588
ISBN 978-1-5108-1960-3

CURRAN ASSOCIATES INC.
proceedings.com